Advances in Intelligent Systems and Computing

Volume 774

Series editor

Janusz Kacprzyk, Polish Academy of Sciences, Warsaw, Poland
e-mail: kacprzyk@ibspan.waw.pl

The series "Advances in Intelligent Systems and Computing" contains publications on theory, applications, and design methods of Intelligent Systems and Intelligent Computing. Virtually all disciplines such as engineering, natural sciences, computer and information science, ICT, economics, business, e-commerce, environment, healthcare, life science are covered. The list of topics spans all the areas of modern intelligent systems and computing such as: computational intelligence, soft computing including neural networks, fuzzy systems, evolutionary computing and the fusion of these paradigms, social intelligence, ambient intelligence, computational neuroscience, artificial life, virtual worlds and society, cognitive science and systems, Perception and Vision, DNA and immune based systems, self-organizing and adaptive systems, e-Learning and teaching, human-centered and human-centric computing, recommender systems, intelligent control, robotics and mechatronics including human-machine teaming, knowledge-based paradigms, learning paradigms, machine ethics, intelligent data analysis, knowledge management, intelligent agents, intelligent decision making and support, intelligent network security, trust management, interactive entertainment, Web intelligence and multimedia.

The publications within "Advances in Intelligent Systems and Computing" are primarily proceedings of important conferences, symposia and congresses. They cover significant recent developments in the field, both of a foundational and applicable character. An important characteristic feature of the series is the short publication time and world-wide distribution. This permits a rapid and broad dissemination of research results.

More information about this series at http://www.springer.com/series/11156

Shuichi Fukuda
Editor

Advances in Affective and Pleasurable Design

Proceedings of the AHFE 2018 International
Conference on Affective and Pleasurable
Design, July 21–25, 2018,
Loews Sapphire Falls Resort at Universal Studios,
Orlando, Florida, USA

 Springer

Editor
Shuichi Fukuda
System Design and Management
Keio University
Tokyo, Japan

ISSN 2194-5357   ISSN 2194-5365   (electronic)
Advances in Intelligent Systems and Computing
ISBN 978-3-319-94943-7   ISBN 978-3-319-94944-4   (eBook)
https://doi.org/10.1007/978-3-319-94944-4

Library of Congress Control Number: 2018947435

Printed on acid-free paper

This Springer imprint is published by the registered company Springer International Publishing AG part of Springer Nature
The registered company address is: Gewerbestrasse 11, 6330 Cham, Switzerland

Advances in Human Factors
and Ergonomics 2018

AHFE 2018 Series Editors

Tareq Z. Ahram, Florida, USA
Waldemar Karwowski, Florida, USA

9th International Conference on Applied Human Factors and Ergonomics and the Affiliated Conferences

Proceedings of the AHFE 2018 International Conference on Affective and Pleasurable Design, held on July 21–25, 2018, in Loews Sapphire Falls Resort at Universal Studios, Orlando, Florida, USA

Advances in Affective and Pleasurable Design	Shuichi Fukuda
Advances in Neuroergonomics and Cognitive Engineering	Hasan Ayaz and Lukasz Mazur
Advances in Design for Inclusion	Giuseppe Di Bucchianico
Advances in Ergonomics in Design	Francisco Rebelo and Marcelo M. Soares
Advances in Human Error, Reliability, Resilience, and Performance	Ronald L. Boring
Advances in Human Factors and Ergonomics in Healthcare and Medical Devices	Nancy J. Lightner
Advances in Human Factors in Simulation and Modeling	Daniel N. Cassenti
Advances in Human Factors and Systems Interaction	Isabel L. Nunes
Advances in Human Factors in Cybersecurity	Tareq Z. Ahram and Denise Nicholson
Advances in Human Factors, Business Management and Society	Jussi Ilari Kantola, Salman Nazir and Tibor Barath
Advances in Human Factors in Robots and Unmanned Systems	Jessie Chen
Advances in Human Factors in Training, Education, and Learning Sciences	Salman Nazir, Anna-Maria Teperi and Aleksandra Polak-Sopińska
Advances in Human Aspects of Transportation	Neville Stanton

(continued)

(continued)

Advances in Artificial Intelligence, Software and Systems Engineering	*Tareq Z. Ahram*
Advances in Human Factors, Sustainable Urban Planning and Infrastructure	*Jerzy Charytonowicz and Christianne Falcão*
Advances in Physical Ergonomics & Human Factors	*Ravindra S. Goonetilleke and Waldemar Karwowski*
Advances in Interdisciplinary Practice in Industrial Design	*WonJoon Chung and Cliff Sungsoo Shin*
Advances in Safety Management and Human Factors	*Pedro Miguel Ferreira Martins Arezes*
Advances in Social and Occupational Ergonomics	*Richard H. M. Goossens*
Advances in Manufacturing, Production Management and Process Control	*Waldemar Karwowski, Stefan Trzcielinski, Beata Mrugalska, Massimo Di Nicolantonio and Emilio Rossi*
Advances in Usability, User Experience and Assistive Technology	*Tareq Z. Ahram and Christianne Falcão*
Advances in Human Factors in Wearable Technologies and Game Design	*Tareq Z. Ahram*
Advances in Human Factors in Communication of Design	*Amic G. Ho*

Preface

This book focuses on a positive emotional approach in product, service, and system design and emphasizes aesthetics and enjoyment in user experience. This book provides dissemination and exchange of scientific information on the theoretical and practical areas of affective and pleasurable design for research experts and industry practitioners from multidisciplinary backgrounds, including industrial designers, emotion designer, ethnographers, human–computer interaction researchers, human factors engineers, interaction designers, mobile product designers, and vehicle system designers.

This book is organized into five sections which focus on the following subjects:

1. Affective and Emotional Aspects of Design
2. Affective Design in Healthcare
3. Sensory Engineering and Emotional Design
4. Kansei Engineering and Product Design
5. Affective Value and Kawaii Engineering

Sections 1 through 3 of this book cover new approaches in affective and pleasurable design with emphasis on emotional engineering and product development. Sections 4 and 5 focus on material and design issues using Kansei Engineering and Kawaii Engineering for user behavior in design process product, service, and system development, human interface, emotional aspect in UX, and methodological issues in design and development. The overall structure of this book is organized to move from special interests in design, design and development issues, to novel approaches for emotional design.

All papers in this book were either reviewed or contributed by the members of editorial board. For this, I would like to appreciate the board members listed below:

Améziane Aoussat, France
Sangwoo Bahn, Korea
Lin-Lin Chen, Taiwan
Kwangsu Cho, Korea

Sooshin Choi, USA
Denis A. Coelho, Portugal
Oya Demirbilek, Australia
Magnus Feil, USA
Andy Freivalds, USA
Qin Gao, China
Ravi Goonetilleke, Hong Kong
Brian Henson, UK
Amic G. Ho, Hong Kong
Wonil Hwang, Korea
Yong Gu Ji, Korea
Eui-Chul Jung, Korea
Jieun Kim, Korea
Kyungdoh Kim, Korea
Kentaro Kotani, Japan
Stéphanie Minel, France
Kazunari Morimoto, Japan
Michiko Ohkura, Japan
Taezoon Park, Korea
P. L. Patrick Rau, China
Simon Schutte, Sweden
Dosun Shin, USA
Anders Warell, Sweden
Myung Hwan Yun, Korea

This book is the first step in covering diverse topics including design and development of practices in affective and pleasurable design. I hope this book is informative and helpful for the researchers and practitioners in developing more emotional products, services, and systems.

July 2018   Shuichi Fukuda

Contents

Affective Design in Healthcare

Sensory Engineering and Emotional Design

Kansei Engineering and Product Design

Affective and Emotional Aspects of Design

Increasing Importance of Emotion in the Connected Society

Shuichi Fukuda[(⊠)]

System Design and Management Research Institute,
Keio University, 4-1-1, Hiyoshi, Kohoku-ku, Yokohama 223-8526, Japan
shufukuda@gmail.com

Abstract. Emotion will increase its importance in the Connected Society due to the following reasons.

(1) Traditional engineering has been final-product- and individual product-focused, with emphasis on product value. But in the Connected Society, products must be designed, developed and operated as a team. Rational and optimizing approaches are no more applicable and we have to look for emotional satisfaction, as Simon pointed out.
(2) Product Quality is improved to the full extent so customers cannot recognize. We need something other than just product quality to explore a new market. Processes will create new values.
(3) Customers are now looking for mental satisfaction instead of material satisfaction. Thus, their expectations and requirements are quickly personalized. Therefore, we must consider how we can satisfy their human needs. Engineering should move toward much more self-motivated and self-determined and we have to satisfy their need for growth.

Keywords: Product team · Emotional satisfaction · Process value
Self-motivated · Self-determined · Need for growth · Emotional messages

1 Introduction

Up to now, we have been focusing our attention on individual products and we would like to have better quality and better performing products. But in the Connected Society, products are connected and work as a team. Thus, their team performance becomes more important than those of individual products. And how a team responds to our expectations will play a very important role.

In other words, our expectations vary from time to time and from case to case and how a product team responds to such changes will bring us more joy than the individual product excellence in quality or performance.

Thus, the connected society may be said that it is very similar to the team sports. We enjoy them not by observing the excellence of individual players, but how they play

© Springer International Publishing AG, part of Springer Nature 2019
S. Fukuda (Ed.): AHFE 2018, AISC 774, pp. 3–13, 2019.
https://doi.org/10.1007/978-3-319-94944-4_1

together strategically in response to the changing situations. Thus, in the Connected Society, what we should focus upon is how we can provide emotional satisfaction to the customers by paying attention to what they expect from the team and its team play.

2 Engineering: Yesterday and Today

Yesterday, our world is small and closed. It was a Closed World (Fig. 1). And there were changes, but these changes are smooth, i.e., mathematically differentiable, so that we could foresee the future and engineers could estimate the operating conditions.

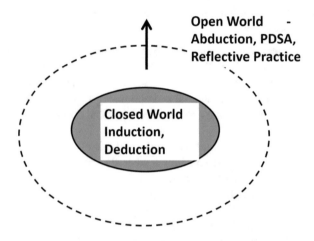

Fig. 1. Closed World and Open World

Our traditional engineering has been final-product focused. In other words, engineering up to now has been none other than product development.

As the world is closed and bounded and it is small, the number of degrees of freedom is not so large therefore we could design and produce products rationally and we could optimize them.

Herbert Simon pointed out [1, 2] that if the number of variables is not too many, we can apply rational approaches and optimize. But if the number of variables becomes very large, our rationality is bounded so we cannot optimize anymore. Instead, we should take emotion into consideration. We have no other choice but to look for a satisfying solution. He called such a result as satisficing (satisfy + suffice).

Today, our world is rapidly expanding, and now it becomes an Open World (Fig. 1) Therefore, changes take place very frequently and extensively. And what makes the problem difficult is that these changes are sharp, so they cannot be mathematically differentiable. Thus, nobody can predict the future. Environments and situations vary from moment to moment. And in the case of engineering, only users know what is happening now. Therefore, engineering today are moving rapidly toward user-centered and emotional satisfaction becomes the new keyword instead of optimization, which has been considered most important up to now.

3 Controllable World

Although our world has kept on expanding, we could have applied rational approaches. Why? This is because we extracted feature points which follow the rational rules.

Let us illustrate this by taking example of identifying the name of a river. A river is flowing, and its behavior is changing all the time. So, if we focus our attention to its flow, we cannot identify the name of a river. But we identify its name by looking for something like trees or mountains which do not change or change slowly. Thus, we can identify the name of a river (Fig. 2).

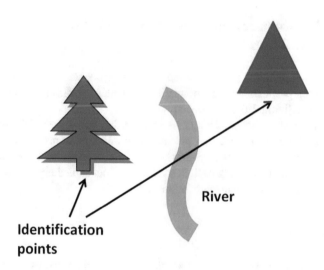

Fig. 2. Identification of the name of a river

We used the same approach to control. For example, in the case of welding, molten pool is just like a river. It is changing all the time. But instead of focusing our attention to molten pool, we find controllable points which surround it. By controlling these points, we can control it rationally. As engineering world is expanding so widely and quickly, we cannot apply rational approaches in a straightforward manner as we did in the past. But by finding such controllable points, we can control things which are otherwise unpredictable and uncontrollable. Thus, we succeeded in expanding controllability beyond rational boundary. Once such a frame of problem-solving is established, then we can solve the problem rationally and optimize (Fig. 3).

But we should note that such an approach was feasible, because the environments and situations do not change appreciably. This approach was possible, because the surroundings of the target do not change appreciably so that we can apply rational approaches, although the target itself alone cannot be rationally treated.

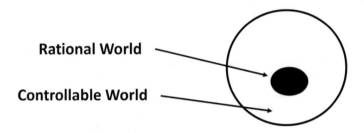

Fig. 3. Rational and Controllable World

4 Emerging Engineering: From Individual Product to Product Team

Up to now, our engineering has focused on final products. We have been making efforts to develop products with better quality and better performance. But these efforts are focused on individual final products.

IoT has changed the scene. Things are now getting connected. Not only human to human, but machine to machine and machine to human. Now products work together as a team. Therefore, we must consider how we can manage and operate them as a team. We have been considering each individual product separately, but what becomes important now is the team management and operation.

5   11 Best, Best 11

Knute Rockne, famous American football player and coached told us a very interesting and important word. He said "Best team cannot be made up by 11 best players. Best team can be made by 11 players who work together as a team and who respond flexibly and adaptively to the changes".

In the old days, situations did not change appreciably from game to game. So, each player has his position in the team formation and he is expected to play his role in the position to the best. Then, that would lead the team to victory. But today, games change so much from moment to moment so that adaptability becomes more crucial. And the formation is no more fixed. It changes continuously to adapt to the situations.

In other words, the formation yesterday was tree-structured, and the structure does not change. Although efficient, it cannot adapt to fast-changing situations, because a tree has only one output node (Fig. 4).

In order to adapt flexibly and adaptively to the changing situations, the formation must be a network (Fig. 5). A tree has only one output node, but in the case of a network, any node can be an output node. Therefore, a network is more adaptable. Further, in the frequently and extensively changing situations today, a network itself must be adaptable. So, its structure changes from situation to situation.

But we must remember that to make a network itself adaptable, a node must possess many links because nodes are called for to team up in the best way with other nodes to

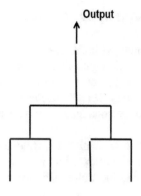

Fig. 4. Tree

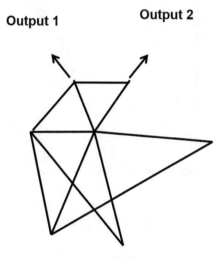

Fig. 5. Network

cope with the situation. The node with a small number of links does not satisfy this requirement.

In the old days, a player who can play best in his position is called a best player, but today, what is expected from him is not such superb individual capabilities, but how many different capabilities he has to enable team working with as many different partners as possible becomes important. In other words, players were going deeper and deeper in the old days, but today they are expected to go wider and wider.

6 Emotional Satisfaction: Why It Becomes Important

As discussed earlier, we cannot optimize and focus our attention on final product performance alone anymore in this Open World where changes take place frequently and extensively. As Simon pointed out, we are forced to change our goal from optimization to emotional satisfaction. Thus, it becomes more and more important with increasing diversification and personalization to consider how we can satisfy our customers emotionally.

Emotional Satisfaction becomes important from another aspect. We have been paying tremendous efforts to make our product work better. Thus, the quality curve of our products is now approaching the ceiling (Fig. 6).

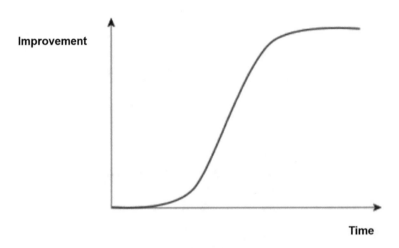

Fig. 6. Quality curve

Weber and Fechner proposed the following law.

$$\frac{\Delta R}{R} = \text{constant}$$

Unless the increment does not increase proportionally to the stimulus level, we cannot recognize the increase. To take a simple example, if a man with a loud voice raised his voice a little, nobody recognizes that he raised his voice. But a man with a small voice raised his voice a little, many people become aware that he raised his voice.

When the quality is improving at the lower level of this curve, customers easily recognize that the quality is improving. But as the quality comes closer to the ceiling, it is very difficult for customers to recognize the improvement. And the task itself of improving becomes increasingly difficult. Therefore, apart from the discussion of Bounded Rationality, we need to look for something else other than traditional function-based approach to expand our market. Thus, the importance of emotional satisfaction is increasing rapidly.

7 Functional Excellence and Emotional Satisfaction

We must take note that functional excellence can be evaluated objectively, But emotional satisfaction is very much subjective. Even if the product does not perform best from the standpoint of engineers, customers may feel emotionally satisfied if their expectations are met. We must remember that one reason why product requirements are diversifying is customers' expectations are becoming more and more personalized so that their requirements are widely diversifying.

Abraham Maslow proposed a hierarchy of human needs [3] (Fig. 7).

Fig. 7. Maslow's hierarchy of human needs

At the base level, people look for material satisfaction. They would like to have better products. But as they go up, their needs change from material to mental and at the top, people would like to actualize themselves. They would like to show how capable they are.

In fact, mountain climbers are a good example. They look for a difficult route, even though there is a much easier one. And when they succeed, they would like to challenge the more difficult route. Challenge is the core and mainspring of all human activities. Games are another typical example. Players of any level can enjoy the game their way, but once they succeed, they challenge for more difficult one.

Thus, the importance of emotional satisfaction is increasing rapidly.

8 From Product Value to Process Value

This change of needs also brings the change in value evaluation. Traditional engineering has focused on a final product, so product value has been considered most important. If a product works better with better quality, that was most valuable.

But when our needs are heading for mental satisfaction, we should remember that the process yields value as much as or sometimes more than product value. In fact, mountain climbing, gaming, etc. are typical examples of how processes yield great value. Thus, we must bear in mind that engineering tomorrow must consider how we can develop process values in addition to product values.

9   Expectation Management

Although the word "Voice of Customers" is well known, we should remember that this is the voice of customers about the product which is already developed. i.e., It is nothing other than the feedback. Its way of thinking is completely based on the B to C basis.

But what is needed today is C to B way of thinking. It is completely the other way around. We should consider what products, and what is becoming more important is, what processes we can develop to meet the expectations of customers. This is Expectation Management and AI would help us a great deal. It would help us detect a group of customers with similar expectations.

Thus, engineering tomorrow will be going the other way, i.e., from B2C to C2B. And how products and processes satisfy the expectations of customers will be the most importance factor for value creation. Again, it is very subjective.

As things are getting connected and products work as a team, we should pay our attention to how a team of products works together to meet the expectation of customers.

To take soccer for example, the fans used to enjoy the beautiful play of each player and winning the game was most important. But today, fans come to the stadium to enjoy how strategically a team plays to respond to the quickly changing situations. When unexpected game strategy comes up, the fans get excited and that is what attracts them to come all the way to the stadium. They enjoy the game because their expectations are betrayed in a beautiful and truly emotionally satisfying manner. Winning the game is still important, but this surprise is much more enjoyable than just a victory. In fact, this is the essence of a game. Why is gamification getting attention these days? It is the same, no matter whether it is a sport or an amusement. Customers enjoy the processes.

10   Importance of Self-determination: Two-Sided Roles of Customers

Deci and Ryan proposed Self-Determination Theory [4]. They pointed out that there are two kinds of motivations, one is intrinsic, the other extrinsic. Even if the job is the same, we feel much happier when we do it out of our own intrinsic motivation than when we do it based on extrinsic motivation. We feel happier when the job is self-determined.

We should remember that today only users know what is happening right now and they are the ones who have to make decisions how to respond to the changes. Thus, adaptability is completely left to their decision-making capabilities.

On the other hand, in the case of traditional engineering, decision-making is completely at the hands of engineers and customers are supposed only to receive a product and use it. There were almost no decision-making tasks on the part of a user how to use a product. But today, customers have two roles. One is the traditional role of a user and the other is how to adapt products, now not a single product, but a team of products, to the changing situations.

To describe it in another way, we can create value not only by focusing our attention to product aspects, but also to processes and by letting our customers feel that they are performing these two roles perfectly. Then, that will provide them with a sense of great achievement and fulfilment.

11   Need for Growth

Another point pointed out by Deci and Ryan, which is quite important, is we have a need for growth. We are born to keep on growing. But traditional engineering forced us to repeat the same way, especially in hardware field.

But if we compare hardware and software developments, we understand easily how software development takes into consideration this issue of our need for growth.

Hardware is developed to satisfy the design specifications which are required for a final product. Thus, hardware development is a one time task. Its development finishes when we deliver a final product to a customer (Fig. 8).

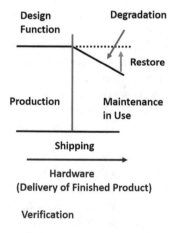

Fig. 8.   Hardware development

However, in the case of software development, it introduces continuous proto-typing approach. First, basic functions are provided to the customers and once they get accustomed to the system, then it is upgraded a little. Thus, step by step, functions grow with customers. Customers feel their expectations are truly taken into consideration in the development process and as the systems grow with them, their need for growth is also satisfied. And as these upgrades reflect the needs of customers, they feel they made the decision and it satisfied their need for self-determination. Thus, software devel-opment truly makes the most of psychology to increase value. Of course, the system is valuable as a product, but what is more important and what attracts customers is its clever strategy based on psychology (Fig. 9).

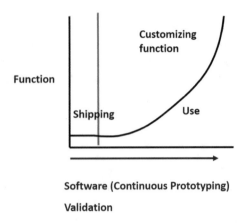

Fig. 9. Software developments

12 From Silent to Communicative Modularization

Today, modularization is attracting wide attention to cope with rapidly increasing diversification. But these modularizations are driven by the idea that if we modularize things, we can put them together as we like and if we can develop a common platform, then we can reduce cost considerably. These motives are producer-centric. And these modules are silent. They do not communicate with each other.

But today things are getting connected and they communicate with each other. Each module can communication with other modules and they can find out what would be the best strategy to respond to the current situation. Products communicate with each other and they can communicate with users to come up with the best satisfying solution.

This is exactly what is being done on the pitch in soccer. Players communicate with each other with voice and with heart to find out the best satisfying strategy to overcome the current situation. Their formation or network changes from moment to moment or from task to task. Each player is a module in industry terms. A user is now a true "playmaker" in the game. He can enjoy making a game.

13 Object-Oriented Programming to IoT and to Symbiotic Engineering

Finally, I would like to point out that most IoT discussions focus their attention on hardware products, but if we recall that in software field, OOP (Object-oriented Programming) also attracted wide attention in the early 1960s. Alan Kay, who is one of the founders of OOP developed Smalltalk during the 1970s. It is interesting to note that Objects, Messages and Methods constitute Smalltalk. And what Kay emphasized is the importance of messaging. He told us that Smalltalk was created as the language to underpin the "new world" of computing exemplified by "human–computer symbiosis" [5].

Kevin Ashton started IoT in 1999, about 20 years later. Ashton was an expert in RFID and he was interested in re-organizing supply chains, which was motivated by his former work at Procter and Gamble. It is interesting that Ashton preferred to call IoT "Internet *for* Things", not "Internet *of* Things" and both Kay and Ashton emphasized the importance of messaging between Objects (OOP) and between Things (IoT).

In fact, it may be safe to say that Objects and Things are basically the same concept. Things (Objects) send out messages and communicate with other Things (Objects) to make a better decision to cope with the quickly changing situations. Although in Small talk, only human-computer symbiosis is taken up, but in the Connected Society, Symbiosis in much broader sense must be established.

Toward this goal, Symbiotic Engineering must be developed and it will lead us to happy life in harmony with the environment, products and society.

14   Closing Remark

The Connected Society will bring us Symbiotic Life. We can live happily together with environments, humans, products and society. To achieve this goal, engineering should move in this direction, where process values become very important and communication will play a crucial role. We must start first from considering the human nature. In other words, we should consider from our inside first and then study how we can harmonize with the outside.

We must consider keeping our customers engaged all the way through the whole product lifecycle from product development to its disposal. And we should let our customers make decisions as much as possible. That would satisfy their needs for self-determination and growth and would bring them great emotional satisfaction, which is the ultimate value in the Open World.

References

1. Simon, H.A.: Administrative Behavior. The Free Press, New York (1976)
2. Maslow, A.H.: A theory of human motivation. Psychol. Rev. **50**(4), 370–396 (1943)
3. Ryan, R.M., Deci, E.L.: Self-Determination Theory: Basic Psychological Needs in Motivation, Development and Wellness. Guilford Publishing, New York (2017)
4. http://gagne.homedns.org/ ~ tgagne/contrib/EarlyHistoryST.html

Playable Cities for Children?

Anton Nijholt[1,2(✉)]

[1] Human Media Interaction, University of Twente,
PO Box 217, 7500 AE Enschede, The Netherlands
anijholt@cs.utwente.nl
[2] Imagineering Institute, Johor Bahru, Iskandar, Malaysia

Abstract. The concept of playable cities was introduced in the city of Bristol (UK) in 2012. While in smart cities the focus is on efficiency, whether it is on city governance, energy consumption, waste management, communication or transport, in playable cities the focus is on making cities more attractive for its inhabitants and visitors by introducing entertaining, playful and humorous interactive installations in the urban environment. In this paper we look at existing playable city projects, look at their characteristics and see whether their 'playability' can be experienced by others than those who are qualified as the 'smart people' that are usually seen as the 'users' of these playful installations. We mention some criticisms on these projects and also compare the playable cities initiatives with initiatives that aim at making cities more child-friendly.

Keywords: Playable cities · Smart cities · Inclusive smart cities
Child-friendly cities

1 Introduction

We have become familiar with the notion of smart cities. In smart cities the focus is on efficiency, monitoring, and control. These issues are embraced by city authorities and large information technology companies are now offering comprehensive systems to make smart cities possible. These systems offer solutions for city governance, traffic control, communication and transport, use of energy, waste management, and many other problems that are becoming bugbears, not only for city authorities and local politicians, but also for citizens.

In smart cities there can be initiatives to solve and attack problems by at least integrating and reason about information that is already available in the many separate databases that are maintained, locally, regionally, or nationally. But this is not the focus of this paper. Another aspect of a smart city is that it has sensors and actuators that provide computers with real-time information about citizens' activities. That is, this is not, for example, about people's water or energy consumption, but it is about their activities and presence at home, in public spaces, while recreating, or while being at work in their workplaces. Tracking of people's activities is done with sensors. We can have miniature cameras, microphones and other sensors embedded in street furniture, in workplaces and home environments. Intelligent computer vision, audio, and computer reasoning software allow the interpretation of what is monitored. This information can be fed into databases for integration with other information and for off-line

© Springer International Publishing AG, part of Springer Nature 2019
S. Fukuda (Ed.): AHFE 2018, AISC 774, pp. 14–20, 2019.
https://doi.org/10.1007/978-3-319-94944-4_2

analysis and producing results that can guide city management decisions. And, of course, there is also the possibility to have local communities profit from this information. Local communities can be given access to some of the sensors and actuators that collect information or control their environment. Cheap sensor, actuator and computing technology can also allow them to design and implement applications that are tuned to their specific needs, for example make their community safer or offer their children safe play facilities.

Large companies that offer 'solutions' to city authorities who want their city to be part of the smart city movement don't focus on the needs of children, elderly, low income people, disabled persons, or local communities in their proposals to city authorities. Their role is to provide software systems that help to collect and integrate data and turn it into information with the aim to make a city more efficient by more informed or even automatic decision making and machine-like city control. In addition, monitoring city events, for example, unusual heavy traffic or a traffic congestion, can also initiate actuators that manage the traffic in real-time in such a way that problems decrease. Smart City plans for US cities include lampposts that are equipped with audio-visual sensors that can detect problems when pedestrians cross a street, but also detect whether in their close proximity a gun has been fired and then warn the police. In Wellington, New Zealand, 'Safe City' sensor technology has been employed to detect begging, rough sleeping and drug abuse on the streets and warn the police when necessary.

Hence, smart city technology can also be used to make cities more livable for its inhabitants, including children, elderly, and disabled persons. What role can be played by the intelligent sensors and actuators that become embedded in the urban environment for recreational and leisure activities? We can talk about urban games or gamification of our daily city activities, and it is also possible to investigate how sensors and actuators embedded in our environments and our own bodily wearable technology can collaborate to have applications that address recreational and leisure needs. Or, in the case of children, needs that deal with their psychosocial and cognitive developments that are not necessarily explicitly addressed but that follow from outdoor play activity.

2 Playable Cities

Smart sensors and actuators are becoming part of our urban environments. They are embedded in our environments and when they become part of the Internet of Things, they can exchange information and make decisions based on real-time events and information that has been collected before. Rather than using such information for making city processes more efficient, we can also look at the possibility to employ this digital technology to introduce playful and humorous interactions, events and products in the digitally enhanced city world.

2.1 The Playable City Contest

Since 2012, with support from the British Council, in the city of Bristol a Playable Cities contest is organized[1] [1]. Design studios, and research institutes and groups are invited to propose playful additions to urban environments. It is quite obvious that proposals need to make use of smart technology. So, sensors and actuators that are already available in the urban environment can be used, project-specific sensors and actuators can be introduced, buildings, statues, street furniture, or whatever is already present in public spaces can be used, and, of course, apart from technology that detects and interprets our behavior, our own smart phones or other wearables can be used in a playful interaction with a public playful installation.

Playable City contest projects introduced so far do not make use of sensors and actuators that are already embedded in the smart city. They are newly implemented in order to make such playful additions possible. And, moreover, after a short period in which city dwellers are supposed to enjoy them, they are again removed from the city.

Playful additions that have been introduced are smart street furniture that acts as chatbots, lampposts that record shadows of passersby and project confusing shadows on the streets, projections of animated and moving animals on streets and walls that invite passersby to play with them, chase them or compete with them (for example, see whether you can jump as high as a kangaroo projected on the wall of a subway station), or a pedestrian traffic light that invites you to smile while you are waiting for the green light. Here is the complete list of award winning designs:

- 2013 Hello Lamp Post (fake interaction using a mobile phone and SMS messages with lampposts and other street furniture)
- 2014 Shadowing (infrared cameras attached to lampposts capture shadows of passersby and project one of these shadows when someone else passes)
- 2015 Urbanimals (jumping and crawling virtual animals are projected on walls and floors, inviting passersby to follow or imitate them)
- 2016 Stop Smile Stroll (initially announced as a disco experience when waiting before and during crossing a street, its actual implementation amounted to showing some 'painted' animations on your camera-captured face. These animations are shown on a display attached to a pedestrian traffic light pole)
- 2017 Star Light, Star Bright (implemented in Oxford; star constellations that are visible above Oxford are mapped on step-activated light installation, people have to work together to turn on a whole constellation).

2.2 Criticisms on Bristol's Playable City Vision

Although initially the contest seemed to be an open competition with 'idealistic', artistic and scientific aims, nowadays it has become a commercial enterprise and with shortlists and winners whose proposals have a high chance of commercial exploitation in as many cities as possible. This is a usual trend. Where initially the development of ideas about a smart city also included ideas about a 'creative city' that makes use and

---

[1] https://www.playablecity.com/.

stimulates the creativity of its population, for city authorities, fed by large ICT companies, the smart city view emphasizes efficiency, monitoring and control. As mentioned in [2] "… what if some smart initiative which started out as publicly funded and with social inclusion as goal, become overtaking by private sector concerns whose goal becomes purely profit-making?" That is, community interests are overtaken by developer's interest.

There are exceptions, for example the already mentioned 2015 winning Urbanimals installation designed by the Polish LAX laboratory and that required both advanced scientific and artistic creativity. But generally, using technology that was already available decades ago, the projects invite you to perform dance movements in the street or require you to smile in order to get directions from a signpost in the street. Pedestrian crossings and bus stops seem to be among the favorite aims of playable city designers and award assigning juries. From the Playable Cities website announcing the 2016 "Stop Smile Stroll" winner: "Stop Smile Stroll will transform a pedestrian crossing into a 30-s opportunity for sharing a moment of magic. It is a playful intervention at pedestrian crossings that brings strangers together for a moment of shared fun, breaking the mundane 'stop and walk' routine." The playability that is aimed at is meant to have pedestrians docilely accept their role of having to dance and smile on command, while traffic is left out of control. As it turned out, what was ultimately implemented were some isolated pedestrian traffic signs equipped with a camera and a display, where the camera captured the facial expression of the pedestrian and the display showed some funny animations on the captured face, not that different from what is offered by commercial photobooths that let you manipulate your picture.

The contest projects have drawn various kinds of criticism. Jesse Bert, Chairman of the Smart Cities Council, in 2013[2]: "Think of all the value they could have created. Directions to the nearest hospital. The arrival time of the next bus. The location of the closet public restroom. Suggestions on nearby sites and attractions. A way to report potholes and other problems. The resource "grade" of buildings, so residents can see which ones are energy-wasters. And on and on. Instead they have programmed in oh-so-cute play conversations with no value. On top of that, they are doing it via yesterday's SMS text messaging technology instead of tomorrow's smart phone platform. The one bright spot? The UK apparently has so much smart city funding that its cities can afford to flush money down the toilet." And, as a follow-up in 2014[3]: I don't have anything against the idea of "playable" cities… But can't we invent smart cities with "playable" features that actually add value to people's lives?"

We can add the more general observations of Feargus O' Sullivan [3] writing in the City Lab journal: "Playable interventions don't democratize cities. To do that, you'd have to fight far more serious threats than crosswalks that aren't fun enough. Your probable first move would be to battle the privatization and control of supposedly public spaces…", and, "Of course, there is one group for whom this stuff often works: children. The infantilizing frolics of playable infrastructure must be great if you're an actual infant." But, unfortunately, with a few exceptions, what is offered by the

---

[2] https://smartcitiescouncil.com/article/dumb-way-get-smart-bristol-launches-pointless-city-game.

[3] https://smartcitiescouncil.com/article/remember-talking-lamp-posts-meet-years-playable-city-winner.

shortlisted and award-winning designs, does not address playability for children, rather it is some kind of 'yuppy playability', playability for the happy and creative few, and playability that disappears from a city after a few months of installation. And Phineas Harper, deputy director of the UK Architecture Foundation, in 2017[4]: "These installations advance a trend of interactive public art commissions which claim to make the city "playable" and thus child-friendly. In reality these installations embody adult ideas of playfulness but fail to reflect the actual behaviour of children. They are like Thomas Heatherwick's rolling bridge – delightful but frivolous one-off follies primarily for adults' pleasure."

In [4] we wrote in the Wall Street Journal about "How to Make Cities More Fun." and we positively commented on the many playable city creative ideas. But the comments we received were not positive. One of the comments was that it made pedestrian crossings a perfect place for pick-pockets. More interesting was the following comment: "Are we really presumed by the political and corporate powers that be to be so void of our own interesting internal mental activity, that we need to be engaged by some contrived banality before we cross the street? … Or is it really more sinister than that? … that what we're being conditioned to do is to no longer rely on our own internal processes, but do become totally dependent on the stimulation of computerized devices."

The Playable City projects that have been introduced so far usually address persons that are familiar with advanced digital technology, persons that can move around freely without physical restrictions and can show bodily gestures, facial expressions, and emotions. However, many potential users or inhabitants of smart environments do not belong to this selective group of smart technology users and because of their age, physical characteristics or disabilities are excluded of being a 'full' participant of playful activities offered by smart urban environments. Recently we see initiatives [5–7] that provide visions for an inclusive accessible urban future, where inclusive includes older people, persons with disabilities [8] and children (child-friendly smart cities [9]). We still have to wait for similar visions from the "Playable City" initiative.

3   Playable Cities for All?

Children, let alone elderly or disabled persons, are not taken into account in the "Playable City" initiative. They are not present at openings when town authorities, cultural policy makers and designers give 'acte de présence'. Neither do we see them during evenings and nights when it is dark enough for some of the installations to be used or standing on tiptoe to become visible for a camera. Projects do not invite collaboration or social interaction. Projects are not co-created with potential users and there is no evaluation taking into account users' opinions. Projects are briefly available and then disappear from the streets.

---

[4] https://www.dezeen.com/2017/03/28/phineas-harper-opinion-children-playgrounds-play-predetermined-adult-designers/.

In cities but also in rural areas children disappear from the streets. Well, not really, they can be found in the backseats of cars playing with their smartphones and going from one children's location to the other. Streets are owned by cars, drivers have to be kept happy and public spaces or playgrounds do not allow children's initiatives that go beyond the risk-free pre-canned play and recreation facilities that they offer and have to be obeyed. No tinkering, no self-made soapbox cars, no rope jumping or hopscotch play, no ball games with self-invented rules or other self-structuring games. Many studies mention the positive associations of unsupervised play and independent mobility with the development of social skills (establishing social relations), safety skills and decision making skills or more generally, psychosocial (emotional), physical, and cognitive skills. In [10] it is mentioned that in most developed countries in the previous decades there has been a dramatic decrease in independent mobility of children (walking, cycling, or public transport). In studies positive associations are reported with gains in daily physical activity [11]. There are of course obvious reasons for the decline in independent mobility and unsupervised play: fear of child abduction, stranger danger, street violence, and careless drivers. If there are no activities and facilities in the own neighborhood, it leads to longer travel distances.

There are many initiatives to make cities more child-friendly and improve the infrastructure with the aim to achieve the benefits in child development that are mentioned above. There is the European 'Child in the City' initiative (https://www. childinthecity.org/) with its seminars and conferences on children's play in urban environments and its focus on childrens' needs in urban planning [9] and childrens' role as fellow urban citizens and as co-researchers of the city. Especially in the United Kingdom many projects can be mentioned that aim at making streets safer for children, reducing traffic, pedestrianising city parts and do not take the view that outdoor children play can only take place in fenced and hazard-proofing playground with equipment that has a prescribed way of play.

4 Conclusions

The playable city discussion has been dominated by examples from the yearly "Playable Cities" contest. We surveyed some of the possible criticisms on these projects and looked at the possibility that community interests is overtaken by developer's interest. This seems especially true when we look at what playable cities can mean for children and the many serious attempts to make cities playable again for children with the help of smart technology.

References

1. Nijholt, A.: Towards playful and playable cities. In: Nijholt, A. (ed.) Playable Cities: The City as a Digital Playground. Gaming Media and Social Effects, pp. 1–20. Springer, Singapore (2017)
2. Hollands, R.G.: Will the real smart city please stand up? City 12(3), 303–320 (2008)

3. O'Sullivan, F.: The Problem with 'Playable' Cities. https://www.citylab.com/design/2016/11/playable-cities-projects-crosswalk-party/506528/
4. Nijholt, A.: How to make cities more fun. Wall Street J. (2017) https://www.wsj.com/articles/how-to-make-cities-more-fun-1496163790
5. Korngold, D., Lemos, M., Rohwer, M.: Smart Cities for All: A Vision for an Inclusive, Accessible Urban Future. BSR report (2016). http://smartcities4all.org/wp-content/uploads/2017/06/Smart-Cities-for-All-A-Vision-for-an-Inclusive-Accessible-Urban-Futur...-min.pdf
6. Thurston, J., Pineda, V.: Smart Cities for All. A Global Strategy for Digital Inclusion Proposed by G3ict and World Enabled. Smart Cities for All Report (2016). http://www.g3ict.org/download/p/fileId_1040/productId_350
7. Fietkau, J.: The case for including senior citizens in the playable city. In: Proceedings of the International Conference on Web Intelligence (WI 2017), pp. 1072–1075. ACM, New York (2017). https://doi.org/10.1145/3106426.310
8. de Oliveira Neto, J.S., Kofuji, S.T.: Inclusive smart city: an exploratory study. In: Antona, M., Stephanidis, C. (eds.) Proceedings (Part II) Universal Access in Human-Computer Interaction. Interaction Techniques and Environments: 10th International Conference, UAHCI 2016, pp. 456–465. Springer, Cham (2016). https://doi.org/10.1007/978-3-319-40244-4_44
9. De Visscher, S.: The subjectification of the child friendly city. In: 8th Child in the City Conference, Ghent, Belgium (2016)
10. Shaw, B., Bicket, M., Elliott, B., Fagan-Watson, B., Mocca, E., Hillman, M.: Children's independent mobility: an international comparison and recommendations for action. Policy Studies Institute, London (2015)
11. Schoeppe, S., Duncan, M.J., Badland, H.M., Oliver, M., Browne, M.: Associations between children's independent mobility and physical activity. BMC Public Health **14**, 91 (2014)

Emotional Crowdsourcing Tool Design for Product Development: A Case Study Using Local Crowds

Shengfeng Qin[1(✉)] and Yuxuan Zhou[2]

[1] School of Design, Northumbria University,
Newcastle upon Tyne NE1 8ST, UK
sheng-feng.qin@northumbria.ac.uk
[2] The Design Lab, University of California San Diego,
9500 Gilman Dr, La Jolla, CA 92093, USA
yuz417@ucsd.edu

Abstract. Crowdsourcing is a useful tool for new product development and business innovation. To maximize its utility, regarding a crowdsourcing tool as an everyday thing used by crowds, in this paper, we integrate emotion and emotional design into a human-centred design process with local crowds and showcase an emotional crowdsourcing platform (tool) design for product development. In order to transform emotional design principles into the emotional crowdsourcing tool design, making it useful, usable and pleasurable, we first prototype a suitable emotion/emotional design process, and then identify a set of emotions and corresponding design features applicable for a crowdsourcing platform design and gain better understanding of common key concerns affecting the users' satisfaction of a platform. Finally, we identify a set of operational toolkits with emotional design features.

Keywords: Emotional design · Crowdsourcing · Platform design
Product development · Emotional design features · Human-centred design

1 Introduction

Demand for mass personalization of products now drives innovation of product design and development. One of the key enabling technologies is crowdsourcing-based digital platform for product design and development. However, existing crowdsourcing platforms for product design have some barriers for design and manufacturing businesses to adopt them in practice [1]. The main barriers are (1) effective involvement with crowd-workers and field experts, (2) the designers' trust of problem-solving ability and quality on a platform, and (3) the users' enjoyment of using such a platform

The authors would like to thank the financial support from the UK Leverhumle Trust's International Academic Fellowship project: IAF-2016-037 and the great supports from Professor Donald Norman, academic colleagues and supporting staff, researchers and the project participants at the Design Lab at UCSD.

(the users here refer to both the designers and crowd-workers). These factors together cause the overall satisfaction problem to a platform. Therefore, how to design and make a crowdsourcing platform (tool) appealing to both the business (the designer) and the crowd workers is a great challenge.

In this research, we regard a crowdsourcing tool as an everyday product used by both designers and crowds. Design of everyday product was previously focused on its usability, aiming to making the product useful and usable [2]. Since Norman [3] set forth the emotional design theory and principles, the design of everyday product has emphasized on emotional design for moving beyond usability to fun and pleasure. Norman's emotional design builds on the three interconnected layers: visceral design, behavioral design and reflective design. It is believed that Norman's emotional design theory can be applied to product and service design. However, there is no explicit emotional design process/method to guide emotional design. Now the question is how to transfer the emotional design theory into design practice [4], for example, a crowdsourcing-based new product development platform design in terms of its interface, key functions, tools, and interactions?

Emotion [5] is generally defined in terms of subjective experiences or feelings, goal-directed behaviors (attack, flight), expressive behavior (smiling, snarling), and physiological arousal (heart rate increases, sweating). Emotion is linked to motivation [5] and as a readout mechanism associated with motivation. Buck in [6] structures emotions into biological emotions (such as love and bonding) and higher-level emotions: Social emotions (such as pride), cognitive emotions (such as surprise, interest and curiosity) and moral emotions (such as feelings of distributive and retributive justice).

Closely related to emotional design, designing emotions [7] and designing pleasurable products [8] are premier references in design research. Desmet in [9] proposed a multi-layered model of product emotions including surprise emotion (such as surprise, amazement), instrumental emotion (e.g. disappointment, satisfactory), aesthetic emotions (such as disgust, attracted to), social emotions (e.g. indignation, admiration) and interest emotions (e.g. boredom, fascination) and developed the PrEmo tool [10] for measuring product emotion. In general, emotion as part of user experience, emotional design in practice is also associated with design for fun [11] and design for user experience (UX) [12] in the human computer interaction research field [13].

Since there exists no clear emotional design process for conducting emotional design practice but certain difficulties in quantitatively assessing emotions [14, 15] and emotional design [16], we have used general human-centered design processes and methods in our case study and conducted a qualitative study on our emotional crowdsourcing tool design. To gain a better understanding of the problem and identify possible solutions, we use the layout design of a kitchen bar with virtual crowdsourcing platform in the case study.

In this case study, we use local crowds in a co-design process, at the idea/concept development stages, they are acted as designers to conduct designers-driven emotion design, at the evaluation stages, they are acted as users to conduct user-driven emotional design. The relationships between the emotion design and emotional design can be found in [17].

The goal of this study is to integrate the designers-driven emotion design and the users-driven emotional design principles [17] into a human centered design process, apply it to a crowdsourcing tool design and hope to develop an emotional design process. The key contributions are following:

- Prototyping a suitable emotion/emotional design process
- Identifying a set of emotions and corresponding design features applicable for a crowdsourcing platform design
- Understanding of common key concerns affecting the users' satisfaction of a platform
- Identifying a set of operational toolkits with emotional design features.

2 Related Work

In order to make a crowdsourcing platform appealing to use, the related work in crowdsourcing field includes: gamification, pricing, incentives, notifications, engagement improvement and task recommendations.

Gamification. Both gamification and crowdsourcing mechanisms share common features in engaging users to participate and contribute, thus, gamification [18] was used in crowdsourcing as engagement techniques for human rights organizations. Aiming to improve performance of campaigning systems of human rights organization, Zeineddine in [18] investigated psychological motivations behind user's engagement. The study provided system design insights in not only crowdsourcing but also interactive social platforms.

Researchers [19] from the University of Amsterdam and IBM Research team combined gamification techniques and crowdsourcing to create an engaging game "*Dr. Detective*" with a gold standard in medical text. The research proposed a design for a gamified crowdsourcing workflow to create an annotated version of medical text. Results from a pilot study have confirmed the usefulness of this gamified crowdsourcing model in medical field. While researchers [20] from the University of Bath argued that to support fairness in collaborative online systems, visual representations like meters in an online game can enable basic inferences about contributions and fairness. This finding is constructive for designing collaboration mechanisms in terms of gamification on the crowdsourcing platform.

Pricing. The work in [21] studied pricing mechanisms for crowdsourcing markets, which introduced the basic framework of designing mechanisms for crowdsourcing market. To maximize tasks and minimize payments, the researchers came up with the platform Mechanical Perk (MPerk) on top of Amazon's Mechanical Turk (MTurk), which provides a good basis for designing pricing mechanisms for crowdsourcing markets that can be even further extended. Others [22] developed an efficient and truthful pricing mechanism (TruTeam) for team formation in crowdsourcing markets. Among four different incentive mechanisms: profitability, individual rationality, computational efficiency and truthfulness, simulations confirm that the superior mechanism is an efficient and truthful pricing mechanism for crowdsourcing markets.

Incentives. The problems in designing effective incentives are discussed in [23]. While existing research have used incentives to improve participation, this research demonstrates the choice of incentives may introduce a bias against different participants demographically. This study suggests utilizing rewards to target desired participants and using various incentives to improve participation diversity. It is noticed [24] that incentives of workers, which are not aligned with requesters, are affecting the quality of crowdsourcing work. To address this issue, researchers in [24] introduced approval voting coupling with incentive-compatible compensation mechanism. The incentive mechanism is proven through the preliminary experiment conducted on Amazon Mechanical Turk. It is also found in [25] that per-task payments on crowdsourcing platforms reduce productivity, although paid crowdsourcing marketplaces are gaining popularity. Among three tested incentive approaches including methods of bulk, coupons, and material goods, the study found that task completion rate increases when participants get paid. Material incentives generally decrease participation over time. Therefore, it is proposed to apply alternating payment mechanisms, which can lead to more work output and higher earnings for workers.

Notifications. The study [26] indicated that information notification system of crowdsourcing is potentially an important tool for developing nations to utilize. However, current limitations such as unstable electronic communication infrastructure are hindering the information providing in real-time notification system in developing countries. To overcome this issue, the study proposed to use SMS as a reliable method to deliver real-time message. The prototype of this idea used prediction algorithm of user behaviors to ensure a better reward and engaging platform.

It is also noticed in [27] that designing effective performance feedback notification systems can stimulate content contribution. Feedback framed either pro-socially or pro-self has positive effect on content contributors, with a gender different response to competitively framed feedback. This finding provides implication on designing notification system of crowdsourcing platform.

Engagement Improvement. In [28], authors introduced *taste-matching* and *taste-grokking*, two crowdsourcing approaches that are designed to capture personal preferences. Over the testing of generic users, both approaches showed improvement in engagement despite different advantages and drawbacks depending on complexity and variability of the task. The problem of how to engage users' interests besides extrinsic rewards is also studied [29]. The research results suggest positive relationships between user's motivation and engagement to a task, and between user's interest to the topic and their motivation. The results provide a preliminary design guideline for a crowdsourcing environment. The idea of intrinsic benefits is essential to helper when the contribution is voluntary. On the other hand, organizations who seek help can increase task attractions by framing the issue in an engaging way that may seem more interesting to the contributors.

Task Recommendation. The importance of having task recommendation in crowdsourcing systems was discussed in [30]. The research results show that combining worker performance history and task searching history can better reflect accurate recommendations that suit worker's interests. Creation of multi-label classification and taxonomy [31] is also important to task recommendation. By tracking the potential helper's location, the location-based recommendation technique [32] can direct people to nearby regions where help is needed and direct their attention based on previous task history for better task recommendation.

In general, there lacks emotional designs approach in crowdsourcing platform design. Here, we focus on the crowdsourcing platform design issues with a new lens of emotional design in the context of supporting new product development.

3   Research Method Design

We follow the human-centered design principles and apply the research-through-design approach to design our investigation processes. The research activities were carried out through 4 workshops and 2 focus groups over 10 weeks. These activities follow the double-diamond design process model in the case study.

Case Selection. At the preparation stage, first, we used a group discussion among the researchers in the Design Lab, University of California, San Diego to decide to convert the current Design Lab into a kitchen bar as a business proposition and outsourcing the kitchen layout design as a case study. The reasons for choosing a kitchen bar design for the case study has twofold. One is that the primary functions, appliance and furniture associated with a kitchen bar are easy to understand for participants. The other is that the workshops and focus group studies will be conducted in the Design Lab thus the participants are easy to refer to it during their research. In addition, based on the case selection, we designed and produced a mini-project briefing document including the following information: the project lead researcher, the Project title-A Case Study of Transferring Norman's Emotional Design Theory into a Crowdsourcing Platform Design, the project background on emotional design and crowdsourcing, the project key purposes/activities and the project time line.

The case study was to redesign the current Design Lab as a Kitchen bar and create a new layout design for the kitchen bar in the Atkinson Hall at UCSD campus. Dimensions could be measured by crowd researchers/participants. Assuming a kitchen layout design project manager can access a crowdsourcing platform and the platform managers to help carry out the kitchen bar layout design from the crowds (both individual crowds and business crowds-company users).

Participants as Crowd. Next, we recruited the workshops/case study participants. There were around 150 research assistants registered on the Design Lab and potentially available for the project. We then issued a call for project participants via email to all of them with the project briefing document. We received 25 replies to express their interests in participation. We then used Doodle to vote and identify most common available time. Based on the most available time, the project selected 17 workshop participants. They are all undergraduates (about half of them are male), 12 majoring in cognitive Science-HCI, 1 in Communication, 2 in Math/Computer Science, 1 in management and Business Studies and 1 in BioEngineering/BioInformatics. We also recruited an Interior Designer and design researcher to participate the workshops. So in total, we had 18 participants in our workshops.

Co-creation. At the second stage-workshop studies, before a workshop started, we published the workshop activity plan with main questions to answer and related background and research information. The workshop activity plan was published as Google Docs for all participants to share, co-creation and co-editing. After the workshop, the outcomes from the workshop were summarized by the Lead researcher and included in the next consecutive workshop activity plan. Therefore, next workshop activities are well connected to previous studies. Workshop activities include a workshop briefing, brainstorming, group discussions, post-it and voting, and prototyping and simulations.

Use Project Stage-Gates Model as Guiding. During this second stage, we spent four workshops to explore (1) what are emotions that could be embedded in a crowdsourcing platform (2) what are key problems along a project development line and (3) what are key functions/tools needed for solving the problems. Most of participants know crowdsourcing systems such as Amazon MTurk, and Human Intelligence Tasks or HITs based problem-solving styles. Arguably, this HITs-based crowdsourcing tool is not very popular in product design field because it lacks supports for a design project development as a whole. While our emotional crowdsourcing tool is expected to support activities along a whole design project development process, thus, in our research, we use the project stage-gates model to guide our explorations with six stages: Discovery, Scoping, Business Case, Development, Test and Launch.

In the above studies, crowds worked as the crowdsourcing platform designers to conduct *emotion design* as discussed in [17].

Simulation. At the third stage, we conducted two focus-group-based simulation studies. Each focus group had 7 participants (picked from the previous workshop participants). In the first focus group, each participant was assigned a role working with the crowdsourcing tool, and as role players, participants simulated key activities for his/her role along a product design project process based on the six stages on the Stage-Gate model. The purpose of this Focus group was to produce a list of key functions/tools needed. The second focus group simulated a kitchen bar layout design via a prototype crowdsourcing tool to verify, refine and update the key functions/tools list.

In the last focus group, first, the key functions/tools were ranked based on the participants' voting. Then the 6-3-5 brainWriting method was used to create ideas to develop emotional design features for 6 top ranked operational tools and finally, the participants voted for best ideas.

In the simulation studies, crowds worked as the end users of the virtual platform to conduct *emotional design* as discussed in [17].

4 Experiments and Results

Emotional Design Process. We used the workshop 1 for participants to understand business needs/user needs at the top stage level of new product development. Research activities include (1) project briefing, (2) Norman's TED talk on Emotional Design (on youtube), (3) demonstration of three crowdsourcing systems: Slack, CrowdSpring and MTurk, (4) human-centered design principles, and (5) emotional design features. The above activities helped the participants to understand different role players such as business requester, crowdsourcing platform and the platform managers, the crowds and helpable companies (business crowds). After the above, we used open discussion, brainstorming, group discussion, post-it and card-sorting to identify an emotional design process for guiding the following-up research activities.

The research results show that the emotional design process has two parts A and B. Part A is about *emotion design* to identify good design experience and emotions from learning and researching and in turn identify good design features. The steps include (1) collect ideas of what good experiences have being used in similar services (learning from experience) (2) identify what design features which enables delivering/provoking good experience, and (3) find out how we can apply such features into our design. Part B is about *emotional design* to apply a human centered design process to a pleasurable UI/UX design for computer supported collaborative work. The key steps are: (1) need finding, user interviews, survey, observation, personas + storyboarding (2) design with low-fi and hi-fi-prototypes, (3) user (usability) testing with observation, and (4) more iterations incorporated with user feedback.

Pleasures, Emotions and Design Features for Consideration. In the second workshop, Participants followed the emotion design process to first identify the pleasures and emotions which could be synthesized in a crowdsourcing tool and then mapped them onto four levels of pleasures by using Designing Pleasurable Product as reference. The goal is to make the tool useful, usable and delightful. The results from discussion, brainstorming and co-editing, are shown in Table 1. Emotions and their associated design features are shown in Table 2.

Table 1. Pleasures and exemplar design features

Expected pleasures	Exemplar design features
Physio-pleasure: sights, sounds, smells, taste, and touch	Colors, shift function, aesthetically pleasing layouts, flat design, minimalism
Soci-pleasure (derived from interaction with others). It combines aspects of both behavioral and reflective design	Facebook/Slack reactions, Facebook mutual friends/friend suggestions, real time notifications, Twitter mentions, Facebook/Instagram likes, tagging in memes, posting on social media
Psycho-pleasure (deals with people's reactions and psychological state during the use of products at the behavioral level)	Social Media reactions and likes, share features, security of data, context-relevant things, discounts/reward system, daily check-in reward system, Attention-holding data stream
Ideo-pleasure (signify the value judgements of their owner at the reflective level)	Rating systems, mutual feedback (Uber/Lyft), people you may know, Invite/Referral based systems, secret groups/events, and premium accounts on websites

Table 2. Emotions and design features

Emotions	Associated design features
Fun	Humor, emoji communication
Happy, proud	Personalization, intonation, reward system
Respect/love	Passion
Surprise/joy	Easter eggs, storytelling

Common Key Concerns from Both Designers and Crowd-Workers. The third workshop was used to identify typical application scenarios/questions with reference to a design project manager (business requester) might ask for help during a layout design on a simulated platform. The exploration followed the project phase-gate model with mapping to the kitchen design project process model.

We first introduced a kitchen design or in general interior design process and key concerns. Some reference books and kitchen design websites were briefed beforehand. Then the participants were asked either work individually or as a group to discuss, co-create, co-edit and co-work on the Table 3. As a result, it shows the possible key questions on a mapped stage model: S-scoping, B-building business case, D-development, E-evaluation, and P-product launch. We also ask the participants to indicate who are qualified for answering questions, the answers show that qualified crowds have different profiles such as professional knowledge and experience.

Table 3. Typical questions

Stages	Key questions look like
S	With reference to the context information, what are key functions the product/system should have? what are estimated costs of developing this product? how many staff are required to run the system/product? could you please vote for top answers for the above questions?
B	Based on the project information, what is your schematic design solution? how many functions are designed in your solution? how long does it take to develop into a full design solution? how much does it cost? could you please vote for top schematic design solutions and comment on your rankings?
D	With the schematic design of the product and project information, what is your full design solution? what are your pickings of appliance? what are information/emotional design embedded? what are your design highlights? how much/long does it need to...? could you please vote for top design solutions and comment on your rankings?
E	Based on the crowd voting from the prior stage, select 2–3 full designs for evaluation including expert reviewing/design reviewing, what is your ranking and comments on each design solution?
P	With new product information, for launching the product with the key product features, how to best organize an onsite launch event? what are best social media platforms for a launch? how to do it?

The research results show that designing a product design crowdsourcing platform needs to refer to its design context in term of process, key design issues, knowledge and skills required, standards and key references, etc. This in turn explains why general-purposed crowdsourcing platforms-based HITs are not suitable for product design.

Key Functions/Operational Tools Needed. From the key questions listed in Table 3 can be seen that at each stage, key questions are asked with different orders, therefore, the key questions are context-sensitive and progressive. This kind of features is well mapped to the phase-gate model. Between two phases, there is a gate for decision such as voting or ranking solutions.

At the workshop 4, we simulated crowdsourcing working scenarios based on key questions and possible answers and identified key tools required to support the design project development in an emotional design process. We discussed and agreed on the general relationship between UX, design for fun, emotional design, gamification, and emoji communication. And in general, we believe their relationships are

- *Positive emotion* → *positive mood* → *positive experience*
- *Gamification* → *fun* → *promoting emotion* → *experience*

We divided participants into 4 groups, each representing a role player then we used group discussion, co-creation, group brainstorming, card sorting, voting and co-editing for most useful design tools/functions. The result is a list of functions and tools for performing specific tasks. Users are classified as A—platform administrators, B-business requesters, and C-Crowds.

Evaluation of Key Operational Tools and Updating. We conducted a focus group study to observe and record crowdsourcing-based design activities on a simulated crowdsourcing platform with the identified tools in Table 4 to refine and update key tools/functions. We used two tables with a splitter to simulate a virtual working environment (without face-to-face communication) between team members. Participants are 8 people: A-the platform administrator, B-Business requester (or a design project manager), O-an observer for the HCI test (or evaluation), C1 and C2 are two team members working on the Task 1, C3 and C4 on the Task 2, while C5 is a single participant working on T1. Task 1 is to design a kitchen bar from existing layout. Task 2 is to design a kitchen bar from a blank layout.

We used two types of cards to carry out all communications on the crowdsourcing platform. On a card, a co-worker in a team will write down information he/she want to pass and indicate what communication tool/function he/she will use (a White card is for a traditional tool to be used) while a Pink card is for an emotional tool to be used). Tools provided for use include floor plan with existing layout design, blank floor plan, a set of White communication/tool cards, a set of Pink communication/tool cards, pencils and pens, rulers, crayons/color coding labels, scissors and glue.

After analyzing information on the communication cards, we obtained the updated tools (See Table 5) by integrating results from Table 4 and the focus group study.

Note that it is important to have a team-building tool and provide a team message board. A private team share space and a team message board (either the platform provided or social media platform) are used for discussion and co-design.

Table 4. Key functions and tools

Users	Functions/tools
A + B + C	Key facts demo/application stories, Pricing policies, Normal legal/engagement terms, IP Policies, Online forum for feedback/improvement, Messaging board for crowd questions, Facebook feeding on participation activities
B	Registration tool/profiling tool, Pricing tool, Expert reviewing tool
A	Registration tool
B + C -> A	Communication tools: email + phone/social media platforms
B + C	Introduction to the platform, Image + file uploading
A + B	Team tool-flexible control of crowd profiles
C	Pricing tool, Voting tool/Voting rewards, Registration tool/profiling tool
Back-End	Task evaluation, Task recommendation

Table 5. Updated key functions/tools

Functions/tools	Possible realization methods
Introduction to the platform	Animation
Key facts demo	Interactive video, cartoon/graphics
IP Policies	Reference templates
Online forum/Messaging board	Project message board (PMB)
Social media tools	Private message
Pricing tool	3rd party tool
Review tool	Pitching tool, sharing tool
Registration tool/profiling tool	Intelligent business/crowd profile building
Team-up tool	Team-building tool
Team collaboration	Team message board
Work-sharing, annotating	Document sharing and co-editing tool
Voting tool with rewards	Action recognition tool
Rewarding tool	Expectation estimation tool
Task recommendation	Crowd evaluation tool and task recommendation tool
Task evaluation	Task decomposition and evaluation tool

Key Operational Tools with Emotional Design Features. In the second focus group, we first co-voted on Google Doc for most important tools which should have emotional design features. The result is shown in the Table 6 with higher voting items.

Table 6. Rated functions/tools with emotional design

Functions/tools	Need emotion features
Basic communication tools (email + phone)	*****
Online forum: message & discussion	***
Messaging board for crowdsourcing questions	*****
Team-building tool	***
Work-sharing & annotating	***
Task recommendation	***

5 Discussion and Conclusion

This paper firstly works through a case study using human-centered design and UX design as guides to develop an integrated emotion and emotional design process and identify common question types for a product design project. Secondly, by simulating real design scenarios on a virtual crowdsourcing platform, it identifies the key functions/tools and associated emotional design features. The research results imply that the design process for a general purposed crowdsourcing platform is different from the platform for a product design with requirements on contextual understanding and emotional design. The integrated emotion and emotional design within a human-centred design process sheds light on this direction.

In summary, this paper demonstrates (1) what are design processes to follow for emotional design, (2) what are emotions and corresponding design features could be used for a crowdsourcing platform to support a product design, and (3) what are common interaction tools which should have emotional design features. Through the case study, we have answered the above questions and gained better understanding of a crowdsourcing platform for product design.

References

1. Qin, S.F., Van der Velde, D., Chatzakis, E., McStea, T., Smith, N.: Exploring barriers and opportunities in adopting crowdsourcing based new product development in manufacturing SMEs. Chin. J. Mech. Eng. **29**(6), 1052–1066 (2016)
2. Norman, D.: The Design of Everyday Things. The MIT Press, Cambridge (1998)
3. Norman, D.: Emotional Design-Why We Love (or Hate) Everyday Things. Basic Books, New York (2004)
4. Qin, S.F.: Transforming emotional design principles into innovative product development platform design. In: CSCW 17 Workshop on Theory Transfers? Social Theory and CSCW Research. ACM, Portland (2017)
5. Buck, R.: Emotion-A Biosocial Synthesis. Cambridge University Press, Cambridge (2014)
6. Buck, R.: Prime theory: an integrated view of motivation and emotion. Psychol. Rev. **92**(3), 389–413 (1985)
7. Desmet, P.: Designing Emotions. Delft University of Technology, Delft (2002)
8. Jordan, P.W.: Designing Pleasurable Products: An Introduction to the New Human Factors. Taylor & Francis, London (2000)
9. Desmet, P.: A multilayed model of product emotions. Des. J. **6**(2), 4–13 (2003)
10. Desmet, P.: Designing emotions. Des. J. **6**(2), 60–62 (2003)
11. Shneiderman, B.: Designing for fun: how can we design user interfaces to be more fun? Interaction **11**, 48–50 (2004)
12. Treder, M.: Beyond wireframing: the real-life UX design process. Smashing Magazine
13. Marcus, A.: Emotion commotion. ACM Interact. **10**, 28–34 (2003)
14. Shashidhar, G., Koolagudi, K., Sreenivasa, R.: Emotion recognition from speech: a review. Int. J. Speech Technol. **15**, 99–117 (2012)
15. Dukes, D., Clément, F., Audrin, C., Mortillaro, M.: Looking beyond the static face in emotion recognition: the informative case of interest. Vis. Cogn. **25**(4–6), 575–588 (2017)

16. Desmet, P., Overbeeke, K., Tax, S.: Designing product with added emotional value: development and application of an approach for research through design. Des. J. **4**(1), 32–47 (2001)
17. Ho, A.G., Siu, K.W.M.: Emotion design, emotional design, emotionalize design: a review on their relationships from a new perspective. Des. J. **15**(1), 9–32 (2012)
18. Zeineddine, I.: Gamification and crowdsourcing as engagement techniques for human right organizations. University of Gothenburg thesis (2012)
19. Dumitrache, A., Aroyo, L., Welty, C., Sips, R.J., Levas, A.: Dr. Detective. https://research. vu.nl/ws/files/1134293/ISWC2013.pdf
20. Kelly, R., Watts, L., Payne, S.J.: Can visualization of contributions support fairness in collaboration? Findings from meters in an online game. In: Proceedings of ACM CSCW 2016, pp. 664–678. ACM, San Francisco (2016)
21. Singer, Y., Mittal, M.: Pricing mechanisms for crowdsourcing markets. In: Proceedings of ACM WWW 2013, pp. 1157–1166. ACM, Rio de Janeiro (2013)
22. Liu, Q., Luo, T., Tang, R., Bressan, S.: An efficient and truthful pricing mechanism for team formation in crowdsourcing markets. In: Proceedings of IEEE International Conference on Communication - IEEE ICC 2015, pp. 567–572. IEEE Xplore, London (2015)
23. Hsieh, G.; Knocielnik, R.: You get who you pay for: the impact of incentives on participation bias. In: Proceedings of ACM CSCW 2016, pp. 823–835. ACM, San Francisco (2016)
24. Shah, N.B., Zhou, D., Peres, Y.: Approval voting and incentives in crowdsourcing. In: Proceedings of the 32nd International Conference on Machine Learning, vol. 37, pp. 10–19. ACM, Lille (2015)
25. Ikeda, K., Bernstein, M.S.: Pay it backward: per-task payments on crowdsourcing platforms reduce productivity. In: Proceedings of CHI 2016, pp. 4111–4121. ACM, San Jose (2016)
26. Singh, A., Li, Y., Sun, Y., Sun, Q.: An intelligent mobile crowdsourcing information notification system for developing countries. In: Xin-lin, H. (ed.) Machine Learning and Intelligent Communications, MLICOM 2016. LNICST, pp. 139–149. Springer, Cham (2017)
27. Huang, N., Gu, B., Burtch, G., Hong, Y., Liang, C., Wang, K., Fu, D., Yang, B., Lan, W.: Designing Effective performance feedback notification systems to stimulate content contribution: evidence from a crowdsourcing recipe platform. In: Proceedings of the 50th Hawaii International Conference on System Sciences, pp. 1453–1462. Hawaii (2017)
28. Organisciak, P., Teevan, J., Dumais, S., Miller R.C., Kalai, A.T.: A crowd of your own: crowdsourcing for on-demand personalization. In: Proceedings of the Second AAAI Conference on Human Computation and Crowdsourcing, pp. 192–200. AAAI (2014)
29. de Vreede, T., de Vreede, G.J., Reiter-Palmon, R.: Antecedents of engagement in community-based crowdsourcing. In: Proceedings of the 50th Hawaii International Conference on System Sciences, pp. 761–770. Hawaii (2017)
30. Yuen, M.C., King, I., Leung, K.S.: Task recommendation in crowdsourcing systems. In: Proceedings of CrowdKDD 2012, pp. 22–26. ACM, Beijing (2012)
31. Bragg, J., Weld, D.S.: Crowdsourcing multi-label classification for taxonomy creation. In: Proceedings of the First AAAI Conference on Human Computation and Crowdsourcing (HCOMP 2013), pp. 25–33. AAAI (2013)
32. Kim, Y., Harburg, E., Azrla, S., Gerber, E., Gergle, D., Zhang, H.: Enabling physical crowdsourcing on-the-go with context-sensitive notifications. In: Human Computation and Crowdsourcing: Works in Progress Abstracts: An Adjunct to the Proceedings of the Third AAAI Conference on Human Computation and Crowdsourcing, pp. 14–15. AAAI (2015)

Shanghai Shikumen Cultural and Creative Product Design Based on Design Thinking

Qianqian Wu, Zhang Zhang[✉], and Li Xu

East China University of Science and Technology,
Shanghai 200237, People's Republic of China
873616093@qq.com, zhangzhang@ecust.edu.cn,
yate9999@hotmail.com

Abstract. The main purpose of this paper is to design cultural and creative product in globalization by using the method of design thinking on Shanghai Shikumen culture. Nowadays culture is a core element in enhancing the city's attractiveness, competitiveness, influence and soft power. As an international metropolis that plays an important role in the Belt and Road initiative, Shanghai urgently needs to develop its cultural industries and promotes the development of cultural creativity. Shikumen, as a traditional residential area combining Western and Chinese elements in Shanghai, is a microcosm of Shanghai modern history and culture. Shikumen remains in the memory of the old Shanghai people. And it is also a model of East-West cultural integration. Shanghai Shikumen culture is studied in this paper. The paper consisted of the following steps: (1) Exploring the history of Shikumen origin by means of literature review, and refining its intrinsic culture and emotional elements. (2) Defining the target population by making field visits and on-site interviews. (3) Designing related cultural and creative product making use of Design Thinking based on the real needs of consumers. And eventually Shikumen culture can be promoted by making related product design. A more distinctive Shanghai cultural image can be created. And it can also help facilitate the cultural exchanges between East and West.

Keywords: Shanghai Shikumen · Cultural and creative product design
Cultural elements · Design thinking

1 Background

Since the 21st century, with the development of technology and the improvement of people's living standards, people's consumption level is continuously increasing. Consumptive structure also produced tremendous change. The new economy with the keywords of "innovation", "knowledge" and "culture" came into being. There developed a "consumer-led economy" which is represented by new technology, new industries and new formats. Under this background, creative industries arise and grow rapidly. A more complete definition of "creative industries" was put forward in the UK 2001 Creative Industries Mapping Document. It defined the creative industries as "those industries which have their origin in individual creativity, skill and talent and which have a potential for wealth and job creation through the generation and exploitation of

© Springer International Publishing AG, part of Springer Nature 2019
S. Fukuda (Ed.): AHFE 2018, AISC 774, pp. 33–40, 2019.
https://doi.org/10.1007/978-3-319-94944-4_4

intellectual property" [1]. Now it is generally believed that cultural and creative industries are equivalent to creative industries. Cultural and creative industries not only have their unique cultural values, but also contribute to economic growth. It plays an important role in the country's economic restructuring. Cultural and creative products convert intangible assets into tangible assets through pricing and sales, promoting economic and employment growth, creating economic value for the cultural and creative industries.

Under the globalize background, culture has become one of the important factors in establishing the national image and enhancing the country's influence. Chinese President Xi Jinping put forward that we should strengthen cultural confidence, promoting the prosperity and prosperity of socialist culture in 19th CPC National Congress. China must vigorously develop its cultural and creative industries and enhance its soft power. As one of the six major international super cities, Shanghai is an important junction of the "Belt and Road" and the Yangtze River Economic Belt [2]. It urgently needs to develop its cultural industry and enhance its urban attractiveness. In December 2017, the CPC Shanghai Municipal Committee and the Shanghai Municipal People's Government promulgated "50 Articles of Shanghai Culture and Creation" focusing on the development of Shanghai's modern cultural and creative industries. It is expected that in the next five years, the added value of cultural and creative industries in the city will account for the proportion of the city's total output value 18% or so. By 2035, a cultural and creative industry center with international influence will be built in an all-round way [3].

Shanghai has a unique "Shanghai School Culture" which integrating Chinese and Western culture. It even formed a unique residential in the course of historical development: Shikumen. It was the basic residential of modern Shanghai citizens. Since the opening in Shanghai, Shikumen has carried generations of memories. It is not only a microcosm of Shanghai's modern history, but also a symbol of modern cities. Therefore, it can provide good material for the development of Shanghai cultural and creative product design. But first of all we need to understand the historical origin of Shikumen and clearly define the cultural foundation of the Shikumen architectural features. This paper aims to apply Shikumen culture elements to the design of cultural and creative products with appropriate form of expression based on the full understanding of Shikumen. Create Shikumen creative design with Shanghai characteristics.

2   Introduction About Shanghai Shikumen

2.1   History of Shanghai Shikumen

Shanghai Shikumen originated in the mid-19th century. In September 1853, Shanghai knife association happened. Since 1860, the Taiping troops successively took over Suzhou and Hangzhou. Refugees in cities and towns such as Shanghai, Jiangsu and Zhejiang avoided eviction of the concession and sought asylum. Refugees in places like Shanghai and Jiangsu had to move into the concession to seek asylum. With the dramatic increase in the number of people in concession, there was a serious shortage of housing. Concession traders took the opportunity to build a large number of simple wooden houses and hired them to refugees for profit. In order to make full use of the limited land to create more benefits, the wooden houses are generally arranged in a row of European terraced houses. However, it is extremely unsafe because wood can easily lead to a firehouse. On the basis of retaining the original arrangement and layout, the wooden houses changed into Brick houses. These brick houses are the prototype of Shanghai Shikumen. Scholars generally agree that Xingrenli, built in 1876, is the oldest Shikumen house.

The uniqueness of Shikumen house is mainly reflected in the stone doors with different styles. Why do we call it Shikumen? There are different versions. There is a view that the ancient imperial palace has five doors and princes' palace has three doors, both the palace's outermost door, called "kumen". So we call this type of residence "Shikumen" ("stone" is pronounced "shi" in China) [4]. Another view is that in Shanghai Dialect, the package or bundle other things are pronounced "ku" [5]. Therefore, houses with stone-framed are commonly known as "Shikumen". It is the main living space of modern Shanghai citizens. According to statistics, in 1945 Shanghai Shikumen accounted for 70% of the total residential area of the city.

There are three main stages in the development of Shikumen: Early Shikumen (1870–1908), late Shikumen and Shikumen alleys (1909–1930) and garden alleys (1930–) etc. With the development of history, Shikumen has been influenced by Western culture in the interior layout and decoration of buildings, which gradually showing the integration of Chinese and Western cultures.

2.2 Cultural Elements in Shikumen

The layout of a separate Shikumen house absorbed some characteristics of traditional houses in China. It is often built into two-story buildings with symmetrical layout based on the central axis, which is similar to the layout of traditional houses in China. Walls in front and behind are up to 5 m, forming an almost enclosed space. Residents can still feel quiet although Shikumen built in downtown. Due to historical reasons for its

Fig. 1. Similarities between Shikumen and the arrangements of European terraced houses

formation, Shikumen adopted the arrangements of European terraced houses, which is also the prototype of Shikumen alleys. You can find the similarities between two arrangements in Fig. 1.

Shikumen outward appearance, decoration, etc. are affected by the impact of Chinese and Western cultures, too. Early Shikumen residential decoration is similar to the decoration of traditional southern residential, mainly decorated with animal & plant.

While the later Shikumen influenced by Western architectural culture, door decoration began to use a large number of Western classical elements such as parallel circles, semi-circles, bow-type circle [6]. Moulding also changed from complex to simple. However, Shikumen decoration in the process of learning from Western decorative elements, made some man-made changes. As shown in Fig. 2, the pure semicircle and acute triangle arch volumes in Western classical architecture have been transformed into the shape of non-semicircular and obtuse triangle in Shikumen. It can be seen that Western culture pursuing individuality, like to use full shape on architectural decoration. However, China pursued "normality", symmetry and evenness in its culture of "normality". So the shapes such as non-semicircle (less than semicircle) and obtuse-angled triangles were used on Shikumen decoration. Shikumen door is decorated with a large number of horizontal lines and diamond on decoration, showing balance and simplicity. To some extent it is inspired by Chinese characters [7]. This further proves Shikumen is a combination of Chinese and Western culture.

Doors of
Western classical architecture | Shikumen

Fig. 2. Different doors in Western classical architecture and Shikumen.

2.3   Emotional Factors in Shikumen

Japanese literature master Okuno Takeo mentioned the concept of "original scenery" in the "original scenery in the text", that is, things formed and fixed in a deep sense inadvertently, maintaining the visual pleasure brought by the original appearance of things, creating comfortable and warm spaces in daily life without destroying nature [8]. The vitality of the city is closely linked with the vitality of the streets and the people. Just as the unique "Nongtang" culture of Shikumen has its vitality. Shikumen appeared along with Nongtang ("alleys" is pronounced "Nongtang" in China, refers to the passage of public transport space) [9]. Anyi Wang in "Everlasting Regret" fully drew the old Shanghai Nongtang style. Therefore, Shikumen not only represents the culture of Shanghai, but also is a carrier of old Shanghai people's memory.

On one hand, Shikumen Nongtang is a place for public recreation. On the other hand, it is an important trading space for traders. Residents enjoy the cool nights in the Nongtang, play chess or chat with each other. Children play all kinds of mini games. Merchants streamed into Nongtang, selling goods at their stalls from morning till night. The stalls that selling picture-story book and stalls selling special snacks were numerous. Figure 3 can restore part of the sale of traders trading.

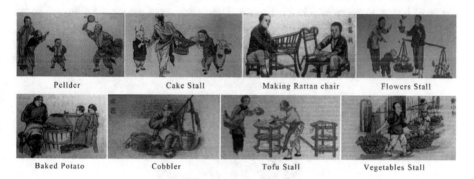

| Pellder | Cake Stall | Making Rattan chair | Flowers Stall |
| Baked Potato | Cobbler | Tofu Stall | Vegetables Stall |

Fig. 3. Traders in Shikumen Nongtang.

3 Shikumen Cultural and Creative Product Design

Tim Brown, IDEO's CEO, applies design thinking into design and defines it as "a discipline that uses the designer's sensibility and methods to match people's needs with what is techno-logically feasible and what a viable business strategy can convert into customer value and market opportunity" [10]. The main steps in the application of design thinking are "Empathize", "Define", "Ideate", "Prototype" and "Test". This paper also applies the relevant method of design thinking to design Shikumen creative products based on the full research.

3.1 Empathize and Define

This article first distributed network questionnaire to know whether people understand Shikumen culture well, their purchase intention, frequency and channels of cultural and creative products. A total of 97 questionnaires was distributed. The target population mainly concentrated in 18–30-year-old higher education population. After analyzing the relevant data, we can draw the following conclusions: More than 80% do not know much about the Shikumen culture in Shanghai. The main channels of understanding are the network, television, typical products of the market (e.g. Shanghai Shikumen wine). In response to Shikumen's cultural development, 71% supported the development of related cultural and creative industries. Meanwhile, about 68% also pointed out the need to raise publicity awareness and increase cultural awareness. About 95% of people purchase cultural and creative products occasionally. Therefore, although the public support the development of cultural and creative industries, they are lack of purchasing power. This reflects the lack of good cultural and creative products in the market. When asked about the product preferences, they specifically pointed out the product's price ratio, practical functions and cultural information.

At the same time, we interviewed a few local college students in Shanghai, and two residents who had previously lived in Shikumen. Those young college students do not know much about Shikumen, which reflects that there are not a lot of policies in promoting Shikumen cultural development. The two residents who once lived in Shikumen are most impressed by Shikumen Nongtang culture. All of them constitute the most beautiful memories in their mind. They also mentioned that Shikumen in the present is no longer in the past. Some Shikumen protection projects such as Tianzifang and Xintiandi are commercialize. Neighborhood relations are no longer as close as they used to be. Based on the above investigation, this paper sorted out the main needs of cultural and creative design as indicated in Fig. 4.

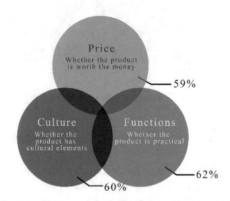

Fig. 4. The main needs of cultural and creative design.

3.2   Practice-Ideate and Prototype

The mind map, as pictured in Fig. 5, showing the main features of Shanghai Shikumen based on the data from previous literature searches and research interviews. We can clearly identify several important features of Shanghai Shikumen: the integration of Chinese and Western culture, unique nongtang culture showing harmonious interpersonal relationship and doors with different forms. And these are also the key points to creative design.

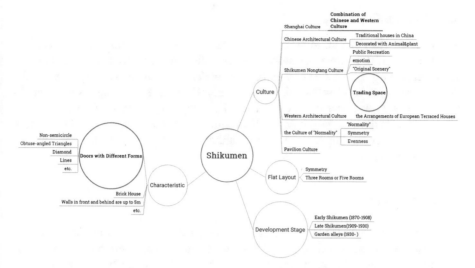

Fig. 5. The mind map of Shikumen

As space is limited, this article mainly shows a design solution. As shown in Fig. 6, based on the compatibility of Shikumen with Chinese and Western cultures, we use western chocolate to represent the Chinese Shikumen features.

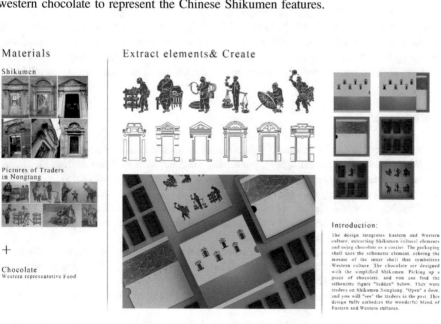

Fig. 6. Shikumen chocolate design

3.3 The Feedback-Test

Show the design program to testers and record the testers' reactions. The testers said that the design program is creative and can reflect the key features of Shikumen. However, there are still some problems: Product design should be more systematic; the concept conveyed to consumers is not clear enough; packaging design is too complicated. There is no better consideration of the complexity in processing and so on. Combined with the above suggestions, there still has great potentials for making further progress. This paper hopes to make the further improvement in Shikumen creative design in the future.

4   Conclusion

This paper delves into the cultural elements and humanistic connotation of Shikumen, trace back to the historical origin of Shikumen and balances cultural elements with modern consumer products, trying to discuss the creative design based on Shikumen culture. The purpose of this article is to make more people understand the historical background and cultural connotation of Shanghai Shikumen. Provide some positive impact for the development and protection of Shikumen. Establish a more distinctive Shanghai city impression.

Acknowledgments. We thank the financial supported by Shanghai Summit Discipline in Design - "一带一路"地域特色产品研发大师工作室 - DC17013. Thanks to everyone who provided valuable suggestions and feedbacks during the writing of this paper, especially Dr. Zhang Zhang and Dr. Li Xu. At the same time, We would like to thank Xiaoyong Ding and Shenbao Yu who committed to Shikumen model making and Shikumen culture promotion.

References

1. https://www.gov.uk/government/publications/creative-industries-mapping-documents-2001
2. Wang, H.: Research on Shanghai's positioning and function in the "One Belt and One Road" and Yangtze river economic belt construction. J. Sci. Dev. **03**, 92–98 (2015)
3. Opinions on Speeding up Innovative Development of Cultural and Creative Industries in Our City. http://www.sh.xinhuanet.com/2017-12/15/c_136827333.htm?from=timeline
4. Zhou, H.: Viewing the characteristics of Shanghai School Houses from the development of Shikumen. J. Cent. Chin. Buil. **01**, 124–127 (1997)
5. Yin, J.C.: Folk Custom in Shanghai: Luwan. Shanghai Culture Press, Shanghai (2008)
6. Xiao, P.: Research on the art of building decoration in architecture. J. Chongqing Univ. (Soc. Sci. Ed.) **04**, 17–21 (2005)
7. Gao, S.: The influence of the golden mean on Shikumen decoration style. J. Archit. Art **02**, 114–116 (2017)
8. Li, X., Guo, Q.: Explore "Original Scenery" on the traditional settlement street. J. Chin. Constr. **02**, 35–37 (2008)
9. Chen X.C.: Understand "Shanghai" through Shikumen Nongtang culture. South Central University for Nationalities (2009)
10. Brown T.C.: Design thinking. Harvard Business Review, June 2008

Research on User's Perceptual Preference of Automobile Styling

Zheng-tang Tan[1], Yi Zhu[2(✉)], and Jiang-hong Zhao[3]

[1] State Key Laboratory of Advanced Design and Manufacture for Vehicle Body,
Hunan University, Changsha, China
410922003@qq.com
[2] School of Art & Design, Guangdong University of Technology,
Guangzhou, China
zhuyi.1983@qq.com
[3] School of Design, Hunan University, Changsha, China
zhaomak@126.com

Abstract. Study the user's perceptual preference of automobile design, help the designer to make clear the reason for positive emotion from the user. Based on the case of electric vehicle styling design from Hunan University, the research regards the design scheme as an alternative plan of "emotional language" and "visual features". A questionnaire survey was conducted to study the perception preferences of different users. The research used both complex network and perceived matching methods, comparing each two alternatives, and suggested that the model of user's perception is an organized process of associated expression and the preference perception is made up of two key factors called "perception center" and "perception connection". The research constructs the prototype design evaluation system, makes each design information of different categories has a visual expression according to user's personalized emotional preference, and helps designers to understand the user's emotional information intuitively.

Keywords: Automobile styling · Perceptual preferences · Emotion activating
Perception center · Perception connection

1 Introduction

Emotion is a physiological response to a person's action on the outside world, which is determined by needs and expectations. When such needs and expectations are met, they produce a pleasant, loving feeling, whereas, conversely, distress and disgust. With the improvement of the product market, the user's attention to the product changes from the material functional requirements to the aesthetic and emotional needs. At present, emotional design is widely used as the theoretical basis for designers to manipulate product feelings to arouse product emotions and to study user's perception information. In his book on emotional design, Norman argues that "products must be attractive, effective, comprehensible, enjoyable and interesting" [1]. That is to say, by designing the emotion to the product, stimulating the user's emotional experience, satisfying the user's emotional

© Springer International Publishing AG, part of Springer Nature 2019
S. Fukuda (Ed.): AHFE 2018, AISC 774, pp. 41–52, 2019.
https://doi.org/10.1007/978-3-319-94944-4_5

psychological needs. The most direct manifestation of this requirement is the preference of users in the face of the same type of product. However, emotion is the inner, the thorough individual phenomenon, and the user's preference perception may be affected by the individual's perception of "difference" or part of the "missing". Further, the user's emotional perception of the product is "weakened", which leads to the deviation between the emotional information obtained by users and the expectations of the design team. In the early stage of design, the association and difference of the perception preferences of each role resulted in the complexity in the decision-making process, and the design convergence stage lacked effective information support. This paper attempts to analyze user perception preference association mode. The visual expression of information through the complex network helps the design team intuitively understand the user's perceptual preferences for the scheme. Improve the perception of products by modifying the alternatives, and enhance the "emotional" information exchange and interaction between the characters. On this basis, the paper summarizes the activation mode of the user's emotional information, and puts forward the key factors for the establishment of user's perception preference. Based on the UNITY engine, multi-role perception information evaluation design auxiliary system is developed, implemented in the 3D environment for the user and the design scheme's 360-degree view review interaction, to explore the user's more comprehensive emotional preference information and assist design decisions.

2 Discussion

2.1 Perceptual Preference of Automobile Styling

In the field of psychological research, preference is a tendentious emotional experience. The user who generates perceptual preference, which means obtaining satisfactory perception, is easier to make choice [2]. In product design, perceptual preferences perform as the degree of preference of various perceptual factors such as the shape, color and surface mechanism of the product while its essence is the intersection of multiple preferences spaces composed of emotional factors, with the features of complexity, fuzziness and multi-dimension [3]. By now, most researches are based on consumers' behavioral preferences, focusing on the relationship between product features and preference perception [4]. Less attention is paid to the formation mechanism and key influencing factors of perceptual preference in the design field. From the user's perception of the electric car product modeling behavior, this paper investigated the performance of the electric automobile modeling about perceptual preference, and analysis the key factors to form electric preference perception on the basis, which tries to to help designers understand how electric cars to form positive emotional experience from the perspective of preference.

The modern design trend has developed to user-centered emotional design [5]. Successful product design can use appropriate design language to create emotional information that can be conveyed and perceived by users. Automobile is a typical high technology, high emotional product. In this sense, automobile design is a perceptual process of emotional empowerment and emotional interpretation. In the automotive design strategy, Renault proposed to design a series of products with human and automobile emotional interaction from the perspective of humanized emotional needs. Banguo, the former design director of BMW, stressed that the user's understanding of car styling is essentially a communication between the shape and the user's emotions [5]. Schmitt used eye tracker to find that in different emotional environments, user's perception of automobile modeling will change [6]. Windhager et al.'s research shows that the perception of the front face of a car is similar to that of face perception, and it is a large proportion in automobile perception [7]. Zhao said that the perception of automobile styling is divided into three levels: the volume, the shape and the graph, and the perception order between the user and the designer is the opposite [8].

In the research of modeling design, traditional automobile modeling perception has the features of integrity, typicality and hierarchy [9]. However, due to the change of the internal structure of electric car styling, the attitude and proportion of the car body are changed, so that the way that users and designers perceive the shape is also changing. Taking the electric car front face as an example, the change of power source determines that it no longer needs the traditional modeling feature of the air grille, so the electric car model is more flexible than the traditional one. Due to the various unique modeling features of different electric vehicle models, the typical design elements of electric vehicle modeling have not formed uniform specifications. At the same time, these design elements also become the user's focus on modeling perception.

The complexity of the modeling perception of electric vehicles makes users and designers have significant differences in their understanding of the modeling of electric vehicles and how they are described. This kind of perceptual preference derived from modeling understanding and perception "absence" inevitably influences the process and evaluation of modeling design. Therefore, it is difficult to find a suitable way to analyze user's perception preference for electric vehicles.

2.2   The Application of Complex Networks in the Field of Design

In recent years, complex networks have been widely used to analyze the relationships among objects in the system. A network is a non-empty set of elements (nodes) and a set of unordered pairs (called edges) of these elements (the application of complex networks in the field of management). The nodes can represent various types of network elements, while edges can represent the relationships among the relational elements. In this paper, network nodes represent the "affective semantics" or "modeling features" contained in the design modeling. The connection between nodes represents the relationship between adjacent nodes. Taking electric vehicle program as an example, the "novelty" in its "affective semantics" is reflected in the "visual features" of "side window" and "head-light". On the basis of network definition, it can be understood that the "novel" and "side window" and "headlights" have connected edges. In the network definition, the number of edges of a node becomes degree, so it can be said that the degree of "novelty" is 2 in this modeling network. The network among the limited edge of the nodes is called the connectivity of the network. The "window" is connected to the "headlights" by connecting with the "novel" side. When "novel" has a large number of connections, it can reflect the centrality of the complex network, so the node is called "center node". The clustering, which is closely related to the nodes, reflects more homogeneous elements in the nodes, and the loose group shows more heterogeneous elements.

User perception information is complex and non-linear in relation to each other. The concept of "robustness", "vulnerability", "center node" and "connectivity" in network features can be used in the analysis of design research. Zhu Yi in his doctoral thesis put forward the concept of complex networks in small world, think network connection exists in modeling design system, and put forward that "the center node and associated elements handled well, it could form a style mutation" [10]. Word mapping is used to discuss the "robustness" and connectivity between the brands, engineering and visual features of a PUMA derivative product [11]. This paper constructs a new method to analyze the relationship between "conceptual semantics" and "visual elements" in user's perception information by means of complex network.

2.3   Communication Channels Between "Affective Semantics" and "Visual Features"

The understanding of automobile modeling activities includes a variety of aesthetic features and the processing of language concepts. It involves the cognitive relationship between users, designers and other roles. In the early stage of automobile design, the information provided by decision-makers is often uncertain and unclear. The policy makers communicate the intangible "emotional concept" to the designer by means of semantic meaning. The designer reinterprets the "decision semantics" and gives it a new interpretation, which diverts into "design semantics". Design, on the other hand, is done in a "given condition" environment. The so-called "given condition" is based on the premise that the decision maker puts forward, and the design behavior can only adapt to the limits set by these "given conditions". From electric model project, the designers often defines "decision-making semantic" as design basis, through the transformation of the benchmark form the detail design semantics, selective use of decision-making information architecture design goal. The two constitute "affective semantics" that connect them to the design scheme, thereby conveying an alternative to the concept of carrying emotions. This design process can be seen as a process by which designers constantly adjust the alternatives and limit the "visual features" to "affective semantics".

The emotional cognition of automobile modeling is the mutual transformation and communication between visual form and language form, involving multi-role cognitive relations. Danhua Zhao pointed out that the emotional activation in the process of modeling design should include the designer put the emotion by the features and the concept of language form, at the same time also includes emotion on the basis of the concept of "language" of users, and with "visual features" related to the perception of mutual emotional experience of concept matching process. OyaDemirbilek points out that the "language" as a tool to convey the "meaning" of the product can not only stimulate the styling of the design, but also help the product to communicate with users on the emotional level [12]. In conclusion, the "language" is the natural medium of emotional information, and forms a pair of mutual influence, mutual activation complex with the product modeling which mainly include visual features, and the mutual transformation and communication is just the emotional reaction of the design.

This study uses the method of perceptual matching to regard the design team that the designer and the manager in the automobile design process as the giver of emotional information. As shown in Fig. 1: team using the "decision semantics" and "design semantics" build "affective semantics" as a medium of emotional information, on this to construct the corresponding "visual features", constitute an alternatives including both sentimental emotion and form. On the other hand, users, as receivers of emotional information, form a perceptual preference for automobile modeling alternatives through the matching and activation of alternative schemes and their own emotional needs. Among them, the difficulty of emotional design is how to effectively activate the latent emotional information of users. In the theoretical environment, the mapping between features and the emotion information described in view of the emotional imagery is fixed, so what model will make emotional information activation is a problem worthy of studying. This paper takes "language" and "features", as a way for users to activate emotional awareness through user modeling experiment study of emotional information, and try to in the case of the practice, study of user's emotional activation patterns in the electric vehicle design through the perspective of multiple emotional information, with the purpose of analyzing user's perceived preference for alternatives.

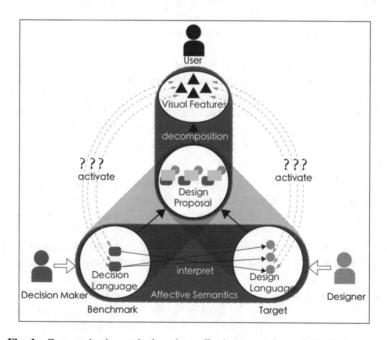

Fig. 1. Communication paths based on affective semantics and visual features.

3 The User's Electric Car "Modeling Feature" and "Emotional Activation" Questionnaire Survey

3.1 Experimental Sample Processing

The electric vehicles shape perception experiment of user's research two problems: the difference expression of electric vehicle modeling perception of two typical users and the key factors of user's preference perception.

Subjects Selected. The experiment was divided into two groups, expert users with automobile design background and young users without relevant experience are selected 30 for each group. Both types of users are typical: the former is rich in design experience, and can sense the feature information of various levels of car modeling. The ordinary user belongs to the group of potential buyers of the new product, and more subjective while evaluating and interpreting the modeling information.

Sample Processing. Processing of sample pictures. This case is based on the evaluation of three design schemes of autonomous electric vehicle renderings in the automobile body from the State Key Laboratory of Advanced Design and Manufacture for Vehicle Body, Hunan University. The sample is based on the actual review of the renderings (two front and back perspectives and a side view). Effect of alternatives to the graph is reproduction, according to the actual proportion sketch plan, appears in the gray-scale images of black and white line box, avoiding any interference from color, light, shadow, drawing techniques and other Non-modeling sensory information on the subjects. Study invited 10 car designers, asked them to display the visual features from the three electric schemes in outline, main features, additional features which could be easily perceived [13], then take those visual features with a Identification rate less than 90% as the non-perceived ones and eliminate Finally, the visual features of each alternative are obtained.

Questionnaire Design. The "modeling features" were split into the front contour, the headlights and other seven universal visual elements and the 4–6 visual elements feature of different schemes. Divided the "concept language" into "decision-making language" put forward by the decision makers, such as novel, comfortable, and "design language" put forward by the designers of their own divergence, for example to explain the "comfort" and put forward the "younger", to explain "novelty" and put forward the "sense of science and technology" and so on. The experiment finally selected 11–14 visual features, 2 decision semantics and 3 design semantics, and constructed the correlation diagram. As shown in the figure, the left side is the questionnaire of scheme 2, and the left part of the questionnaire is the affective semantics, and the other visual features are also divided by scheme 2. When a (emotional) information is thought to be associated with the perceived information, and the connection is the cause of the preferred solution, the two are wired. For example, the user thinks that "novelty" and "taillight" has some relation, and love such design expression, and the two are connected. When the user completed the questionnaire, the designer made a statistical analysis of the collected relevant data, and constructed the "relational degree triangle matrix", which is to fill 1 in the matrix corresponding to the two elements, and do not fill in when there is no correlation. As shown in Fig. 2 from left to right, there are the three electric vehicle design schemes, the questionnaire of electric vehicle scheme 2 and its data statistics.

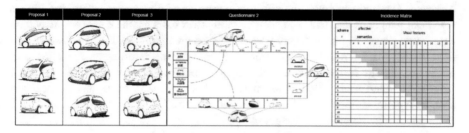

Fig. 2. Design of user's modeling perception experiment.

3.2 Experimental Data Processing

A total of 60 valid samples were received, and the correlation information was 1415, which was 63% for "expert users" and 37% for "ordinary users". The "visual features" and "concept language" of the scheme are seen as the nodes in the network, and the connected nodes are connected by edges, and the activation information network diagram of the scheme can be traced [11]. For all the sample activation relation, below 20% was seen as irrelevant, while more than 20% was seen as for a certain correlation.

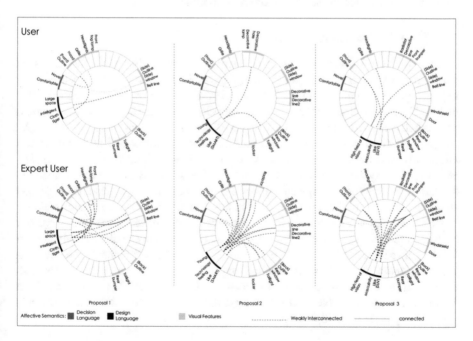

Fig. 3. The emotional activation mode of ordinary users and expert users.

Among them, express ones between 20–50% with dashed lines, while ones more than 50% with solid lines, said the initial line thickness as a unit, when the correlation is increased by 10%, the line increases 50%, form a network as shown.

Shown as Fig. 3, the red region is the decision language and yellow is the design language. The two constructs the affective semantics of the three schemes, and blue is the visual feature of the scheme. The upper and lower rows are the emotional activation association diagrams for "ordinary users" and "expert users". As Fig. 3 shows, due to the difference of knowledge background and interpreting information modeling purpose, two classes of users emotional activation patterns also reflects the significant difference: expert user activation associated with significantly more than the ordinary user, its type in addition to the "language" and "features" activation patterns (86.75%), at the same time there is the "language" and "language" (4.5%), "features" and "features" activation patterns (8.75%), make the emotional information in the "concept language" and "visual elements" in the flow to each other, mutual activation, forming high connectivity of network intertwined. However, the activation correlation of "ordinary user" is low, which is independent and scattered, and its activation mode is limited to the single feature mode (100%).

It is worth noting that the study found that in 3 schemes, users were more inclined to the correlation between "design semantics" and "visual features" rather than "decision semantics" and "visual features". As shown in the contrast between red and yellow area, the correlation of yellow area is significantly more than that of red. Analysis thinks this is due to the designer's "design semantic" is the interpretation of the "decision semantic", with more details that can be got, suggestive of expression, which makes it easier for users to produce a correlation such as above.

3.3 Experimental Results Analysis

Analysis the user for alternative emotional activation patterns is to grasp the sense from the perspective of multiple emotional information preference formation mechanism of the key factors [14], so as to help the designer to compare two schemes of sensory information. This research takes 60 user data through Gephi software to create a better sense of each scheme preference correlation network diagram, as shown in Fig. 4. Through a complicated network diagram, we can make a large number of different categories of information according to user's personalized perception preference for decision makers and designers pay close attention to its emphasis of intuitive. As can be seen in the figure, the "novel" emotional information in the decision semantics is rich, and different design elements are used in different schemes. On the contrary, the emotional information of "comfort" is relatively single, which is only reflected in the correlation of the side contour, back contour and large space in scheme 1. This is because the electric car body is short, the interior space is compact, which is far away from being comfortable. In scheme 1, the design semantics makes the comfort into a large space, which increases the internal space by moving the c column backward. In scheme 1, the "comfort" "big space" and "side profile" constitute the stable network structure of the triangle, so that the comfort can be perceived by users in multiple angles.

Comprehensive the above analysis, this study proposed the user's perception are embodied in the associated expression of an organized process, its cognitive preferences reflected in the network diagram for solution "perception center" and "perception connection" of two key factors, as follows.

Perception Center. Seen in the red region of Fig. 4. In each scheme, one perceptive information which is picked most times is selected to define a concept language, which is called the central feeling of the user's perception preference of the scheme. As shown in Fig. 4-A, the user's "perception center" is novel, with six edges connected to it. "Visual features" such as side outline, side window and so on make the correlation with "concept language". This correlation makes the scheme reflect the strong emotional demand information, and the user also has a rich emotional activation of the information and visual features.

Perceptual Connection. Seen in the blue region of Fig. 4. It means that the user activates the network through a series of "language" and "features", and connects the design information of the whole alternative scheme to enable the user to circulate the sensory information of the overall shape. As shown in Fig. 4-A, the "side outline" connects the two network Spaces formed by "novel" and "big space" to become the connection point of perceptual information flow. On the contrary, the emotional demand of "comfort" in Fig. 4-C is not reflected in the circulating network, the rear light, rear window, side door and other visual features are scattered nodes. This kind of modeling information is not activated by the user and belongs to the missing part of perception information. The scheme shows a weak "perception connection", where the scattered node information needs to be the key modification part of the designer in the next design process.

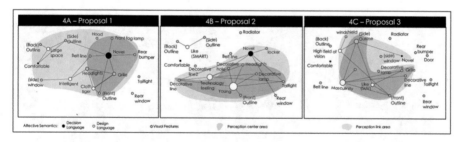

Fig. 4. The net and graphic expression of perceptual preference of alternatives.

4   System Construction and Research Summary

4.1   Prototype Construction of User Evaluation System Based on Perception Preference

The meaning of the design decision is the information organization with the strategy of "image", and the decision-making process of the scheme is based on the comparison of the alternatives. Through the complex network expression of user perception information, it can intuitively establish the user's perception of the scheme and the missing part of the scheme. This can help decision makers and designers to find the solution at the center of the nodes, analysis the correlation degree between the final product and initial concept, to compare the user's connectivity solutions, which enables the decision-maker to obtain the user's complete emotional information, in order to make decision makers more intuitive choice of alternatives. This scheme - based decision - making model also enables designers to clearly design problems and improve design quality.

Based on this purpose, the interface and process design of multi-role perception information evaluation system was studied, and the prototype development of the system was completed in October 2017. As shown in Fig. 5, this is a key interface for the process and data display used by users in the system. It is used in the iterative evaluation of electric vehicle design of Hunan University to prove the usability and effectiveness of the software. In the case, because the project is an exploratory innovation design, the design input in the style is not clear. The research group obtained 15 managers and 50 users, and asked them to count the emotional information expectation semantics of electric vehicles, and got the 8 highest affective semantics such as "dynamic", "fashion", "smart" and so on. In the project file upload interface, the designer imported the scheme into the system and completed the input to the affective semantics of the scheme. The following figure shows the user interface. First, the user enters the system, and the affective semantics are selected and expected based on the observation of the overall shape. Then, the design features are evaluated and evaluated according to the single emotion vocabulary. When the calibration is completed, the system will form the network of the modeling features and affective semantics. When the number of people is evaluated, more reliable role perception information can be obtained to achieve the purpose of assisting design iteration.

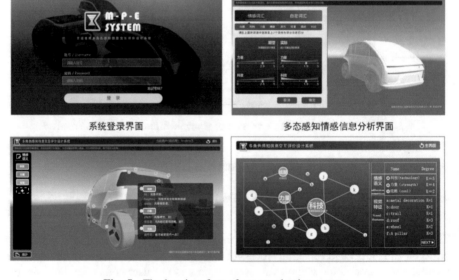

系统登录界面   多态感知情感信息分析界面

Fig. 5. The key interface of user evaluation system.

5 Conclusion

From the perspective of product modeling, this paper analyzes the difference of perception from different users in the product modeling, to study the mode of emotional activation between the "conceptual language" of modeling and the "visual features" of user's perception. Through the research method of network graph, two key factors, "perception center" and "perceptual connection" in perceptual preference are clarified. The method has "clarity" and "guidance", so that the inconsistent design scheme can visualize the user's multi-dimensional perception preference. Through the concept of "language" and "visual features" in the emotional space information exchange, we can timely found the shortage of the modeling design, makes the design concept has a clearer directivity, which provides information support for designers to modify and analyze the scheme.

Acknowledgments. We would like to thank National Nature Science Foundation of China (51605154), Humanities and Social Sciences Foundation of Ministry of Education of China, 2017, (17YJCZH275), Philosophy and Social Science Foundation of the Guangdong Province, China, 2017, (GD16XYS33), for providing this research with financial support.

References

1. Norman, D.A.: Emotional Design: Why We Love (or Hate) Everyday Things. Basic Books (2005)
2. Hong, Z.: A Research to the Influences of Operating Product on the Preference for Form Factors – A Case Study of USB Flash Drive. Tatung University, Taiwan (2008)
3. Zhuang, M.: Evaluation model towards emotional design factors of attractive and pleasurable products (2008)
4. Lee, J.H., Chang, M.L.: Stimulating designers' creativity based on a creative evolutionary system and collective intelligence in product design. Int. J. Ind. Ergon. **40**(3), 295–305 (2010)
5. Jiang, W.: Personalities of the Automobile - A Study on Building Personalities in Car Design Process. University of Success, Taiwan (2011)
6. Schmitt, R., Köhler, M., Durá, J.V., et al.: Objectifying user attention and emotion evoked by relevant perceived product components **3**(2), 315–324 (2013)
7. Zhao, D.: A car styling-based study: the designer's intension and user's interpretation. Hunan University, Changsha (2013)
8. Zhen, W., Tan, Z.: Research on vehicle modeling features based on holistic cognition. Packaging Eng. **34**(24), 51–54 (2013)
9. Zhao, D.: A car styling-based study, the design intention and interpretation. China Youth Press, Beijing (2014)
10. Zhu, Y.: A study on design complexity and design computing. Hunan University, Changsha (2014)
11. Rasoulifar, G.: Integrated Design Process of Branded Products (2014)

12. Demirbilek, O., Sener, B.: Product design, semantics and emotional response. Ergonomics **46**(13–14), 1346–1360 (2002)
13. Zhao, D., Zhao, J.: Automobile form feature and feature line. Packaging Eng. **28**(3), 115–117 (2007)
14. Hsiao, S.W., Wang, H.P.: Applying the semantic transformation method to product form design. Des. Stud. **19**(3), 309–330 (1998)

Modeling Subjective Temperature Using Physiological Index

Tatsuya Amano[1]([⊠]), Takashi Sakamoto[2], Toru-nakata[3],
and Toshikazu Kato[4]

[1] Graduate School of Chuo University,
1-13-27 Kasuga, Bunkyo-ku, Tokyo 112-8551, Japan
al3.6rx3@g.chuo-u.ac.jp
[2] National Institute of Advanced Industrial Science and Technology, AIST
Central-2, 1-1-1 Umezono, Tsukuba, Ibaraki 305-8568, Japan
takashi-sakamoto@aist.go.jp
[3] National Institute of Advanced Industrial Science and Technology, AIST,
2-4-7 Aomi, Koto-ku, Tokyo 135-0064, Japan
toru-nakata@aist.go.jp
[4] Chuo University, 1-13-27 Kasuga, Bunkyo-ku, Tokyo 112-8551, Japan
kato@indsys.chuo-u.ac.jp

Abstract. It is known that comfort is affected by various factors such as sensible temperature, air quality, light, sound, and floor area. Among them, the sensible temperature has a great influence on comfort and has been studied extensively. A new standard effective temperature has resulted from a study of the typical quantification of sensible temperature. However, the model has two problems: The first is that individual differences cannot be considered. Second, there are factors that cannot be considered among the factors for sensible temperature. In this research, subjective temperature is predicted using individual physiological data. By doing this, we expect to see the individual differences that could not be observed using the existing method. Moreover, there is a possibility that subjective temperature can be predicted without considering many factors.

Keywords: Physiological index · Subjective temperature

1 Introduction

1.1 Background

It has been asserted that 47% of 701 occupations will be replaced in the United States because progress in AI has been rapid [1]. For this reason, creative thinking is required in modern society. Creative thinking is often a result of abundant knowledge and also, a comfortable state of the individual [2]. Comfort is a result of various factors such as sensible temperature, air quality, light, sound, and area.

Among them, sensible temperature has a significant influence on the comfort level of an individual, and several extant studies have focused on it.

© Springer International Publishing AG, part of Springer Nature 2019
S. Fukuda (Ed.): AHFE 2018, AISC 774, pp. 53–59, 2019.
https://doi.org/10.1007/978-3-319-94944-4_6

1.2 Previous Research

The standard new effective temperature (SET*) is one measure of sensible temperature. SET* can consider not only the temperature and humidity, but also the influence of heat radiation, wind, clothing parameters, and working conditions. Therefore, by using SET*, it is possible to predict the temperature experienced by a person more accurately than by existing methods. In fact, this indicator is reliable, and it is the preferred standard of the American Society of Heating, Refrigerating, and Air-Conditioning Engineers (ASHRAE).

1.3 Problem

However, the above model has two problems: The first is that individual differences cannot be considered. Some people feel hot in the same space, while others feel cold. Second, there are factors that cannot be considered among the factors for sensible temperature. For example, in addition to temperature, factors such as humidity, wind speed, radiant heat, external factors of SET*, season, menstrual cycle, diurnal rhythm, race, sex, physique, body tissue, posture, acclimatization, acclimation, housing environment, bathing, nutrition, medication, and conditions of insomnia. have been found to have an effect [3].

Furthermore, in recent years, it has been found that oxygen concentration [4], uneven environment [5], lighting [6, 7], color [8, 9], and color temperature [10] affect the subjective temperature.

In addition, it is known that the subjective temperature sensed by humans varies depending on the site, even with the same heat level [11]. With so many factors that are not taken into consideration in SET*, the subjective temperature cannot be predicted accurately. Owing to the sheer number of factors, it is also difficult to develop accurate models.

1.4 Purpose and Approach of This Research

Therefore, in this study, we consider the individual differences. Moreover, we aim to predict the sensible temperature using limited factors. In this research, subjective temperature is predicted using individual physiological data. By doing so, we believe that we can observe the individual differences that could not be accounted for using the existing methods. Moreover, there is a possibility that subjective temperature can be predicted without considering many factors, such as temperature, humidity, radiant heat, wind speed, light, sound, and partial heat stimulation.

2 Previous Research

Gagge et al. announced the new effective temperature (ET*) as an indoor bodily sensation index in 1971 [1].

ET* evaluates the subjective temperature using the thermal equilibrium equation of the human body, considering the temperature, humidity, radiation heat, and wind speed as variables.

However, because the amount of activity and clothing amount are fixed for calculation, there are cases wherein the ET* cannot correctly predict the subjective temperature.

Therefore, SET*, which can consider the air flow, activity amount, and clothing amount, was developed by Gagge et al. [13].

As a result, multiple heating elements could be represented as a single hot and cold element. The calculation formula is shown below.

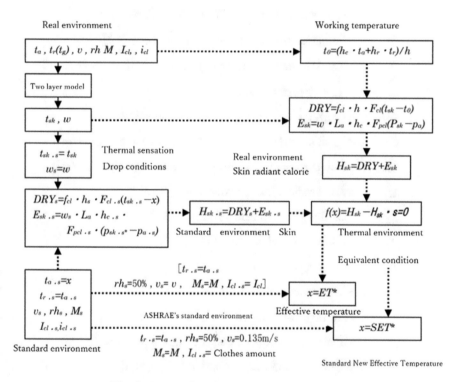

Fig. 1. Process flow for calculation of SET* (14)

t_a : temperature[°C]

rh : Relative humidity[%]

pa : Water vapor partial pressure[kPa]

t_g : Globe temperature[°C]

t_r : Average radiation temperature[°C]

v : wind speed[m/s]

M : Metabolic rate[met]

hc : Convective thermal conductivity[W/(m² · °C)]

hr : Radiation thermal conductivity[W/(m² · °C)]

h : Total thermal conductivity[W/(m² · °C)]

Icl : Real clothes amount[col]

icl : Clothing vapor transmission efficiency[ND]

DRY : Sensible surface sensible heat radiation amount [W/m²]

Esk : Skin surface evaporative emission heat quantity [W/m²]

Hs : Total radiant heat of skin surface[W/m²]

Fcl : Clothing heat transfer efficiency [ND]

$Fcpl$: Clothing heat transfer efficiency [ND]

Fcl : Clothing area increase factor [ND]

w : Skin wet area rate [ND]

t_{sk} : Average skin temperature [°C]

t_a : Standard environment equivalent temperature (x=,°C)

psk^* : Saturated water vapor content in T_{SK} [kPa]

pa^* : Saturated water vapor content at t_a °C[kPa]

Fig. 1. (*continued*)

However, with SET*, it is impossible to consider individual differences. There are other thermal factors that cannot be considered further (Fig. 1).

3   Experimental Method

3.1   Outline of Experiment

The experiment involved the following steps:

- Change the subjective temperature of subjects and acquire physiological index data at that time.
- Analyze physiological data, and model subjective temperature.
- Using a comparison with existing models, clarify the accuracy of the new model.

The subjective temperature of subjects is varied by considering temperature, radiant heat, and wind. We plan to acquire capillary thickness, blood flow information, body temperature, amount of perspiration, heart rate, and myopotential around the shoulder of an individual.

3.2   Reasons for Adopting Physiological Data

We measure the following values as physiological data.

1. Capillary thickness, blood flow, blood pressure.
2. When the body temperature drops, the capillaries contract to prevent heat from escaping. However, when body temperature rises, heat is released by expansion of the capillary, which might indicate a correlation with subjective temperature.

3. Body temperature

It is thought that the thermal equilibrium of the human body affects human thermal sensation. Because thermal equilibrium affects body temperature, we believe that measuring body temperature will eventually help in predicting subjective temperature.

4. Amount of perspiration

Owing to increase in the temperature and humidity, not only do people feel hot, but the amount of perspiration also increases [15]. Thus, it is highly probable that the amount of perspiration correlates with the subjective temperature of a person.

5. Heart rate

It is thought that coldness and heat affect people in the form of thermal stress. Therefore, by using the heartbeat as a stress index, subjective temperature can be predicted.

6. Myoelectric potential around the shoulder

It is said that women produce less heat because of less muscle mass and they are less protected against cold [16]. When feeling cold, humans are thought to produce heat by activating muscles to maintain body temperature. Therefore, by measuring the myoelectric potential and estimating the amount of activity of the muscle, the person's subjective temperature can be predicted. The shoulder was predicted to be most susceptible to thermal effects, and was considered as the region of measurement.

4   Future Work

A model tailored to each person must be created. Some people can easily be modeled by using blood flow information, and others can be easily modeled by using perspiration amount. Therefore, more-accurate predictions can be made by developing model expressions for people.

By approaching each person based on that model, we intend to develop a space in which everyone can be comfortable. As individual approaches, light, sound, and partial stimulation are considered.

This study revealed that light and sound can change a person's temperature perception of radiant heat. Light can stimulate directly only to the target. There are also many studies of directional loudspeakers regarding sound [17], and it is possible to present stimulus only to the target. In other words, it is possible to control the subjective temperature by stimulating an individual by using light or sound. In addition, Wada et al. conducted an experiment to change the overall subjective temperature by local stimulation [11]. Therefore, we believe that a system can be realized using which everyone can be comfortable, using a small heater or cooler as a wearable device and combining it with the model created in this study.

5   Conclusions

In this study, we demonstrated that a specific thermal stimulus could be perceived as being warmer by providing additional light stimulus and sound stimulus. However, this finding was subjective, and we encountered subjects who were not able to perceive the change in the sound and light stimulus. We surmise that the awareness of the participant regarding change in the light and sound stimuli could influence his or her result. It is also possible that the selection of the stimulus was not appropriate. It is impossible to judge whether the subject was able to perceive the stimuli change using the current method; hence, we plan to group subjects together in future experiments.

The purpose of the experiment conducted herein was to enable the radiant heat perceived by the participants as more warm by using the stimulation of sound and light. It was clear that the thermal sensation was influenced by light and sound stimulations. We intend to evaluate the effects of light and sound accurately in further studies.

Acknowledgments. I am deeply grateful to the members of the Human Media Engineering Laboratory of the Faculty of Science and Technology, Chuo University, and the Kansei Robotics Research Center, who are enthusiastic about research discussions and cooperation for experiments. This work was partially supported by JSPS KAKENHI grants, "Research on Sensitivity Symbiosis Mechanism within Groups in Real Space/Information Space" (No. 25240043) and TISE Research Grant of Chuo University, "KANSEI Robotics Environment."

References

1. Frey, C.B., Osborne, M.A.: The Future of Employment: How Susceptible Are Jobs to Computerisation? Oxford Martin School Working Paper (2013)
2. Yoshida, M.: Fluctuation measurement and comfort evaluation of EEG. J. Japan Acoust. Soc. **46** (1990)
3. Yamasaki, K., Nojiri, K., Yokoi, M., Ishibashi, K., Higuchi, S., Maeda, T.: Comparative study of susceptibility to the heat and the cold on themore gulatory responses in Japanese female adults. Jpn. J. Physiol. Anthropol. **11**(1), 13–20 (2006)
4. Yamashita, K., Matsuo, J., Tochihara, Y.: Skin and forearm blood flow response during oxygen inhalation in different thermal conditions of the body. Jpn. J. Physiol. Anthropol., **9**(1) (2004)
5. Matsuo, J., Murayama, T., Tochihara, Y.: Influence of vertical temperature difference on physiological/psychological responses. Jpn. J. Physiol. Anthropol. **9**(2), 60–61 (2004)
6. Kusano, Y.: Physiological psychological effects under neutral temperature range by LED light irradiation in winter and summer.Jpn. J. Physiol. Anthropol. **3**, 142 (2003)
7. Morita, T., Tokura, H.: The influence of different wavelengths of light on human biological rhythms. Appl. Human Sci. **17**(3), 91–96 (1998)
8. Kim, S.H., Tokura, H.: Cloth color preference under the influences of menstrual cycle. Appl. Hum. Sci **16**(4), 149–151 (1997)
9. Inoue, Y.: Experimental study on evaluation of composite condition between room temperature and surface color part 1: evaluation of preferred surface color in winter. Jpn. J. Physiol. Anthropol. **8**(2), 94–95 (2003)
10. Iseki, T., Yasukouchi, A.: The influence of difference in color temperature of illumination during isothermal exposure and after exposure on body temperature regulation. Jpn. J. Physiol. Anthropol. **5**(2), 72–73 (2000)
11. Wada, S., Morita, T., Kinoshita, S., Yoshida, A., Shimazaki, Y.: Effects of local stimulation on thermal sensation of human body part. Trans. Soc. Heat. Air-conditioning Sanit. Eng. Jpn. (2012)
12. Gagge, A.P., Stolwijk, J.A.J., Nishi, Y.: An effective temperature scale based on a simple model of human physiological regulatory response. ASHRAE Trans. **77**(1), 247–262 (1971)
13. Gagge, A.P. Nishi, Y., Gonzalez, R.R.: Standard effective temperature − a single temperature index of temperature sensation and thermal discomfort. In: Proceedings of The CIB Commission W 45 (Human Requirements)Symposium, Thermal Comfort and Moderate Heat Stress, Building Research Station, London, September 1972, pp. 229–250. HMSO, 1973)
14. Adaptation measures for Heisei Island phenomenon in Heisei 24 year and survey work on heat island countermeasure after earthquake. https://www.env.go.jp/air/report/h25-02/index.html
15. Zhang, J.F.: A basic study on design of comfortable clothing in a hot environment – physiological and psychological responses of human body when unclothed and when wearing perforated film clothing. Academic Repository of BUNNKA GAKUEN (2014)
16. Tamura, T.: Cold sensitivity and clothing: from the standpoint of clothing physiology. Academic Repository of BUNNKA GAKUEN (1999)
17. Enomoto, S., Ise, S.: A proposal of the directional speaker system based on the boundary surface control principle. Trans. Inst. Electron. Inf. Commun. Eng. A. **87**(4), 431–438 (2004)

On Poster Design in the Subway Public Space in China Based on the Regional Culture

Yejun Yuan and Huanxiang Yuan[(✉)]

Industrial Design Department, School of Mechanical Engineering,
Huazhong University of Science and Technology, Wuhan 430074, Hubei, China
290229745@qq.com, 1850744624@qq.com

Abstract. At present, when urbanization is going on swiftly and on a large scale in Mainland China, subway has actually become an important vehicle for propagating a city's cultural philosophy and manifesting regional customs owning to the use of various forms of public art. Looking back through the city subways in the world, many of them are excellent manifestations of historical features and representations of modern innovations owing to the use of various forms of public art. How to make posters, one of the forms of public art, culturally loaded when constructing distinct and unique subway public space in cities is what this paper intends to explore, and accordingly, a three-step procedure is proposed: collecting regional cultural elements, making creative designs and evaluating designs.

Keywords: Poster design · China subway stations · Subway public space
Regional culture

1 Introduction

The construction of the first subway line in Mainland China began on July 1, 1965 in Beijing, and opened on October 1, 1969. Then came the second subway line (first in Tianjin) on December 28, 1984, and the third line (first in Shanghai) on May 28, 1993. Until 2006, the development of the metro construction in China was slow and circuitous. With the arrival of the New millennium, a new wave of construction of subways has been ushered in with the cities ever growing fast and the country's economic capability being improved a lot. To this day, 31 cities headed by Beijing, Shanghai, Guangzhou and Shenzhen have built 128 subway lines, with a metro milage of 3881.77 km, and it is estimated that by 2050, more than 300 subway lines will have been constructed in over 200 cities in Mainland China [1].

With the ever-growing development of metro construction in Mainland China, one problem appears, that is, the loss of identity for cities. As has been mentioned by Qinzhong Ma in his book, "the urban environment in which we live is not more and more different, but more and more similar... Not only that, the situation is getting worse, with one city copying another city..." [2], while Liangpu Wu also points out in his book that "cities around the world are now facing so-called 'identity crises'. For decades, we were so proud of the huge amount of construction, but suddenly, we found surprisingly that the South and the North were of one kind, and the inside and outside

© Springer International Publishing AG, part of Springer Nature 2019
S. Fukuda (Ed.): AHFE 2018, AISC 774, pp. 60–66, 2019.
https://doi.org/10.1007/978-3-319-94944-4_7

of the city were of one kind, too" [3]. Facilities in the subway and subway space in different cities and even in the same city are constructed as "standard" and give passengers a sense of deja vu.

To solve this "identity crisis" problem, the involvement of regional cultural elements into the public art in the construction of subway stations works. Metro stations are the most common urban space, with the largest flow of passengers, who are recipients of public art works, thus, the integration of regional and cultural elements can not only connect the ground and the underground to form a new underground space, but also make it a good place to display artworks and spread culture. If we take a look at world metro stations, we can find a lot of unique ones, which follow the natural ecological features, reproduce history and customs of the region, and produce a particular space full of artistic atmosphere and humanistic atmosphere, and are good manifestations of well-designed metros based on the regional culture.

Poster, as defined in the Oxford English Dictionary, is a placard displayed in a public place. In the Advertisement Dictionary published by London-based International Textbook Publishing Company, poster means printed advertisements posted on cardboards, walls, large planks or vehicles, or printed advertisements that are otherwise displayed [4]. It is one of the main forms of outdoor advertising, also one of the most widely distributed and largest forms of media in the subway public space. It has a flexible placement, which can be put in a fixed place, such as the escalator side of the station walls, the station hall, the subway passageway, walls, pillars, or on a moving place, such as special trains, subway LED (light emitting diode) digital media, car TV, platform PIS (passenger information system passenger information system), etc. The content of the poster can be as rich and varied as possible, and the main elements are words, pictures, photos, seals, comics or even QR codes, used for promoting special products such as products, services, and spreading cultural customs and cultural heritage.

During 150 years of the world's metro development, subway advertising has been progressing, which gives birth to distinctive urban subway posters. New York Subway posters (see Fig. 1) give people a sense of fashion and modernism. London Underground Posters (see Fig. 2) seem to be a "visual feast" of the British culture; Moscow Metro Posters (see Fig. 3) show a strong ethnic characteristics of Russian culture and arts; there are numerous posters for literary and artistic activities in Paris Metro Stations (see Fig. 4), endowing people a strong feel of Paris art; Shanghai Metro Posters (see Fig. 5) impress people with both an international financial center and a modern metropolis, and Beijing Metro Posters (see Fig. 6) highlight its ancient culture and modern culture.[1]

In short, with global economic integration and cultural globalization, metro has become an important means of transport for people to commute in Beijing, Shanghai, Shenzhen, Gaugnzhou and other major cities in our country, and the largest amount of local advertising medium, while subway posters tend to display geographical features and promote regional culture.

---

[1] Picture Source: http://p.bigbigwork.com/searchnew.htm.

Fig. 1. New York Subway Posters

Fig. 2. London Underground posters

Fig. 3. Moscow Subway Posters

Fig. 4. Paris Subway Posters

Fig. 5. Shanghai Subway Posters

Fig. 6. Beijing Subway Posters

2   Necessity of Integrating Regional Culture into Subway Poster Design

Culture has a complex meaning. Currently, the definition of culture mainly includes two aspects. One is material culture, and the other is spiritual culture, both of which encompass all civilizations created by the mankind. Regional culture, then, refers to all the material and spiritual cultures created by the people in the region, which has dual features: being homogeneous and heterogeneous. Take for example, all the regional cultures in our country are subordinate to Chinese culture, and that is homogeneity, while there are different geographical areas in our country, and different regional cultures breed, and that is heterogeneity. Most of this difference is caused by different geographical environments, while some are attributed to historical reasons. Take Lingnan culture in Guangdong and Jingchu culture in Wuhan, for example. These two cultures belong to the branch of Chinese civilization, but there are essential differences in between. In this sense, the regional culture is the sum of the material and spiritual civilizations that have common characteristics and are formed together by the people in a given region in the long-term historical process [5].

A city of a high concentration of population tends to develop a highly distinctive regional culture. Since urban public space has the responsibility of creating, spreading and inheriting regional culture, subway public space is supposed to carry a greater part of it.

Firstly, subway poster is one of the popular means of publicizing regional culture. Chinese civilization has a long history and has developed to this day with various local cultures derived from it. It is, indeed, the designers' responsibility to design posters that can carry forward Chinese civilization, and various local cultures in their design. People who use subway space are both local urban residents and foreign tourists. Local residents, accustomed to the local life and adaptable to the local culture, are eager to see the reflection and representation of their local culture, while foreign visitors, as spectators, are always looking for "novelty" and "characteristics" in their travel. These two groups of people together set a demand on subway poster design with the local culture elements.

Secondly, subway poster is a useful means of educating people and promoting products. As one of the printed advertisements, it can be divided into two types: public posters and commercial posters. For public posters, they are to spread public welfare information, to orientate people and to educate people, while for commercial posters, they are to advertise the product or service, publicize the commercial brand and attract customers. For either of them, the design can never be separated from the regional culture, nor can it be divorced from the aesthetic taste of the local people. Otherwise, public posters will not play their essential role, and commercial posters will be unpopular to and even boycotted by the local people.

Lastly, subway poster is an effective means of disseminating information. According to Ergonomic, about 80%–90% of the information people obtain is through the visual system [6]. Placard, as a powerful eye-catching element in the subway space, is an explicit part of the culture, which can be observed and perceived numerous times. Even hurried passengers will slow down and take a look at those posters that are well designed and mixed with cultural elements.

3 Ways of Making Subway Posters Culturally Loaded

In subway posters, the information conveyed, the material adopted, and the artistic form used will decide whether they have any local cultural color. To guarantee that posters are designed in a way that cultural elements are integrated, the following four ways would help.

First, be theme-related. This is the most common way of designing posters and also the most intuitive way to express regional culture. For example, in the choice of themes, combine unique folk-customs and landscapes with the creative subject, such as in a commercial subway poster for Zongzi (see Fig. 7), a traditional Chinese rice-pudding, Zongzi used in Huzhou, a place in Zejiang Province which is famous for its Zongzi made special and delicious, is different from Zongzi made in other places. Similarly, Dragon and Lion Dance in Guangzhou (see Fig. 8), Ice Engraving in Harbin (see Fig. 9), with a strong local characteristics, are good elements of regional culture. Again, unique local landscapes such as Qianling in Xi'an, the Bird's Nest in Beijing (see Fig. 10), represent, to a certain extent, the history and present of the regional culture.[2]

Fig. 7. Huzhou Zongzi

Fig. 8. Dragon & Lion Dance in Guangzhou

Fig. 9. Ice Engraving in Harbin

Fig. 10. The Bird's Nest in Beijing

Second, be form-based. When designing subway posters, different forms can be used. For example, in this digital age, posters featuring digital display bring a scientific and technological experience; paper-cut art forms a three-dimensional space in the plane; the use of visual illusion helps to form a real three-dimensional visual space, by linking the background with posters and through perspective distortion and color cover; the introduction of interactive placard display brings a new vision and a new thinking, through the interaction of the audience [7].

Third, be material-based. China has a quite varied landform, which produces different materials. For instance, cities on the Loess Plateau, using loess as a material in poster design, brings a refreshing visual experience to the audience, and can arouse and the audience's resonance at the same time. Similarly, sand on the desert, grass on the prairie, snow on the snowy mountains, litchi trees in the Lingnan Area, etc., are all unique geographical elements, which can help to express and spread the local culture.

Last, be artistic form-based. China' vast land contributes to geographical differences, different materials, differences in language, pronunciation and habits, which, in turn, generated a wide range of art forms. Therefore, the North, the South, the Central Plains and the Lingnan Area in our country all have their unique style of painting, architecture and particular art forms, not to mention the unique art forms of ethnic minorities such as Tibet Region and Xinjiang Region. Thus, in the poster design, these differences must be observed and reflected.

---

[2] Picture Source: http://huaban.com.

Apart from the aspects mentioned above, there are other things that can reflect regional culture, such as color. People in Northern Shaanxi prefer reddish green, so it is effective to use reddish green in subway posters so as to bring the local people a feeling of ebullience, and increase the appealing of the subway posters.

4 Operational Procedures for Designing Culturally Loaded Metro Posters

Subway placard design, to properly reflect regional culture, can not be made merely out of the designer's creative inspiration, or on a whim of the designer. There has to be a systematic design process or procedure to ensure the rationality and success of the design. To achieve it, a three-step procedure is proposed here which includes data collection, creative design and outcome assessment.

First, to collect data of regional cultural elements. The geographical and cultural elements should include major events related to the area, typical architecture, traditional art forms, and folk culture in that area. It is better to collect and organize the complete regional culture to set up a special resource library that can provide designers with maximum convenience and stimulate designers to creatively use these regional cultural elements. The more comprehensively the investigation is done into regional and cultural elements, the more likely the poster design is to serve the purpose.

There are several ways to collect geographical and cultural elements. The first method is reviewing literature. To understand the history and culture of a place, local history is a very good channel, which comprehensively records the books and documents of the natural, social, political, economic and cultural aspects or specific issues of a certain area in a certain period. Of course, comprehensiveness may not be detailed enough. In this case, use local history as an index, get access to library literature, or online book database, such as China National Knowledge Network, to find the relevant geographical and cultural elements recorded and described in detail. The second method is field investigation. Due to the Heritage Conservation Act in China, many historical relics are preserved in the form of cultural relics or stored in historical relic museums. By conducting field investigation, we can get not only accurate information, but also a complete visual and psychological experience. The third method is visiting experts. Designers need to understand history, and they need to know culture. Therefore, to visit professional authority is a shortcut to collect historical and cultural elements.

Second, to conduct creative designs based on regional culture. Inspirations are the sparkles generated by designers after a lot of thinking. Each designer may have their own unique way so that no one method can be applied to all designers. However, there are two common ways that designers can adopt, allowing them to make a breakthrough when they encounter bottlenecks.

The first method is brainstorming. To do brainstorming, a group of 5 to 15 people gather around to create a relaxing atmosphere. In order not to spoil the good atmosphere, attendees do not usually make any negatives comments, but inspire each other to propose more ideas. It makes reasonable use of the excitation mechanism, association response, competition spirit and other psychological factors. Designers inspire each other, and thus open the door to creativity.

The second method is reverse deduction method, which is based on linear logic thinking, and in which results go first followed by conditions that can meet the result, and then conditions are subdivided so that they are operational, and get tested last. On the basis of the realization of the conditions, elicit a suitable design. The third method is reference method. To borrow others' ideas is difference from copying others, in that it occurs based on a profound understanding of the work of others, that is, the re-application of the thinking of others, instead of copying results. When doing this, remember to broaden the mind, and create a unique creative design based on the current situation of the project. In all, there are many creative design methods, and the method which is the most suitable for one person is a good one.

Last, to assess designs systematically. when evaluating creative designs, the most important thing is to ensure objectivity of assessors and comprehensiveness and accuracy of the evaluative dimensions, and to assign an appropriate evaluation weigh. Designers should display their designs and elaborate on their ideas, application of regional culture and the design results by PPT. Employ those having the same identity as the subway crowd to work as judges to rate the design for each dimension, which are supposed to include visual aesthetics, typicality of cultural embodiment, stand-out of the subject designed and globality of the design. On the basis of the above work, make a statistical analysis of each design and choose the one that scores the highest.

5   Conclusion

To summarize, culturally loaded subway poster design is not simply a pile of design elements. Rather, it is a careful choice of the elements of the regional culture, which has to be typical and unique, based on an extensive study of literature, a thoughtful consideration of elements to be used for the design, which has to be creative and innovative, based on the collective efforts of the team, and a systematic evaluation of the design, which has to be objective and accurate, in terms of visual aesthetics, typicality of cultural embodiment, and globalism of the design. This article thus puts forward suggestions above on the ways of making metro posters culturally loaded and operational procedures for designing culturally loaded metro posters, aiming at introducing some rough ideas of my own and inspiring valuable ideas from others for the later construction of subways in Mainland China.

References

1. China Subway. https://baike.baidu.com
2. Ma, Q.: Fundamental Theories on Public Art. Tianjin University Press, Tianjin (2008)
3. Liangpu, W.: Generalized Architecture. Tsinghua University Press, Beijing (2011)
4. Baidu Encyclopedia. https://baike.baidu.com
5. Liu, K.: Regional Culture and Local Studies. Xue Yuan Press, Beijing (2015)
6. Jia, R.: Poster—a form of advertising design language. J. Des. Forum, 37–39 (2000)
7. Cheng, X.: Public Art Design Principles and Creative Performances. China Water Conservancy and Hydropower Press, Beijing (2016)

Research on the Effect of Different Types of OLED Screens on Visual Search Performance and Visual Fatigue

Yuan Lyu[1], Yunhong Zhang[2(✉)], Wei Li[2],
Zhongting Wang[2], and Jiao Li[3]

[1] Elementary Education College of Capital Normal University, Beijing, China
[2] AQSIQ Key Laboratory of Human Factor and Ergonomics,
China National Institute of Standardization, Beijing, China
zhangyh@cnis.gov.cn
[3] Tianjin Normal University, Tianjin, China

Abstract. This study mainly focuses on performance and visual fatigue difference of visual search process with the different designs of OLED mobile phones screens under the situation of normal using. The research simulated light conditions of office environment, and 20 subjects aged 18–40 years old were randomly selected to explore the differences of visual search tasks performance, flicker fusion frequency, visual fatigue subjective perception and EEG under different conditions. The experiments is designed within the subjects and mobile phone screen with independent Blu-ray processing models, dependent variables for tasks completion performance and subjective perception of visual fatigue and comfort. The test results show that, under the confidence level of 95%, the performance of the search tasks in the non-eye mode mobile phone screen condition is superior to that under the eye protection mode. The decrease amplitude of the critical fusion frequency under the condition of the eye-protection mode phone screen was significantly smaller than that of the non-eye-protection mode. There was no apparent fatigue in the completed search tasks under the condition of the eye-protection mode phone screen, but there was apparent fatigue in that of non-eye-protection mode. EEG data showed that there was no difference in search attention and mood index between two different modes of mobile phone. The screen color displayed in the eye protection mode phone screen is better than that of the non-eye mode. On subjective fatigue, there was no significant difference in subjective visual fatigue experience after the same visual search tasks performed under two different cell phone screen conditions. In summary, the color display is better under the screen condition of the eye protection mode, and the fatigue phenomenon is not easy to appear. This research provides data support for the design of smart phone screen.

Keywords: Eye protection mode · Non-eye protection mode · Visual search
Visual fatigue · EEG

© Springer International Publishing AG, part of Springer Nature 2019
S. Fukuda (Ed.): AHFE 2018, AISC 774, pp. 67–75, 2019.
https://doi.org/10.1007/978-3-319-94944-4_8

1   Introduction

Smart phone has become an indispensable part of modern life because of its multi-functions, such as communication, social intercourse, entertainment and consumption and so on. Flat panel display has the characteristics of planarization, light, thin and power saving, which is in line with the inevitable trend of development of the future image display. Among them, as the third generation display technology OLED (Organic light-emitting diode displays), it has the advantages of low cost, small radiation, strong adaptability, energy saving, large visual angle and fast reaction speed, which represents the development direction of the display in the future [1]. "Blue light" refers to the part of visible light that is closest to ultraviolet light. The light frequency in the wavelengths range from 400 nm to 500 nm is slightly lower than ultraviolet rays and it is the most energetic parts of visible light. Blue light of 415–455 nm with the nature of short wavelength and high energy hurts the eyes most in visible lights, called harmful blue light [2]. Each pixel of the OLED screen can be self-luminous, as long as the voltage is input to the electrode, the excitation layer can produce the required color light. The blue light band produced is mainly concentrated in 460 nm, so it is not a harmful blue light. Therefore, OLED screens is generally considered to reduce eye damage. The external performance of visual fatigue is usually used as the subjective index to measure the degree of visual fatigue. Visual fatigue is called ocular fatigue syndrome in clinical, usually manifests as eye pain, distension, burning, foreign body sensation, tears, photophobia, blurred vision, diplopia, dry eyes, and even ocular inflammation and headache and so on, then cannot read [3]. Chengming Chen et al. (2004) conducted a study with comparing the visual fatigue caused by the OLED screen and the LCD screen. The study measured the blink frequency, heart rate variability, and electroencephalogram (αwave, βwave, α/β) of 12 subjects while viewing 2D images, as well as the objective indicators such as critical fusion frequency before and after watching. The results showed that there is a certain significance of EEG beta and alpha/beta, blink frequency, critical fusion frequency index and subjective scale of the subjects, while the EEG alpha and heart rate variability index were not significant. The experiment proves that compared with LCD screen, OLED screen can effectively reduce the visual fatigue caused by viewing screen [4]. But is it better to protect the eyes of OLED color screen after filtering the blue light? This study used the visual search tasks to verify that, through a series of eye jumps (saccades) and gaze (fixation), visual search tasks obtained the information of external stimuli and finding out the specific stimuli in a certain stimulus background with a stronger purpose. Compared with the ordinary viewing behavior, the requirements of more attention and more frequent knuckles will cause a higher degree of visual fatigue [5].

In the case of the same indoor lighting conditions, this study will use the visual search tasks to test the tasks performance, comfort and reduce the visual fatigue of different mobile phones, to verify the effect of OLED color screen after filtering the blue light to the eye protection mode.

2 Method

2.1 Design

This experiment was to test the subjective and objective differences of visual search tasks performance using mobile phone displays controlled by different displays (filtered blue light OLED display and general OLED display) under indoor light uniform conditions. This experiment was a two-factor within-subject design. Two factors were different screen samples and time factors. Samples contain two different conditions: filtered blue light OLED display and general OLED display, respectively. Time factor was before and after reading. Dependent variables were search time, response accuracy, visual fatigue, emotional response and comfort under different conditions of the tasks.

2.2 Participants

Seventeen ordinary right-handed adults from 18 to 35 years old (11 male and 9 female, mean age = 25.4, SD = 2.58) were recruited and paid to participate in the experiment. All of them had 4.8 normal or corrected-to-normal visual acuities and healthy physical conditions, without ophthalmic diseases.

2.3 Materials

The screen resolution of the general OLED display was 2650 * 1440 and its brightness was 366.36 cd. And the screen resolution of the filtered blue light OLED display was 1080 * 1920 and its brightness was 397.93 cd. The size of all the sample phones were 5.5 in.

2.4 Apparatus

The research used Standard logarithmic visual acuity chart developed by the eye hospital of WMU [4], the BD-II-118 type critical fusion frequency, single electrode EEG equipment to record the EEG indicators during the experimental process. The visual fatigue scale developed by James E. Sheedy was used to test the visual fatigue after the tasks, including eye fatigue (such as eye burning, eye pain, eye drain, eye irritation, eye tearing, visual blur, double vision and eye dryness, etc.) [5].

2.5 Procedures

Experiments were conducted in a quiet laboratory that experimental environment illumination value between 500 lx–700 lx. First, participants signed the informed consent and completed a general survey about their demographic information. The experimenter explained the experimental tasks and asked the subjects to practice until subjects could be very skilled in the experimental tasks. The experimental tasks was based on the ISO ISO9241-304 "Human-System Interactions of Human-System Interactions-Part 304: User Performance Test Methods for Electronic Vision Displays", which required subjects to find one target ring from a series of circles and click the left mouse button when the target ring was found. There were 3 blocks and 65 trails in each block. The searching time was about 30 min. The critical fusion frequency test was applied before and after the reading tasks. The subjects need wear EEG to record EEG changes in the entire mobile phone reading tasks process. The test sequence of samples in each subject was balanced by ABBA method. After completing the test of each sample, the subjective interviewed on the comfort and other aspects of all the samples. There was 20 to 30 min rest between each two tests to avoid visual fatigue. After completing the experiment, the subjects would got a certain reward.

2.6   Data Analysis

The changes value of visual search and visual fatigue data were analyzed by IBM SPSS 20 Statistics software (IBM-SPSS Inc. Chicago, IL). The method of repeated-measurement ANOVA analysis was applied to the experiment data to compare the differences of 3 factors from 3 quantitative indicators, namely: search time, response accuracy and subjective report data.

3   Results and Analysis

3.1   Visual Search Performance

The results of visual search tasks test showed that the tasks performance under the condition of general OLED display was better than that under the condition of filtered blue light OLED display. The results showed that the correct rate of tasks was higher under the condition of general OLED display (M = 98.23, SD = 2.08). Tasks completion time was shorter (M = 6618.80, SD = 2202.34). Under the condition of filtered blue light OLED display, the accuracy of tasks completion was lower (M = 94.52, SD = 14.54). Tasks completion time was longer (M = 7802.50, SD = 2000.88). The repeated-measured ANOVA analysis was applied the accuracy and completion time of users under different conditions. The results showed that there was no significant difference in tasks completion accuracy under different conditions (F = 1.231, p > 0.05). There was a significant differences in tasks completion time under different conditions (F = 5.128, p < 0.05). Tasks

Table 1. The analysis results of variance of tasks completion efficiency with repeated measurement under different conditions

	SS	df	MS	F	Sig. (2-tailed)	η^2
Accuracy of tasks completion under different conditions	137.938	1	137.938	1.231	.281	.061
Tasks completion time under different conditions	14011456.900	1	14011456.900	5.128	.035*	.213

Note: "*" indicates that there is a significant difference at the 95% confidence level.

completion time on general OLED display was significantly shorter than that on the filtered blue light OLED display (P < 0.05). Owing to the performance results of visual search tasks, the tasks completion efficiency of users in the general OLED display conditions was higher (Table 1).

3.2 Comparison of Critical Fusion Frequency

Critical fusion frequency was the objective measurement index of visual fatigue. The lower the critical fusion frequency was, the severer the visual fatigue was. The repeated measurement ANOVA was applied to the critical fusion frequency data before and after the visual search tasks on the general OLED display. The results showed that there was a significant difference between them (F = 18.701, p < 0.001). The CFF value of general OLED display after completing visual search tasks (M = 33.61, SD = 2.229) was significantly lower than that of before the tasks (M = 34.56, SD = 2.552) (P < 0.001). There was no significant difference between the CFF data before and after visual search tasks under the filtered blue light OLED display condition (F = 1.405, p > 0.05). The repeated-measured ANOVA analysis was applied to the CFF data of the general OLED display and the filtered blue light OLED display after visual search tasks. The results showed that there was a significant difference between the two conditions (F = 8.397, p < 0.001). The post hoc analysis showed that the CFF data after completing the tasks on the filtered blue light OLED display (M = 34.25, SD = 2.330) was significantly higher than that of the general OLED display(M = 33.61, SD = 2.229). The degree of visual fatigue increased after using the general OLED display to complete the visual search tasks, and the degree of visual fatigue was the same as that before tasks after using the filtered blue light OLED display to complete visual search tasks. That was, after using the filtered blue light OLED display to complete the three groups of visual search tasks, the degree of visual fatigue did not increase significantly, but after using general OLED display, the degree of visual fatigue increased significantly.

The repeated-measured ANOVA analysis was applied the accuracy and completion time of users under different conditions. The results showed the decreasing amplitude of the critical fusion frequency under the tasks conditions of two kinds of mobile phone screens. The results showed that the decreasing amplitude of critical fusion frequency under the general OLED display condition (M = −0.944,SD = 0.976) was larger than under the filtered blue light OLED display (M = −0.305, SD = 1.149), F = 8.397, p < 0.01, = 0.306. The results demonstrated that after using the general OLED display to complete the visual search tasks, the visual fatigue was significantly severer than that of after using the filtered blue light OLED display.

3.3 Subjective Assessment of Visual Fatigue

The results of subjective assessment of visual fatigue, included eye burning, eye pain, eye strain, eye irritation, eye tearing, visual blur, double vision, eye dryness and headache. Higher the value was, severer the visual fatigue was. The means of using the filtered blue light OLED display were lower than using the general OLED display in the eight indicators of eye burning, eye pain, eye strain, eye irritation, eye tearing, visual blur, double vision, eye dryness.

The repeated-measured ANOVA analysis of the visual fatigue after completing the visual search tasks on the two mobile phone screens showed that there was no significant difference in eye burning($F = 0.144$, $p > 0.05$), eye pain ($F = 1.598$, $p > 0.05$), eye strain ($F = 0.123$, $p > 0.05$), eye irritation ($F = 2.547$, $p > 0.05$), eye tearing ($F = 1.098$, $p > 0.05$), visual blur ($F = 1.399$, $p > 0.05$), double vision ($F = 1.049$, $p > 0.05$), eye dryness ($F = 1.240$, $p > 0.05$), and headache ($F = 0.032$, $p > 0.05$). The results showed that there was no significant difference in subjective visual fatigue after completing the visual search tasks under the general OLED display condition and the filtered blue light OLED display condition (Fig. 1 and Table 2).

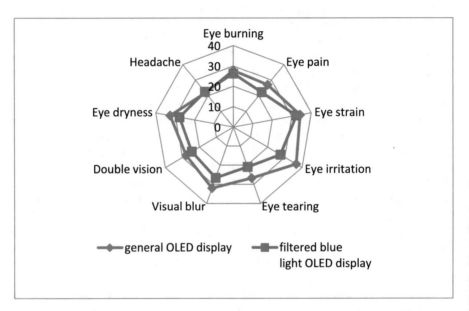

Fig. 1. Comparison of subjective visual fatigue after completing visual search tasks under different conditions

Table 2. Comparison of subjective visual fatigue after completing visual search tasks under different conditions

	SS	df	MS	F	Sig. (2-tailed)	η^2
Eye burning	22.5	1	22.5	0.144	.709	.008
Eye pain	225.63	1	225.63	1.598	.221	.078
Eye strain	36.10	1	36.10	0.123	.729	.006
Eye irritation	837.23	1	837.23	2.547	.127	.118
Eye tearing	336.40	1	336.40	1.098	.308	.055
Visual blur	286.23	1	286.23	1.399	.251	.069
Double vision	119.03	1	119.03	1.049	.319	.052
Eye dryness	235.23	1	235.23	1.240	.279	.061
Headache	3.03	1	3.03	0.032	.861	.002

3.4   The EEG Data

The EEG data included attention index and emotion index. The results showed that the attention index decreased slightly when using the filtered blue light OLED display for the visual search tasks (M = −5.00, SD = 11.71). The attention index rose slightly when using the general OLED display for the visual search tasks (M = 3.55, SD = 12.65). The repeated-measured ANOVA of attention index showed that the variation (post test - pretest) of attention index under the condition of filtered blue light OLED display was smaller than under the condition of general OLED display, and the difference between them was marginally significant (F = 4.343, p = 0.051). Users consumed more attention resources to complete the visual search tasks under the condition of filtered blue light OLED display than under the condition of general OLED display. When using the filtered blue light OLED display and the general OLED display to perform the visual search tasks, the emotion index increased. The rising amplitude of emotion index under the condition of filtered blue light OLED display (M = 5.15, SD = 15.030) was larger than under the condition of general OLED display (M = 6.05, SD = 12.032). The repeated-measured ANOVA of emotion index showed that there was no significant difference in the variation amplitude of emotion index under the condition of filtered blue light OLED display and general OLED display (F = 0.048, p > 0.05) (Table 3).

Table 3. The analysis of subjective visual fatigue caused by completing the visual search tasks under different mobile phone conditions

	SS	df	MS	F	Sig (2-tailed)	η^2
Attention index	731.025	1	731.025	4.343	0.051	0.186
Emotion index	8.100	1	8.100	0.048	0.829	0.003

3.5   The Results of Subjective Perception Scale

The results of the user's evaluation of the comfort and satisfaction of the two mobile phone screens showed that the difference was significant in display color preference (F = 4.411, p = 0.049) between the general OLED display and the filtered blue light OLED display. The preference for the color of filtered blue light OLED display (M = 66.15, SD = 12.816) was significantly higher than that of general OLED display (M = 57.45, SD = 17.503) (p < 0.05). However, there were no significant differences between the filtered blue light OLED display and the general OLED display in terms of overall comfort, difficulty of completing the visual search tasks, comfort of the display color and satisfaction of the overall display effect (see Table 4) (Fig. 2).

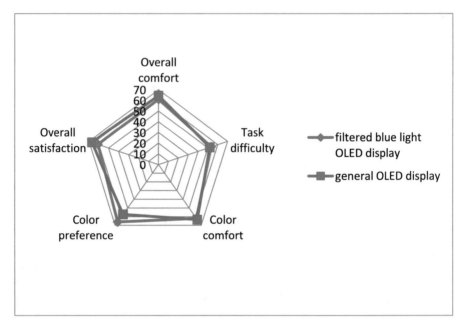

Fig. 2. Comparison of two mobile phone screens on subjective comfort

Table 4. The repeated-measured ANOVA of subjective questionnaires for different mobile phone screens

	SS	df	MS	F value	Sig. (2-tailed)	η^2
Overall comfort	119.025	1	119.025	0.496	.490	0.025
Tasks difficulty	19.600	1	19.600	0.90	.768	0.005
Color comfort	21.025	1	21.025	0.113	.741	0.006
Color preference	756.9	1	756.9	4.411	.049*	0.188
Overall satisfaction	372.1	1	372.1	2.116	.162	0.100

Note: The tagging "*" indicates a significant difference at the 95% confidence level.

In summary, according to the users' evaluation of comfort and satisfaction of the two mobile phone screens, they generally agreed that the overall comfort, tasks difficulty, color comfort and overall satisfaction of the general OLED display and filtered blue light OLED display were both good. But they thought the color experience of the filtered blue light OLED display was superior to the general OLED display.

4   Conclusion and Discussion

The purpose of this study was to compare the experience differences between the general OLED display and the filtered blue light OLED display in the visual search tasks. There were significant differences in the completion time of visual search tasks under the two conditions. The efficiency of visual search tasks under the condition of general OLED display was better than that under the condition of filtered blue light OLED display. At the same time, the decline rate of attention index while using the general OLED display to complete the visual search tasks was significantly lower than that of using the filtered blue light OLED display. These differences may be related with that the definition of the general OLED display was better than that of the filtered blue light OLED display and the edge design made it easier to display the edges. According to the test of visual fatigue, the users of the general OLED display had significant visual fatigue, but there was no significant visual fatigue after completing the tasks by using the filtered blue light OLED display. The results showed that, under the condition of filtered blue light OLED display, the decline rate of critical fusion frequency was significantly lower than which under the general OLED display. In addition, in terms of subjective color preference, the users preferred the color displayed on the filtered blue light OLED display, which may indicate that compared with the general OLED display, the filtered blue light OLED display was more comfortable in its design. It can effectively reduce visual fatigue. But the edge design of the filtered blue light OLED display needs to be improved.

Acknowledgment. The authors would like to gratefully acknowledge the support of the National Key R&D Program of China (2016YFB0401203), and China National Institute of Standardization through the "special funds for the basic R&D undertakings by welfare research institutions" (522018Y-5942, 712016Y-4940).

References

1. Shao, Z., Zheng, X., Yu, C.: OLED in flat panel displays. Chin. J. Liq. Cryst. Disp. **20**(1), 52–56 (2005)
2. Ma, X.: Welcome the arrival of the OLED TV era. View of Househ. Electr. Appl. **5**, 35–37 (2008)
3. Chen, L., Zhao, K.: Research on the cause of visual fatigue and prevention. Sect. Ophthalmol. Foreign Med. Sci. **29**(6), 367–370 (2005)
4. Standard logarithmic visual acuity chart developed by the eye hospital of WMU. People's Medical Publishing House, July 2012
5. Sheedy, J.E, Hayes, J., Engle, J.: Is all asthenopia the same? Optometry Vis. Sci. **80**(11), 732–739 (2003)

Verification of Brain Activity When Watching TV Commercials Using Optical Topography

Haruka Tanida$^{(\boxtimes)}$ and Toshikazu Kato

Chuo University, 1-13-27 Kasuga, Bunkyo-ku, Tokyo 112-8551, Japan
a13.stxe@g.chuo-u.ac.jp

Abstract. Traditionally, the efficacy of TV commercials has been evaluated using surveys. However, studies show that human decision-making can be unreliable, and may not reflect the true preferences of consumers. Because of this, there has been a growing interest in using physiological indices, such as brain activity, to measure consumer responses to advertising. The purpose of this study was to evaluate consumer's brain activity while watching TV commercials. This research is now being verified and we will discuss about the results in the AHFE. We believe that this research can be applied not only to TV commercials, but also to a wide range of video content, including Internet video advertisement, movies, virtual reality, and others.

Keywords: NIRS · Oxygenated hemoglobin · Neuromarketing

1 Introduction

In recent years, due to the spread of smartphones and social media, the video advertisement market is rapidly expanding. Along with diversification of media, advertisers are also developing video advertisements with various specifications. The video advertising market reached 84.2 billion yen, which is 157% of the previous year. The demand for smartphone video advertising has doubled compared to the previous year, and is expected to reach about 70% of the market as a whole. Moreover, it is expected to reach 291.8 billion yen by 2022 [1].

Conventionally, subjective methods such as questionnaires and interviews were often used to evaluate video advertisements. However, there are many uncertainties in human decision-making. The reasons for decision-making and the reasons for decision-making provided by respondents are not necessarily consistent with their original reasons [2]. Petter Johansson and colleagues [3] conducted experiments to investigate the reasons for selection in cases where subjects switched to a different image from the one they had selected as a favorite previously. Their results suggest that people cannot identify clear reasons for their decisions. McClure et al. [4] asked subjects to evaluate the taste of Coca-Cola and Pepsi-Cola. The results showed that there was a difference in taste preference evaluation depending on whether or not the brand name is displayed. This indicates that the evaluation is influenced by the brand name unknowingly even if

© Springer International Publishing AG, part of Springer Nature 2019
S. Fukuda (Ed.): AHFE 2018, AISC 774, pp. 76–81, 2019.
https://doi.org/10.1007/978-3-319-94944-4_9

the subject intends to evaluate the taste of the beverage. Therefore, the results of questionnaires and interviews do not necessarily reflect the preferences of respondents.

Therefore, in addition to subjective assessment, some researchers propose a more reliable evaluation method by considering objective evaluation using a physiological index. One such physiological index is the brain activity. In recent years, neuromarketing, which measures the brain's reactions and explores the processes underlying consumers' preference judgment, has attracted considerable attention. In neuromarketing, it is possible to estimate a preference based on brain activity, which is not based on consumers' conscious processing [5]. As a result, the potential preferences of consumers can be clarified, and this approach is expected to be useful for marketing research on newly marketed products and for the development of new products. In this study, we attempt to estimate the preference of subjects using analysis of brain activity when they are viewing video advertisements.

2 Previous Research

Azehara et al. [6] examined the relationship between the first impression of a television advertisement and the prefrontal cortex. It was found that different parts were activated at the time of viewing a video with a good impression and a video with a bad impression. Based on this result, in this research, we measure the brain activity of the prefrontal cortex related to the impression produced by television advertisements.

Hofer et al. [7] presented a positive image and a negative image to the subjects and measured the brain activity at that time. They observed differences in the brain activities of men and women, and it was shown that women showed greater response to negative images. Gender differences in brain activity can occur not only when watching images but also when watching movies. However, in the study by Azehara et al. [6], gender differences were not considered. In this research, we measure brain activities when watching TV commercials and also verify gender differences.

3 The Present Study

In this research, we measure brain activity data when presenting television advertisements as a visual stimulus and verify the difference between men and women.

3.1 TV Advertisement

The involvement of celebrities to raise awareness of an advertisement is recognized as one of the most effective strategies. In addition to raising awareness, the use of celebrities in advertisements is deeply related to the distribution, pricing, and brand strategy [8]. It can be said that an evaluation of celebrities appearing on TV advertisements is important for advertisement strategy. Furthermore, since this study aims to clarify the difference between males and females, we selected television advertisements with first-person viewpoint video, whose target audience (male or female) is clear. The first-person view images are movies that are filmed from a viewpoint similar to that of the observer. We chose a relatively new TV advertisement that was broadcast after 2013. In addition, since the subjects of this study are students, the age range of the talent was set to 19 to 35 years. Ten television advertisements were selected (5 male talents and 5 female talents) (Table 1).

Table 1. Sex of the talent and product name of the selected TV commercial

№	Talent	Product
1	Male	Instant food
2		Soup
3		Air conditioner
4		Gum
5		Cell Phone
6	Female	Car
7		Alcohol
8		Instant food
9		Alcohol
10		Mobile application

3.2   Control Task

Azehara et al. [6] prepared control tasks to extract only the impression factors of TV advertisements. The control task is a video that combines randomized pixel values of the image of TV advertisements and plays the sound in reverse. They are meaningless pictures having the same physical quantity as TV advertisements. In this research, we also use this method to compare the brain activity when watching television advertisements and the brain activity when watching the control task with the same physical quantity as that of the TV advertisement. By doing so, we believe that we can extract brain activities that respond only to impressions among various factors.

3.3   Measurement of Brain Activity

In this study, the brain activity of a subject during presentation of a visual stimulus was measured using near-infrared spectroscopy (NIRS) (Fig. 1). The NIRS device records the concentration changes in oxy-hemoglobin (Hb) and deoxy-Hb in the cerebral blood flow. The near infrared light irradiated to the head is absorbed, diffused, and approaches the cerebral cortex. Thereafter, it is condensed onto the optical fibers attached on the surface of the head. Because the absorption spectrums of oxy-Hb and deoxy-Hb differ, NIRS can measure changes in the amount of oxy-Hb and deoxy-Hb. Brain activity is also measured using positron-emission tomography (PET) and magnetic resonance imaging (MRI). However, these approaches measure brain activity with the subject's body restrained, and therefore it is difficult to measure brain activities in a natural state. In comparison, NIRS has the advantage that brain activity can be measured non-invasively without placing constraints on the position and orientation of the body.

Fig. 1. NIRS (ETG-4000 Hitachi Medical Co.)

4 Experimental Method

The subjects included 10 right-handed university students (5 female and 5 male students). We presented a TV advertisement for 15 s and control tasks for 15 s around the TV advertisement (Fig. 2). This process was repeated 10 times, because we prepared 10 TV advertisements. The TV advertisements were shown randomly, to avoid the order effect. The erebral blood flow was measured with NIRS at the time of visual stimulus presentation. We used a measurement probe with a 3 × 11 holder and acquired cerebral blood flow data from 52 channels. In accordance with the International 10–20 system, we measured activity in the prefrontal cortex.

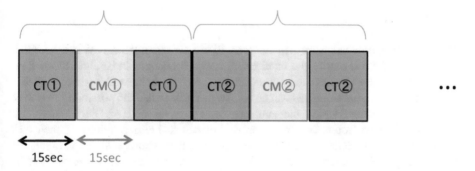

Fig. 2. Initially, the control task is presented for 15 s. Next, we present the TV advertisement for 15 s. Finally we present the control task again.

The above visual stimulus is presented to all subjects and the following four patterns are compared (Fig. 3). 1. Male subjects are watching TV advertisements for male talent. 2. Male subjects are watching TV advertisements for female talent. 3. Female subjects are watching TV advertisements for male talent. 4. Female subjects are watching TV advertisements for female talent.

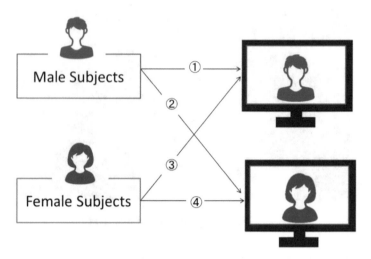

Fig. 3. Four patterns to compare

Yokoyama et al. [9] state that the attention elements of TV advertisement are different by male and female. Men pay attention to the title at the beginning, catch phrases of products, commodity cuts and so on. On the other hand, women pay attention to expression that requires interpretation and sympathy, such as talent facial expressions and imaginary scenes. Therefore, we made the following hypothesis

Hypothesis: Under pattern 3 (Female subjects are watching TV advertisements for male talent), the changes in cerebral blood flow are the largest.

5 Outlook

This research is now being verified and we will discuss about the results in the AHFE.

In this study, we clarify the relationship between changes in cerebral blood flow and the impression of television advertisements, and we believe that a more reliable evaluation of TV advertisements can be achieved. There is also the possibility to clarify potential preferences, which the audience have not yet noticed. Furthermore, by obtaining objective indicators based on differences in brain activities of men and women, it becomes clear which TV advertisements give a strong impression to men or women. Especially, internet video advertisements, which are rapidly expanding in market size, can be targeted and delivered separately for each gender. By distinguishing TV advertisements that suit men and women, it is possible to deliver videos more effectively.

We believe that this research can be applied not only to TV commercials, but also to a wide range of video content, including internet video advertisements, movies, virtual reality, and others.

Acknowledgments. We are deeply grateful to the members of Human Media Engineering Laboratory of the Faculty of Science and Technology, Chuo University, and the Kansei Robotics Research Center, for their participation in research discussions and collaboration in experiments.

This work was partially supported by a JSPS KAKENHI grant, "Research on Sensitivity Symbiosis Mechanism within Groups in Real Space/Information Space" (No. 25240043) and a TISE Research Grant from Chuo University, "KANSEI Robotics Environment".

References

1. Cyber Agent: Market research on domestic video advertisement (2016)
2. Hashimoto, Y., Tsuzuki, T.: Mismatch between interpretation of intention and results in multi-attribute decision-making: the choice blindness paradigm and decision-making by bounded rationality. Rikkyo Psychol. Res. **55**, 45–53 (2013)
3. Johansson, P., Hall, L., Sikstrom, S., Olsson, A.: Failure to detect mismatches between intention and outcome in a simple decision task. Science **310**, 116–119 (2005)
4. McClure, S.M., Li, J., Tomlin, D., Cypert, K.S., Montague, L.M., Montague, P.R.: Neural correlates of behavioral preference for culturally familiar drinks. Neuron **44**, 379–387 (2004)
5. Shibata, T.: Measuring the decision-making process of purchase: from neuroeconomics to neuromarketing. IEICE **96** (2013)
6. Azehara, S.: Analysis between First Impression of Likability and Change of Oxy-Hb in Frontal Cortex during Watching Video, Chuo University Graduate School Master's thesis (2018)
7. Hofer, A.: Gender differences in regional cerebral activity during the perception of emotion. NeuroImage **32**(2), 854–862 (2006)
8. Su, P.J.: The effect of celebrity advertising on advertising communication: literature review and research directions. Waseda Bus. Rev. **44**, 21–37 (2009)
9. Yokoyama, R.: Have Your TV ADS Reached Audience? (2017)

Affective Design in Healthcare

Evaluation of the Effect of the Amount of Information on Cognitive Load by Using a Physiological Index and the Stroop Task

Yushi Hashimoto[✉], Keiichi Watanuki, Kazunori Kaede,
and Keiichi Muramatsu

Graduate School of Science and Engineering, Saitama University,
255 Shimo-okubo, Sakura-ku, Saitama-shi, Saitama 338-8570, Japan
y.hashimoto.687@ms.saitama-u.ac.jp

Abstract. There have been several recent attempts to aid car drivers by providing information on internal and external car environments. The optimal amount of information must be determined to avoid confusion. In this study, the "Stroop task" was used for information processing, and the cognitive load was gradually increased by adding information in stages. We designed and conducted two tasks that originate in the "Stroop task"; these two tasks feature significant differences in cognitive load. We also measured brain activity using near-infrared spectroscopy (NIRS) under the assumption that such activity can be used as an index of cognitive load. Both tasks were associated with increased oxy-hemoglobin levels in the prefrontal area, and the task with a higher cognitive load was associated with a more substantial increase in oxy-hemoglobin; this indicates that oxy-hemoglobin levels may be used as an objective index for the evaluation of information-associated cognitive load.

Keywords: Near-infrared spectroscopy · Cognitive load · Mental work load

1 Introduction

Recently, there have been several attempts to aid car drivers by providing information on internal and external car environments; external information may include traffic signs and signboards, while internal information may include heads-up displays (HUD) and sound navigation. Drivers must be able to recognize and process information from their driving environment such as load condition, degree of congestion, and weather. The driving aids mentioned above could make driving safer and more comfortable, if used appropriately. However, if an excess of information is presented, resulting in mental overload, the driver may become confused, which may potentially become dangerous. Therefore, the optimal amount of information (i.e., that which enhances but does not impede normal driving operation) must be determined. In the present study, "information" refers to visual and audio stimuli, as sight and hearing are the two main senses used to recognize information while driving.

The information processing tasks designed herein originate in the "Stroop task." We add sound information to the original "Stroop task" and gradually increase cognitive

© Springer International Publishing AG, part of Springer Nature 2019
S. Fukuda (Ed.): AHFE 2018, AISC 774, pp. 85–93, 2019.
https://doi.org/10.1007/978-3-319-94944-4_10

load by adding information in stages. We use the term "mental workload" to denote information processing while driving in accordance with previous research [1]. The cognitive load, or mental workload, associated with given tasks is measured; the mental workload indicates workload and work strain following mental activity. Mental workload can be assessed using three different indices: the performance index, the subjective index, and the physiological index. The cognitive load and the task difficulty are evaluated using the correct answer rate as a performance index and the NASA-Task Load Index (NASA-TLX) as a subjective index. Subsequently, the brain activity is measured and used as a physiological index in response to changes in the cognitive load and task difficulty level. Oxy-hemoglobin (OxyHb) in the prefrontal area measured by near-infrared spectroscopy (NIRS) is used as a brain activity indicator. We use OxyHb to index cognitive load variations following changes in the amount of information.

2 Mental Workload

Mental workload is defined as "mental stress" and "mental strain," where mental stress refers to work characteristics and mental strain refers to worker response. Various methods for measuring mental workload have been developed; the NASA-TLX and Subject Workload Assessment Technique (SWAT) are practical examples. In this study, the Japanese version of the NASA-TLX is used, as the subject's native language is Japanese [2]. Mental workload measurement methods can be classified into performance indices, subjective indices, and physiological indices. The performance index covers measurable behavior during work, such as correct answer rate and reaction time. The subjective index gauges subjective symptoms measured via questionnaires. Physiological indices are based upon biological information [2, 3]. We use correct answer rate as a performance index, NASA-TLX as a subjective index, and OxyHb as a physiological index.

Both the mental and physical workloads have temporal aspects such as accumulation, as shown in Fig. 1. Overload1 occurs when the max workload exceeds the worker processing capacity. Overload2 occurs when the accumulated workload exceeds the worker limit. In this study, we consider overload2 affects the physiological index [4].

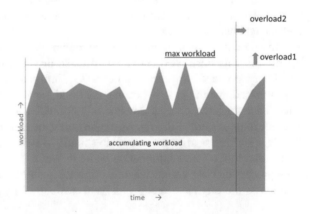

Fig. 1. Workload component (refer from previous research [3]).

3 Measurement of Brain Activity

In this study, the Stroop-based task tests information processing. Relationships between blood flow and solving Stroop tasks has been reported in previous studies [4, 5]; it has been shown that increases in OxyHb occur predominantly in the lower left frontal gyrus. The relation between Stroop task completion and brain function is described from the viewpoint of working memory, and brain activation in the ACC portion has been reported [6]. Other areas are also closely related to working memory, such as the prefrontal dorsal region and the ventromedial region, which are located on the left outer side from the central prefrontal region. In this study, cerebral blood flow measurement is performed on the left outer side of the prefrontal cortex. Because the OxyHb value is relative, the value was standardized using Eq. 1.

$$\text{Oxy}(t)_{z-\text{score}} = \left\{ \text{oxy(t)}_{\text{raw}} - \mu_{\text{rest}} \right\} / \sigma_{\text{rest}}. \tag{1}$$

4 Stroop Task

The Stroop task presents words representing various colors with various character colors, and the subject delivers an answer consisting of the character color. Previous work has shown that answer times are longer in cases in which the character color and the meaning of the word do not coincide with each other (as compared with cases in which the character color matches the meaning of the word). This is because the processing of sensory information (character color) and language information (character meaning) causes a cognitive conflict and causes the Stroop effect to lengthen the reaction time when answering with the character color. For example, a stimulus showing the word "red" written with blue ink indices a higher cognitive load than a stimulus showing "red" written with red ink. In this study, in addition to the original Stroop task presented with font color and character meaning, we designed an experiment that loads auditory speech of color words; this is intended to elucidate the difference in the amount of information between the original Stroop task and the new task and to measure the corresponding change in the mental workload.

Previous studies have reported that it is difficult to memorize the arrangement of the buttons used to answer the Stroop task [7]. We address the problems caused by (1) the answering method (in which button memorization itself becomes a task) and (2) body movement (in which pressing the button affects biometric measurements). For the sake of explanation, the conventional Stroop task is termed "Original," the approach based on the conventional Stroop task involving an optimized answering method is termed "Task 1," and the approach with added sound is named "Task 2" (this task also uses the optimized the answering method). Figure 2 shows an example of an Original answer. In this case, font color answers are affected by the problems mentioned above. Figure 3 shows an example of a Task 1 answer; button 1 represents the correct answer if the character meaning and the text color do not match, while button 2 represents the correct answer if the character meaning and the text color do match. Figure 4 shows an example of a Task 2 answer. Button 1 represents the correct answer if the character meaning, text color, and speech meanings do not match; button 2 is used if two of the three match; and button 3 represents the correct answer when all three stimuli match. That is, buttons further to the right indicate increasing numbers of matched color stimuli, and the button design is simplified.

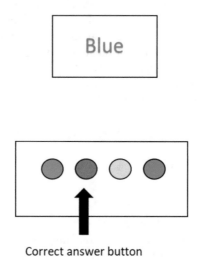

Correct answer button

Fig. 2. An example of an original answer.

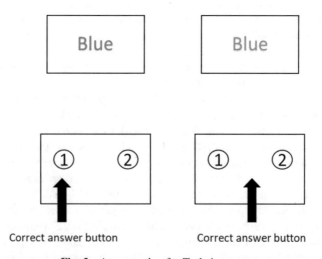

Correct answer button   Correct answer button

Fig. 3. An example of a Task 1 answer.

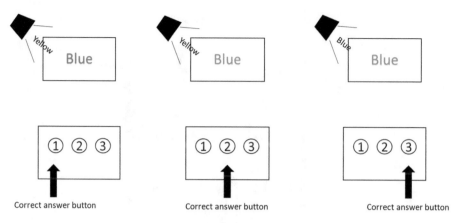

Fig. 4. An example of a Task 2 answer.

5 Experiment

Task 1 and Task 2, described in the previous section, were performed five times each for each subject, and brain activity was measured by NIRS throughout the experiments. The subject responded to NASA-TLX questionnaire for each experiment. During a total task time of 300 s, the stimulation was presented for 5 s 50 times with 1 s of rest between presentations. The subjects were healthy males in their 20 s. The experiment schedule is shown in Fig. 5, and the experimental outline is shown in Fig. 6.

Fig. 5. Task schedule

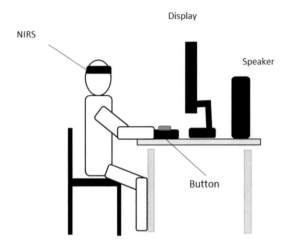

Fig. 6. Experiment environment.

6   Results

Task 1 and Task 2 results were averaged for each subject. Due to mental workload accumulated in the middle and/or final stages of the experiment, overload2 occurred, which is thought to affect cerebral blood flow. Thus, the 300 s ASK time was divided into three equal parts, and the average values of the parts were compared. In Task 2, OxyHb was observed to increase with time in 4 of the 10 subjects. In 5 of the 10 subjects, OxyHb increased from the middle to the end of the experiment after decreasing. Averaging was performed for each of the two groups of subjects. Results of multiple comparisons between rest and each of the three task sections are shown in Figs. 7 and 8. Bonferroni's multiple comparison method was used. The remaining subject displayed no significant reaction between the central prefrontal area and the left outer side. There are subjects for which OxyHb increases are observed during both Task 1 and Task 2, but Task 2 features greater increases; in other experiments, no

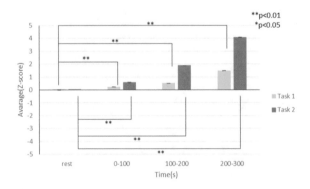

Fig. 7. Comparisons of scores in different thirds of the experiment for NIRS-Group 1.

increases were observed during Task 1, but increases occurred in the second half of the experiment during Task 2.

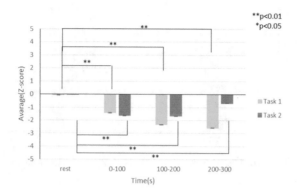

Fig. 8. Comparisons of scores in different thirds of the experiment for NIRS-Group 2.

T-test results between Task 1 and Task 2 for the NASA-TLX index and the correct answer rate are shown in Figs. 9, 10 and 11. Both mental demand and the mean weighted workload score (WWL), which shows the magnitude of the load, increased significantly from Task 1 to Task 2, while the correct answer rate declined significantly. Both results indicate that the workload evidenced by the physiological index is larger in Task 2. Data corresponding with actions, such as button check, were not used in this analysis.

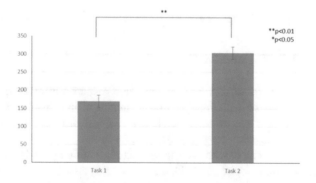

Fig. 9. Comparison of NASA-TLX mental demand between tasks.

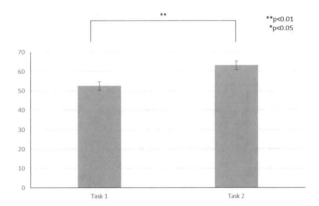

Fig. 10. Comparison of NASA-TLX WWL between tasks.

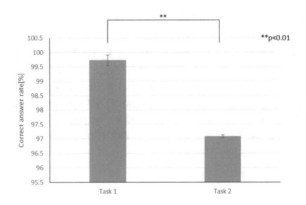

Fig. 11. Comparison of correct answer rates between tasks.

7 Discussion and Conclusion

We measured mental workload via brain activation response in order to evaluate the cognitive loads associated with changes in the amount of information presented. Performance and subjective indices were used to evaluate task difficulty and cognitive load levels during brain activation response measurements.

Subjects were presented with two tasks featuring differences in the amount of information given. Subjective NASA-TLX index results confirmed that mental burden and WWL increased significantly with increasing amount of information. The results also show a difference in load between the two tasks due to mental factors. The correct answer rate, which was used as a performance index, decreased significantly with increasing amount of information, which showed that the difficulty of a task is increased by increasing the amount of information; this result is in agreement with the subjective evaluation. These two results indicate that there was a significant difference in cognitive load between the two tasks.

Regarding the physiological index, subjects fell into two groups. In the task with the larger amount of information (Task 2), subject Group 1 underwent an increase in OxyHb between the middle and final stages. In Group 2, reductions in OxyHb were observed early in the task, while OxyHb increased between the middle and final stages; OxyHb did not increase at the beginning of the task, but did increase after the middle stage. This may be due to an accumulation of mental workload resulting in exceedance of processing limits. In one instance, OxyHb increased even in the task with the least amount of information; however, the OxyHb variation was small compared with the task with a large amount of information. We conclude that the difference in results between the two tasks arose from the differences in cognitive load between the tasks.

References

1. Akihiro, S., Shuji, O.: Trends in Ergonomic Problems in Roads and Traffic Engineering-Introduction of mental workload measurement and NASA-Task Load Index-. Development Public Works Research Institute Monthly Report, No. 561, pp. 9–14 (2000). (in Japanese)
2. Shigeru, H., Naoki, M.: Japanese version of NASA task load index: sensitivity of its workload score to difficulty of three different laboratory tasks. Ergonomics 32(2), 71–79 (1996). (in Japanese)
3. Shigeru, H: Mental Workload Theory and Measurement (2001). (in Japanese)
4. Michihiro, F., Mitsunori, M., Hisatake, Y., Tomoyuki, H.: Discussion of the relation between the cerebral blood flow and reaction time during stroop test. Sci. Eng. Rev. Doshisha Univ. 53(4), 19–24 (2013). (in Japanese)
5. Ryohei, M., Yoshihide, M.: Relationship between stroop task difficulty and brain activation levels − analysis of changes in blood concentrations of oxygenated hemoglobin (Oxy-Hb) based on near infra-red spectroscopy (NIRS). J. Teikyo Heisei Univ. 24(1), 199–203 (2013). (in Japanese)
6. Naoyuki, O.: Prefrontal cortex and working memory. High. Brain Funct. Res. 32(1), 7–13 (2012). (in Japanese)
7. Yongning, S., Yuji, H.: Development and practice effect of a new computer-based stroop/reverse-stroop test. Cognit. Psychol. Res. 9(1), 19–26 (2011). (in Japanese)
8. Schroeter, M.I., Zysset, S., Kupka, T., Kruggel, F., Yves Von Cramon, D.: Near-infraed spectroscopy can detect brain activity during a color-word matching stroop task in an event related design. Hum. Brain Map 17(1), 61–71 (2002)
9. Rasmussen, J.: Information Processing and Human-Machine Interaction, North-Holland, p. 215 (1986)
10. Task Force of the European Society of Cardiology and the North American Society of Pacing and Electrophysiology: Heart rate variability. Standards of measurement, physiological interpretation, and clinical use. Circulation 93(5), 1043–1065 (1996)
11. Daneman, M., Carpenter, P.: Individual differences in working memory and reading. J. Verbal Learn. Verbal Behav. 19, 450–466 (1980)

Examination of the Brain Areas Related to Cognitive Performance During the Stroop Task Using Deep Neural Network

Tomohiro Nishikawa$^{(\boxtimes)}$, Yushi Hashimoto, Kosei Minami,
Keiichi Watanuki, Kazunori Kaede, and Keiichi Muramatsu

Graduate School of Science and Engineering, Saitama University,
255 Shimo-okubo, Sakura-ku, Saitama-shi, Saitama 338-8570, Japan
t.nishikawa.398@ms.saitama-u.ac.jp

Abstract. To examine brain areas related to the cognitive load condition during the Stroop task, we proposed a method using a Deep Neural Network (DNN). We acquired cerebral blood flow data in congruent and incongruent tasks by near-infrared spectroscopy (NIRS) equipped with 22 ch. The data were used to train a DNN, and the influence of each factor on the output was evaluated. Our DNN model consists of independent input layers for each channel of NIRS, as well as fully-connected hidden layers and output layers. Our results suggest that the medial prefrontal cortex (focusing on cognition) and the left inferior frontal gyrus (focusing on language processing) were involved in the cognitive load during the Stroop task. These results in the Stroop task were consistent. Therefore, the proposed method's utility was confirmed.

Keywords: Stroop task · Near-infrared spectroscopy · Cognitive load
Deep Neural Network

1 Introduction

The Stroop task is a popular cognitive task in which the participant is shown a word written in colored letters and asked to name the color. When the letter color disagrees with the meaning of the word (incongruent condition), the participant will typically respond with a delayed discrimination time and an increase in the number of false reactions, compared with when the two attributes are consistent (congruent condition). This phenomenon is known as the Stroop effect or Stroop interference [1].

In the past decade, many studies have been conducted regarding cerebral blood flow on cognitive tasks [2]. To evaluate the cognitive load on the Stroop task, it is necessary to examine the tendency of cerebral blood flow. Generally, statistical methods, such as averaging data, are used to perform this examination. However, cerebral blood flow exhibits large individual differences, and these statistical methods may not be entirely appropriate.

This research aimed to examine the brain area involved in the cognitive load during the Stroop task. First, we measured cerebral blood flow during the congruent and incongruent conditions on the Stroop task. To measure cerebral blood flow, near-infrared

© Springer International Publishing AG, part of Springer Nature 2019
S. Fukuda (Ed.): AHFE 2018, AISC 774, pp. 94–101, 2019.
https://doi.org/10.1007/978-3-319-94944-4_11

spectroscopy (NIRS), which is a neural imaging tool capable of measuring the hemo-dynamic change in the cerebral cortex, was used. Second, we proposed a feature extraction method using a Deep Neural Network (DNN); this is a machine learning algorithm, extended from a neural network, which consists of artificial neurons that model human cranial nerve cells. The learning process comprises iteratively adjusting the synaptic weights of the network to classify data using back propagation training. In typical, machine learning models are "black boxes;" however, we extract features by evaluating the influence of each factor on the output.

2 Near-Infrared Spectroscopy (NIRS)

NIRS is a neural imaging tool, capable of measuring hemodynamic changes in the cerebral cortex [3, 4]. These measurements are performed using a near-infrared light with a wavelength of 700–900 nm, which exhibits a high transmittance through the body and a variable absorbance, depending on the oxygenated state of hemoglobin. The neural activity in the brain is analyzed by measuring changes in oxygenated hemoglobin concentration (oxy-Hb) and deoxygenated hemoglobin concentration (deoxy-Hb), which result from changes in blood flow. In general, oxy-Hb increases in blood when a certain part of the brain is activated [5]. In cognitive tasks such as the Stroop task, cerebral blood flow in the prefrontal cortex (related to cognitive function) is measured, and oxy-Hb is used as an indicator of cerebral activation reaction.

3 Acquisition of Cerebral Blood Flow During the Stroop Task

To acquire and evaluate brain activity during cognitive load, we measured cerebral blood flow during the Stroop task.

3.1 Experimental Procedure

Each participant performed the congruent condition task for 45 s, and the incongruent condition task for 45 s. Rest sections of 45 s were included before and after each task. Each participant named the color by pressing the button at hand. Each participant named the color by pressing the button at hand. There were four letters, indicating each of the following colors: red, blue, yellow, and green; the letters were Japanese. A total of 28 healthy Japanese volunteers participated. The participants' mean age was 22 years (range = 20–26, SD = 1.08 years). All participants exhibited normal color vision and normal linguistic-cognitive development; further, all participants were native Japanese speakers. To assess brain activity, we measured oxy-Hb changes in the prefrontal cortex by wearable optical topography (WOT-220, manufactured by Hitachi High-Tech Solutions Corporation). The measurement position consisted of 22 channels on the prefrontal cortex, based on the International 10–20 system.

3.2   Results of Cerebral Blood Flow Assessment

The changes in oxy-Hb in the prefrontal cortex, averaged among participants, are shown in Fig. 1. The congruent task is indicated by red lines and incongruent task by blue lines. Some channels (ch. 1, 2, 5, 17, 19, 20, 21, and 22) did not work well, depending on interference by hair or other physical barriers. There was no significant difference between congruent and incongruent tasks in terms of oxy-Hb. Individual differences were large because of the large variance. These results suggest that cerebral blood flow exhibits large individual differences, and that statistical analysis, such as averaging, may not be appropriate. Moreover, it is difficult to evaluate the areas of the brain that are related to cognitive load using these methods.

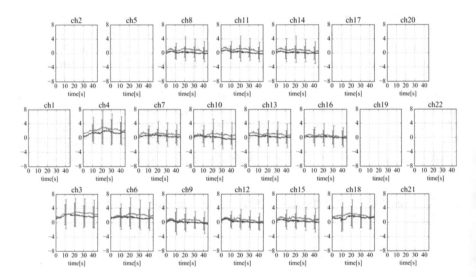

Fig. 1. Changes in oxygenated hemoglobin (oxy-Hb) in the prefrontal cortex. The congruent task is indicated by red lines and incongruent task by blue lines. The arrangement of the graphs reflects the location of the channels on the prefrontal cortex.

4   Method

To examine the brain area involved in the cognitive load during the Stroop task, we proposed a feature extraction method using a DNN.

4.1   Preprocessing

During DNN learning, oxy-Hb data is normalized. By normality testing, it was confirmed that the data followed a normal distribution (Average = 0.94, SD = 2.92). Thus, the significance level was set to 1%, and the value in the adoption area (-5.84 to 7.73) was normalized from 0 to 1. Data outside the adopted area was smoothened to 0 or 1. In addition, the data measurements at 5 Hz were averaged every second.

4.2   Training

Oxy-Hb data was trained by our DNN model, as shown in Fig. 2. Our DNN model structure consists of independent input layers for each channel of NIRS, as well as fully-connected hidden layers and output layers.

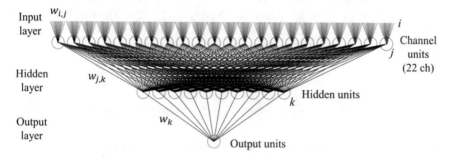

Fig. 2. Our Deep Neural Network (DNN) model consists of independent input layers for each channel of near-infrared spectroscopy, along with fully-connected hidden layers and output layers. DNN trains oxy-Hb data so that the output value approaches 0 for the congruent task and 1 for the incongruent task.

An input layer collects the time series data of each channel by weighting. Assuming that data length is i (= 45), the number of channel unit is j (= 22), input data is $x_{i,j}$, weights is $w_{i,j}$, and the activation function is f, the output of the channel unit j is as follows.

$$y_j = f\left(\Sigma_i x_{i,j} w_{i,j}\right) \tag{1}$$

A hidden layer collects the input from the 22 channel units by weighting, thereby enabling nonlinear classification. Assuming that the number of hidden units is k (= 10) and weights is $w_{j,k}$, the output of the unit k is as follows.

$$y_k = f\left(\Sigma_j y_j w_{j,k}\right) \tag{2}$$

An output layer collects the input from the hidden layer by weighting. Our DNN has one output unit, which is constructed as follows.

$$y = f\left(\Sigma_k y_k w_k\right) \tag{3}$$

The activation function is "Sigmoid", shown in (4). The output of each unit is centered at 0.5, and approaches 0 (if it takes a negative value) or 1 (if it takes a positive value).

$$f(x) = 1 / (1 + \exp(-x)) \tag{4}$$

Our DNN trains oxy-Hb data so that the output value approaches 0 for congruent task and 1 for incongruent task. The number of data entries for training was 56 (28 participants × two tasks). The number of training repeats (epoch) was set to 10000. To reduce overfitting, dropout [6] was used. Dropout was adapted in each layer with ratio = 0.1. For an optimization method, we used Adam [7] with a step size alpha = 0.001, exponential decay rate of the first-order moment beta1 = 0.9, second beta1 = 0.999 and small value for the numerical stability = 1e − 0.8. The weight matrix was initialized with Gaussian samples, each of which has zero mean and a deviation d as calculated in (5). "*In_size*" is number of input units.

$$d = \sqrt{1/in\_size} \tag{5}$$

The model was implemented using the Python library Chainer [8].

4.3   Analysis of the Influence Factor of Each Channel on Output

After training, using weights of the DNN, we analyzed how much each channel unit affect the output, which was classified as an incongruent task. The learned weight expressed the influence on each unit. Thus, focusing on the output layer, using output weights w_k, the influence factor a_k of each hidden unit was defined by the following equation.

$$a_k = f(w_k) \tag{6}$$

Focusing on the hidden layer, using hidden weights $w_{j,k}$, the influence factor a_j of each channel unit was defined by the following equation.

$$a_j = \Sigma_k\{f(w_{j,k})a_k\} \tag{7}$$

Because a_k corresponds to each channel of NIRS, the channel with larger a_k has more features classified within the incongruent task. Focusing on the input layer, using input weights $w_{i,j}$ and a_k, the time series influence factor of each channel $a_{i,j}$ was defined by the following equation. By mapping this, features of oxy-Hb change in each channel were evaluated.

$$a_{i,j} = f(w_{i,j})a_j \tag{8}$$

5   Results

Since the result changes according to the initial value of the model, 100 experiments were conducted and evaluated by averaging. The learning curve was shown in Fig. 3. Mean squared error (MSE) loss, in accordance with the increase of epoch, was reduced. MSE loss of epoch = 10000 was 0.120.

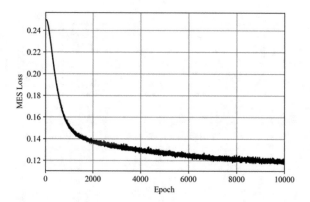

Fig. 3. Learning curve of training mean squared error (MSE) loss. The horizontal axis is the epoch and the vertical axis is the training MSE loss.

The influence factor a_j of each channel unit is shown in Table 1. Notably, ch. 4, 10, 11, 14 and 15's units exhibited a particularly large influence. This suggests that larger features of the incongruent task were included in ch. 4, 10, 11, 14, and 15, within the prefrontal cortex.

Table 1. Influence factor of each channel unit a_j. Larger influence factor is related to greater influence of the output on the incongruent task.

Influence factor aj			
ch.1	0.00	ch.12	2.82
ch.2	0.00	ch.13	2.63
ch.3	2.78	ch.14	3.57
ch.4	4.23	ch.15	3.43
ch.5	0.00	ch.16	0.93
ch.6	0.68	ch.17	0.00
ch.7	0.73	ch.18	0.89
ch.8	1.39	ch.19	0.00
ch.9	2.92	ch.20	0.00
ch.10	3.40	ch.21	0.00
ch.11	3.50	ch.22	0.00

Time series influence factor of each channel $a_{i,j}$ is shown in Fig. 4. In ch. 3, 4, and 10, the influence factor was varied. In ch. 9, 12, and 15, the influence factor was small at the beginning of the task and was large during the task.

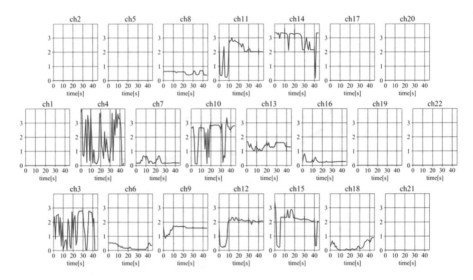

Fig. 4. Time series influence factor of each channel $a_{i,j}$. The arrangement of the graphs reflects the location of the channels on the prefrontal cortex.

6 Discussion

Our results showed that, in ch. 3, 4, and 10, the influence factor was varied. There is a possibility that the DNN trained noise, which might be body motion during the task. The main cause of body motion was button selection by hand. This was a limitation of the experimental setting. However, variation of the influence factor might have a periodic effect on cerebral blood flow, because of cognitive load. Furthermore, in ch. 9, 12, and 15, the influence factor was small at the beginning of the task and large during the task. This means that, at the initiation of the task, there was no feature to indicate difference between congruent and incongruent tasks; this feature appeared during the task. These channels are located in the medial prefrontal cortex (mPFC; focusing on cognition) and the left inferior frontal gyrus (LIFG; focusing on language processing). Therefore, it may be that the influence of language processing on cognitive load during the incongruent task was extracted. Further, cognitive loads may accumulate, so it is reasonable that the influence factor became larger in the latter half of the task. In the Stroop task, the load is suspected to occur during language processing, so the result was reasonable. By applying this analysis method, it will be possible to use NIRS to examine the brain areas involved in tasks within other experiments. In the future, we plan to conduct experiments with tasks that incorporate higher cognitive loads of various types. In addition, some channels (ch. 1, 2, 5, 17, 19, 20, 21, and 22) did not work well when using simple NIRS. To accurately evaluate brain activity, it may be necessary to use NIRS methods that can measure in a wider range.

7 Conclusions

Change in oxy-Hb average among participants exhibited no significant differences between congruent and incongruent tasks. Individual differences were large because of a great degree of variance. This suggested that cerebral blood flow exhibits large individual differences, and that statistical analysis, such as averaging, may not be appropriate. As a new analysis method to examine the brain area involved in the cognitive load during the Stroop task, we proposed a feature extraction method using DNN. The results of our training and analysis suggest that the mPFC (focusing on cognition) and the LIFG (focusing on language processing) were related to cognitive load. By applying our method to other tasks, utility will be confirmed as new analysis method of cerebral blood flow.

References

1. Stroop, J.R.: Studies of interference in serial verbal reactions. J. Exp. Psychol. **18**, 643–662 (1935)
2. Ehlis, A.C., Herrmann, M.J., Wagener, A., Fallgatter, A.J.: Multi-channel near-infrared spectroscopy detects specific inferior-frontal activation during incongruent stroop trials. Biol. Psychol. **69**, 315–331 (2005)
3. Villringer, A., Planck, J., Hock, C., Schleinkofer, L., Dirnagl, U.: Near infrared spectroscopy (NIRS): a new tool to study hemodynamic changes during activation of brain function in human adults. Neurosci. Lett. **154**, 101–104 (1993)
4. Owen-Reece, H., Smith, M., Elwell, C.E., Goldstone, J.C.: Near infrared spectroscopy. Br. J. Anaesth. **82**, 418–426 (1999)
5. Okamoto, M., Dan, H., Shimizu, K., Takeo, K., Amita, T., Oda, I., Konishi, I., Sakamoto, K., Isobe, S., Suzuki, T., Kohyama, K., Dan, I.: Multimodal assessment of cortical activation during apple peeling by NIRS and fMRI. Neuroimage **21**, 1275–1288 (2004)
6. Hinton, G.E., Srivastava, N., Krizhevsky, A., Sutskever, I., Salakhutdinov, R.R.: Improving neural networks by preventing coadaptation of feature detectors. arXiv preprint: arXiv:1207. 0580 (2012)
7. Kingma, D.P., Ba, J.L.: Adam: a method for stochastic optimization. In: 3rd International Conference on Learning Representations, Ithaca, pp. 1–13. arXiv.org (2015)
8. Tokui, S., Oono, K., Hido, S., Clayton, J.: Chainer: a next-generation open source framework for deep learning. In: Workshop on Machine Learning Systems (LearningSys) at NIPS (2015)

An Analysis of a Human-Exoskeleton System for Gait Rehabilitation

Lei Hou$^{(\boxtimes)}$, YiLin Wang, Jing Qiu, Lu Wang, XiaoJuan Zheng, and Hong Cheng

Center for Robot, University of Electronic Science and Technology of China,
No. 2006, Xiyuan Avenue, West Hi-Tech Zone,
Chengdu 611731, Sichuan, China
{007hou, qiujing, hcheng}@uestc.edu.cn,
kellywangsx@126.com, 201622080425@std.uestc.edu.cn,
zxjcfrobot@163.com

Abstract. In this paper, we describe about assessment of our AIDER exoskeleton using objective and subjective methods. The results show that a difference of transferring center of mass between with the exoskeleton walking task and without the exoskeleton walking task. It also shows a difference of transferring center of mass between the SCI patient and the healthy subjects during the waking task. These results reveal that the mechanical structure caused changing in subjects' center of mass, and this may affect subjects' safety and comfort.

Keywords: Exoskeleton · Human-machine system
Objective and subjective methods

1 Introduction

People who suffer from spinal cord injury (SCI) with lower limb paraplegia need to use wheelchair for long time that will cause shoulder pain [1], osteoporosis [2], muscular atrophy [3, 4] and kinds of complications [5]. Therefore, paralysis need rehabilitation training in the whole of his life. The lower extremity exoskeleton is an emerging technology that is a typical system of human-machine integration, which can bring movement into correspondence with the human body, assist human walk, and provide the rehabilitation training for patients with spinal cord injury. Patients passed through the rehabilitation training with exoskeleton, most of them can increase muscle strength and bone density, and improve them the quality of life [6–8]. Therefore, more and more the exoskeleton robot used in the field of medical rehabilitation [9].

In order to design a secure and comfortable human robot system, the effects of the exoskeleton on human needs to analyze [10]. However, most studies focus on exoskeletons' structures, actuators and control methods, and only few studies discuss its safety, comfort and efficiency.

In this paper, a comprehensive assessment of a lower limb exoskeleton was conducted by analysis of kinematic behaviors and vital signals for human robot system on AIDER 3.2 exoskeleton.

© Springer International Publishing AG, part of Springer Nature 2019
S. Fukuda (Ed.): AHFE 2018, AISC 774, pp. 102–108, 2019.
https://doi.org/10.1007/978-3-319-94944-4_12

2   The AIDER Exoskeleton

AIDER (Center for Robotics, China) is a lower limbs wearable exoskeleton robot to help spinal cord injury patients to rehabilitation training that has 8 degrees of freedom distribute 2° in hip joint, 1° in knee joint and 1° in ankle joint. Hip joint and knee joint have the active drive where fit motor in [11]. Ankle joint has passive drive where fit spring in [11]. Moreover, each joint has a band that can tie user to transmit force and moment, and there are four segments had physical interaction with user, distributing in waist, thigh, shank and foot on AIDER. For transmitting large force and moment, four aluminum alloy baffles were installed in the back of two thighs, the front of two shanks. The function of four baffles is that can make the patients with spinal cord injury keep standing straightly when they are walking with AIDER.

AIDER detailed function is that assist human in sitting-to-standing, standing, walking and standing-to-sitting, and AIDER just like human lower extremity as show in Fig. 1, for improving users' security, the robot is equipped with a pair of crutches, the user through the crutch handle button to control the exoskeleton walking and keep his body balanced by crutches. When human activities with wearable exoskeleton, the function of crutches is important [12]. The suitable user of the ADIER is from 160–190 cm height and under 100 kg.

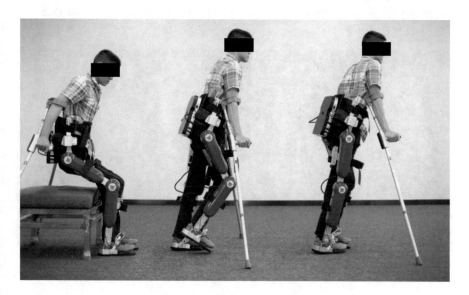

Fig. 1. The AIDER exoskeleton

3   Experimental Method

3.1   Participants

One spinal cord injury patient and 20 healthy male volunteers were participated in this experiment (average age: 23 ± 1 (mean $\pm$ standard deviation)) without any experience in exoskeleton. The average height is 174.3 cm, and the average weight is 63.1 kg. The patient was injured in T12 and L1, spinal cord injured level B and incomplete SCI. None of them reported psychiatric or other related diseases that might affect the outcomes. This study was approved by the Logistics Department for Civilian Ethics Committee of Sichuan Province 81 Rehabilitation Center and the Ethics Committee of Center for Robot, University of Electronics and Science Technology of China. All participants were informed about the experimental procedure and written consent was obtained before the experiment.

3.2   Instrumentation

As Fig. 1 shows, the motion data was collected by Vicon system (Oxford Metrics Limited, American) with 80 Hz sampling frequency. Eight 3D infrared high-speed cameras had been used, they evenly distributed in experimental environment, as Fig. 2 shows. Foot pressure data was collected by BP400600 force plate (AMTI, American), 120 Hz sampling frequency were selected. The OMRON electronic sphygmomanometer HEM-7121 was used to collected blood pressure and heart rate data.

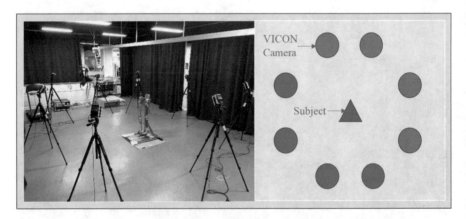

Fig. 2. Experimental environment

3.3   Experimental Design

The experiment consisted of a training session and a data collecting session. First, all subjects were trained with our exoskeleton by perform our training program until they could walk with the exoskeleton with a stable performance independently in the training session. Our training program consists four parts: sitting-to-standing, standing and balancing, walking and standing-to-sitting. Each subject was trained one hour a day. During the training.

During the data collecting session, the 20 healthy subjects performed sitting-to-standing, standing-to-sitting and 3 meter-walk with/without the exoskeleton on the normal indoor flat ground, and the spinal cord injury patient only performed sitting-to-standing, standing-to-sitting and 3 meter-walk with the exoskeleton. Each subject performed 6 times. The gait speed was fixed at constant speed. As Fig. 3 shown, the subject was instructed to start from sitting-to-standing, then stand quietly for 10 s before 3 meter-walking, after walking, they performed standing-to-sitting. Kinematic data was by the force plate system and the motion capture system during the trials. A foot frame ensured that each foot was solely placed on each force plate. The motion capture markers were attached to the lateral malleolus, knee, heel, big toe, the lateral tibia, head and shoulders. Heart rate data and blood pressure were collected before and after each 3 meter-walk trials.

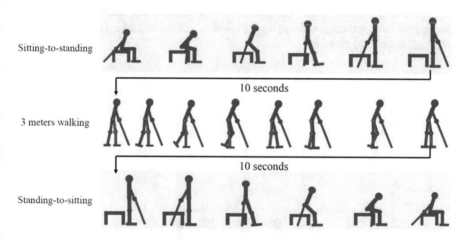

Fig. 3. Time chart of the experiment

3.4   Data Collection and Analysis

The motion data was recorded at 80 Hz with eight 3D high-speed cameras during sitting-to-standing, 3 meter-walk and standing-to-sitting trials. This data was determined the subjects status. The foot pressure data was only recorded during 3 meter-walk at 120 Hz. Both devices were synchronized by the Nexus 2.2.3 software. Performance time data was collected by a sports watch.

The data was divided to 4 blocks: sitting-to-standing, standing and balancing, 3 meter-walk and standing-to-sitting. We analyzed the 3 meter-walk block separately.

4   Results and Discussion

Figures 4 and 5 show the force plate data. Figure 4 shows that healthy subjects' changing in the mess of center during the 3-m walking with/without the exoskeleton. As the figure shown, in case of the walking task with the exoskeleton, the subjects transferred their center of mass smoothly according to fulcrum. For example, when subjects stood on the left leg and swung the right leg, the mess of center changed to left, and vice versa. On the other hand, the subjects changed their center of mass clumsy during the 3 meter-walking task. This is because that the subjects need to use crutches to stand and keep the human-machine system's balance. This an unfamiliar motion for those healthy subjects cause they could not transfer the center of mass smoothly. As Fig. 5 shown, the SCI patient transferred his center of mass clumsy and wider than the healthy subjects' result. The SCI patient does not have enough muscular power in his lower body for keep the human-machine balance, so he must use crutches to keep the human-machine system's balance and greater crutches support was required. Thus, his movement was greater than the others. The result revealed that the mechanical structure caused changing in subjects' center of mass, and this may affect subjects' safety and comfort.

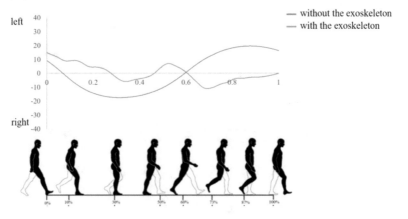

Fig. 4. Transferring the mass of center with/without the exoskeleton

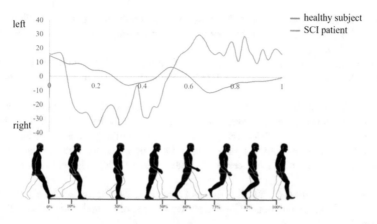

Fig. 5. Transferring the mass of center (Comparison of SCI patient and healthy subjects)

5 Conclusion

In This study, we analyzed transferring the center of mass with our AIDER exoskeleton. It revealed that there was a gait difference between the exoskeleton and human and a difference of transferring center of mass between SCI patient and healthy subjects. In the future further studies shall be conducted to collect and analyze more data from the human-exoskeleton system to optimize the exoskeleton and then improve the users' safety and comfort.

Acknowledgments. This research project is supported by National Key Research and Development Plan (2017YFB1302300).

References

1. Samuelsson, K.A.M., Tropp, H., Gerdle, B.: Shoulder pain and its consequences in paraplegic spinal cord-injured, wheelchair users. Spinal Cord **42**(1), 41 (2004)
2. Sheng-Dan, J., Li-Yang, D., Lei-Shang, J.: Osteoporosis after spinal cord injury. Spinal Cord **36**(12), 822 (1998)
3. Signorile, J.F., Banovac, K.M., Gomez, M., Flipse, D., Caruso, J.F., Lowensteyn, I.: Increased muscle strength in paralyzed patients after spinal cord injury -effect of beta-2 adrenergic agonist-. Arch. Phys. Med. Rehabil. **76**(1), 55–58 (1995)
4. Misawa, A., Shimada, Y., Matsunaga, T., Sato, K.: The effects of therapeutic electric stimulation on acute muscle atrophy in rats after spinal cor injury. Arch. Phys. Med. Rehabil. **82**, 1596–1603 (2001)
5. McKinley, W.O., Gittler, M.S., Kirshblum, S.C., et al.: Spinal cord injury medicine – medical complications after spinal cord injury: identification and management. Am. Acad. Phys. Med. Rehabil. **83**(3), S58–S64 (2002)
6. Karelis, A.D., Carvalho, L.P., Castillo, M.J., Gagnon, D.H., Aubertin-Leheudre, M.: Effect on body composition and bone mineral density of walking with a robotic exoskeleton in adults with chronic. J. Rehabil. Med. **49**, 84–87 (2017)

7. Raab, K., Krakow, K., Tripp, F., Jung, M.: Effects of training with the ReWalk exoskeleton on quality of life in incomplete spinal cord injury: a single case study. Spinal Cord Ser. Cases **2**, 15025 (2016)
8. Kolakowsky-Hayner, S.A., Crew, J., Moran, S., Shah, A.: Safety and feasibility of using the Ekso TM Bionic Exoskeleton to aid ambulation after spinal cord injury. J. Spine, **S4**, 003 (2013)
9. Suzuki, K., Mito, G., Kawamoto, H., Hasegawa, Y., Sankai, Y.: Intention-based walking support for paraplegia patients with robot Suit HAL. Adv. Robot. **21**(12), 1441–1469 (2007)
10. Rathore, A., Wilcox, M., Morgado Ranirez, D.Z.: Quantifying the human-robot interaction forces between a lower limb exoskeleton and healthy users. In: 38th Annual International Conference of the IEEE Engineering in Medicine and Biology Society (EMBC), Orlande, Florida (2016)
11. Chun-Feng, Y., Hong, C., Ye, C.: Design of a wearable sensing system for a lower extremity exoskeleton. In: International Conference on Mechatronics and Automation, Takamatsu, Japan (2017)
12. Hassan, M., Kadone, H., Suzuki, K., Sankai, Y.: Wearable gait measurement system with an instrumented cane for exoskeleton control. Sensors (Basel) **14**(1), 1705–1722 (2014)

Using Social Interaction in Rehabilitation to Improve Stroke Patients Motivation

Fu-Yu Liu[1(✉)] and Chien-Hsu Chen[1,2]

[1] Department of Industrial Design,
National Cheng Kung University, Tainan, Taiwan
asfiksears@gmail.com, chenhsu@mail.ncku.edu.tw
[2] Hierarchical Green-Energy Materials(Hi-GEM) Research Center,
National Cheng Kung University, Tainan, Taiwan

Abstract. Rehabilitation for stroke is important for recovery, but for patients, living their daily life independently after stroke is not an easy task. Difficulty accepting their new condition, facing the changes of roles and self-concept are mentioned as psychological challenges that lead to low social interaction affecting relations around them. Additionally, low levels of emotional support may cause depression and decrease motivation. It has been proven that sufficient social support can influence the training outcome resulting in better performance, and a faster recovery. The aim of this study is (1) to explore the concept of sharing in rehabilitation at the purpose of improving patients' motivation through (2) developing a user-friendly tangible device to train their upper limbs fine motor (ULFM) skills. This study focuses on both the development of the rehabilitation training device, and collection and comparison of data of their interaction with patients and the patient motivation levels.

Keywords: Health care · Stroke rehabilitation · Motivation · Social support
Social sharing

1 Introduction

After suffering through a stroke, an individual's life will be changed suddenly and dramatically, usually with physical disabilities introduced into the patient's daily living [1]. In most cases, following hospitalization and an initial recovery period, patients will have to undergo a lengthy and tedious rehabilitation period in order to recover their bodily functions such as motor movements [2].

The rehabilitation process involves spending numerous hours a week doing exercises in a community setting among other patients and therapists. With the low therapists to patients' ratio, the elderly are usually left on their own to do their prescribed exercises after the initial setting up and introduction of the process. Without constant supervision, where some of the elderly skive or do not adhere to instructions, their recovery progress can be considerably delayed. In addition to the severity of each individual's condition, the rate of recovery differs for each patient, which can encourage or discourage their efforts.

Apart from the mundane repetitive process, the sudden change in physical abilities can lead to patients finding difficulties accepting their self-concept and social role

© Springer International Publishing AG, part of Springer Nature 2019
S. Fukuda (Ed.): AHFE 2018, AISC 774, pp. 109–120, 2019.
https://doi.org/10.1007/978-3-319-94944-4_13

capacity [3], which affects their level of motivation in daily living, especially during rehabilitation when they are reminded of their disabilities. Over time, the negativity of their mindset coupled with the lack of emotional support may lead to Post Stroke Depression(PSD) [4].

Furthermore, keeping in good relations with family, friends during this period can prove to be challenging, requiring additional effort on top of the negativity faced by all parties, particularly for the patient. While rehabilitation mainly involves the physical body of the patient, emotional and social connectedness too play a vital role in the rehabilitation process as they affect the psychological well-being of the patient as well as having the ability to influence their motivational level during rehabilitation [5].

From the numerous forms of social interaction observed, the act of sharing is one that is the simplest and most intuitive form of connecting to others. Sharing one's experiences, success or even one's presence can be sufficient to render emotional support to another that requires it, of which can be further enhanced through mutual reciprocation.

Therefore, our aim of this study is (1) to explore the concept of social support in the form of sharing, in rehabilitation to improve patients' motivation through (2) developing a user-friendly tangible device to train their upper limbs fine motor (ULFM) skills. To investigate the impact of this research direction, this study will focus on both the development of the rehabilitation device for training ULFM skills. Data such as their interaction with the device, motivational levels determined through tracking and monitoring, and feedback from interviews and questionnaires, will be collected and compared for analysis.

2   Background

This section presents the research and understanding of the (1) stroke disease, (2) difficulties and potential obstacles faced in current rehabilitation, (3) social connectedness with its importance and influence in the context of rehabilitation and (4) the application of current technology in rehabilitation.

2.1   Stroke and Rehabilitation

Stroke Rehabilitation. Stroke is a global issue with the number of people suffering from stroke increasing annually [6]. There are two major types of stroke: ischemic strokes (IS) and hemorrhagic stroke (HS) [7]. Depending on the type of stroke and the corresponding area of damaged brain cells, symptoms include: arms and legs paralysis, trouble in speaking and understanding, difficulty in seeing, breathing problems and loss of consciousness [8]. For motor function recovery, strengthening of muscles and prevention of another stroke, stroke survivors often need to undergo a repetitive rehabilitation exercise. Through this training process, their nervous system would be stimulated in the Penumbra, the area surrounding the damaged cells, to rebuild the connection between the brain and corresponding area [9, 10]. The goal of rehabilitation is to return patients back to their normal life and to live as independently as possible.

However, the recovery process is time-consuming and often takes months or even years, with the possibility of patients falling into a depressive state [4].

Problems in Rehabilitation Today. With the amount of patients requiring rehabilitation increasing every year but the amount of therapists remaining relatively unchanged, the ratio between physiotherapists and patients is constantly increasing. Based on rehabilitation center that this study is working with, the therapist-to-patient ratio can exceed 1:15. Patients are often left to their own to use the training equipment without supervision after the initial introduction to their prescription. Rehabilitation helps with motor functions recovery as the constant repetition of an action strengthens the involved muscles while forming muscle memory. However, this same repetition results in the patient perceiving the exercise and rehabilitation in its entirety to be mundane and boring. In addition, results of rehabilitation take an extended period before any significant progress is seen. This in turn can discourage patients from inputting further effort as they feel more subjected to their disability over time. As with most rehabilitation centers as observed, progress tracking of a patient's recovery is done with the functional assessment every 3 to 6 months. It is only during this time when both therapist and the patient will be updated on their progress, if any.

2.2 Psychological Challenges and Motivation

Psychological Challenges. Studies have shown that about one-third of stroke survivors get Post Stroke Depression(PSD) due to the changes in their daily life, such as the need to rely on others for daily activities, and lack of social interaction with others caused by decreased mobility [4]. Depending on the level of depression as well as the effects and progress of their rehabilitation, quality of life and mortality may also be adversely affected [11]. In addition, the common psychological challenge lies with the assumption of self-concept and role capacity [3]. This new condition also affects their ability to perform duties of their original roles in life such as a parent, partner or employee among many others. As such, self-preservation behaviors may be adopted during their struggle to accept their new condition, which could lead to the declination of interactions and human relations with others [12].

Motivation. In contrast to extrinsic motivation with its external rewards, intrinsic motivation refers to behavior that is driven internally, with the motivation to engage in a behavior arising from within the individual because it is naturally satisfying. In accordance to the Self Determination Theory (SDT), a proposition that all human beings have fundamental psychological needs; to be competent, autonomous, and related to others, as shown in Fig. 1 [13]. Motivation for training increases and is sustained when all three needs are satisfied [14]. Additionally, these three are in direct proportion with each other, such as when one's opinion are accepted (autonomy) and recognition received from others, the other two needs of relatedness and competence would be satisfied as well [15]. In this study, we are placing our focus and emphasis on the dimension of relatedness, to bridge the gap between overall motivation and the two aspects of autonomy and competence, in the patients' rehabilitation process.

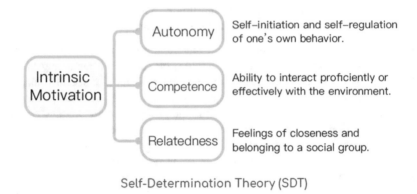

Self-Determination Theory (SDT)

Fig. 1. Three dimension of intrinsic motivation: Autonomy, Competence, Relatedness.

2.3   Social Support and Sharing Behavior

Social Support and Social Isolation. Studies show that receiving adequate emotional and psychological support from family members, peers or therapists is an important factor in promoting rehabilitation exercise [5, 16]. With sufficient support given, depression can be prevented and motivation sustained through the entire training process. Thus, social support performs a vital role in the rehabilitation process. Social isolation differs from loneliness. It refers to the phenomenon that a person who rarely contacts other people including family members and friends, or even avoiding all possibilities of contact with others. Therefore, someone who is socially isolated will not always be lonely and vice versa [17]. However, studies have indicated that suffering from social isolation may increase the risk of mortality by an average of 29% [18]. In addition to the increased risk of mortality, loneliness and social isolation can also lead to coronary artery disease(CAD) and increased risk of stroke [19]. Another study also shows the relation between stroke and social isolation, where stroke survivors are tracked for five years. The results show that lack of social interaction may cause depression and pressure which in turn lead to poor performance in rehabilitation and higher risk of stroke recurrence, myocardial infarction, death etc., [20]. Therefore, awareness for the need of social interaction and prevention of social isolation after stroke may reduce the risk of other complications.

Social Sharing Behavior. Research shows that apart from using cooperation and competition in the context of rehabilitation to boost motivation through social interactions, there exist other forms of achieving social support [3]. From observations conducted in the rehabilitation center, the elderly patients can be seen enjoying the act of sharing. For example, gifting of handcrafted accessories to fellow patients and even therapists or sharing of personal experiences that they recently had. This act of sharing helps boost their level of confidence, which in turn forms a sense of belonging to the center and their peers, which formed the intrinsic motivation as discussed previously. Another study discusses how social sharing helps with interpersonal emotion regulation

[21], where individuals managed to relieve pent-up emotions [22, 23] elevate levels of positivity [24], and even gain emotional support from the mere presence of others [25].

2.4   Technology in Rehabilitation

Due to the need for more engagement in rehabilitation, studies have shown that using technology such as robotics, Augmented Reality(AR), Virtual Reality(VR) and Kinect in the context of rehabilitation, provides an alternative approach on gamifying and providing attractive feedback. Although technology-based therapy offers a good form of motivation, there still exist shortcomings unsuitable for upper limbs fine motor training. In the application of robotics, studies indicate that while they are useful in improving the function and strength of upper limb rehabilitation [26], most are not affordable for the common family. Otherwise, the Kinect is more affordable and applicable to train gross motor skills for mobility in the beginning, especially for patients whose conditions are not severe. However, this technology has not advanced enough to detect fine motor movements such as finger gripping. Another concern is the learning curve and usability of such technology for the elderly. Although the AR and VR technology are new in the field of rehabilitation training and bring along a diverse multitude of possibilities to therapy, the design of the system is not suitable and familiar enough for the elderly. This leads to a steep learning curve and results in frustration [27]. In conclusion, technology and gamification holds the potential to provide a customizable and friendly training platform when utilized appropriately.

3   Design Methods

3.1   Observation

Participants. Patients and two therapists from the rehabilitation center were recruited. One of the therapists is an occupational therapist and the other a physical therapist, both senior and professional in rehabilitation training. They shared specialized training knowledge and practical operation instructions regarding methods to improve patient's ability correctly and problems faced currently. The patients are the elderly in rehabilitation center who require training for recovery such as stroke survivors and Parkinson patients.

Procedures. Data collection and observation was conducted with regards to the contextual inquiry principles [28]. Permission was obtained from the rehabilitation center for the observation and data collecting, inclusive of audio record and digital photos. Prior to the testing, a brief introduction of the research and experiment was given, and the current situation and basic information of the center explained by the therapist. Following which, interviews were held with the therapists to understand more about the obstacles faced during rehabilitation, as well as the demand and difficulties of stroke patients with upper limb disabilities. They gave us an extensive explanation of rehabilitation and stroke disease, which helped us in forming more comprehensive design principles. In addition, the physical therapist also recommended anatomy books to understand movements for fine motor training for deeper analysis.

Analysis. Data from the contextual inquiry were organized and three crucial points from the process were identified and illustrated in Fig. 2. It was common to see that patients were not paying attention on their training as well as performing inaccurately due to the lack of an apparent goal. Additionally, without supervision or accompaniment, they might not exercise voluntarily. However, one thing that caught our attention was that they prefer to interact with the therapists and their peers, sharing their personal life experience during the training. Therefore, the three key points identified from the contextual inquiry are (1) the lack of goals equates the loss of their attention, (2) absence of feedback to visualize the training progress and correction of any inaccurate movements, and (3) appropriate accompaniment and social interaction may raise their motivation.

Lack of Goals   Short of Feedback   Social Interaction

Fig. 2. Three main points derived from contextual inquiry. (1) lack of goals results in loss of attention, (2) absence of feedback to show the training progress accuracy of movement execution, and (3) appropriate social interaction may raise their motivation.

3.2   Development of Fine Motor Training Design

Design Consideration. The design principles for the training device are listed in Table 1 below. According to the theory background and the data analysis from contextual inquiry, the data is organized into four principles. Firstly, activity of daily living (ADL) based training content aim to bridge the gap between training movement and daily activity to ensure that the training outcome can be applied to their daily living. To satisfy all stages of patients, having levels of varying difficulty are considered. Showing the training information provides a convenient and clear visual for the patient to understand their training progress. This also offers therapists and family members with information to adjust the training content and take care of them. Moreover, apart from being attractive and interactive, visual feedback can serve as a guide to correct their movements. Lastly, introducing social engagement and interaction can make rehabilitation seem more enjoyable and fun. Their efforts can be transformed into something tangible, which they can share with others, forming the basis of social interaction.

Table 1. Design principles.

Principles	Correspondence problems
• ADL-based training content	Ensure the training content related to the activities of daily life
• Level of difficulty	To fit different needs of each individual and training stages
• Visual feedback	To visualize the training progress and correct any inaccurate movement
• Social sharing element	Improve motivation through social behavior

Hardware and Software. Both hardware and software development follows the design principles previously discussed. For the physical training part, we developed a tangible device with acrylic cylinders of five varying sizes to train fine motor skills. One display is used for displaying the content and visual feedback. An Arduino mega microcontroller is used to control the input and output. The sizes of acrylic cylinder (with D = 100, 75, 50, 25, 7 mm) were determined from the gripping and grasping behavior based on the common activity of daily living, such as opening a jar, switching radio button, turning door knob etc. Additionally, for the software part, Processing is used to address the data and operating the pixel art component of the system. By using five input cylinders and five corresponding output grids, users can generate a pixel artwork of at least 10*10 grids (depending on the task level) at the end of the training.

Idea Generation. We are aiming to know how much the concept (integrating training and coloring action) and visual feedback will motivate the patients. Below are the concept details of the design and training procedure as shown in Fig. 3. and Table 2.

Steps of the Training

(1) User selects the color based on the goal color of the card (a).
(2) Rotate the cylinder to select a color (d). (Total of 6 colors)
(3) Press the cylinder to fill the box with the selected color (from step (2)), (e).
(4) The corresponding boxes of the grid will be colored.
(5) Repeat step (1) - (4), (g).
(6) Using this device, the user will get their own pixel art at the end of the hand training process (h). The user would need to repeat the action for 20 times and at the end all 100 grids(10*10) on the screen would be colored (h).
(7) User can share the result by sending the image to family member or displaying the artwork on the wall in rehabilitation center (i).

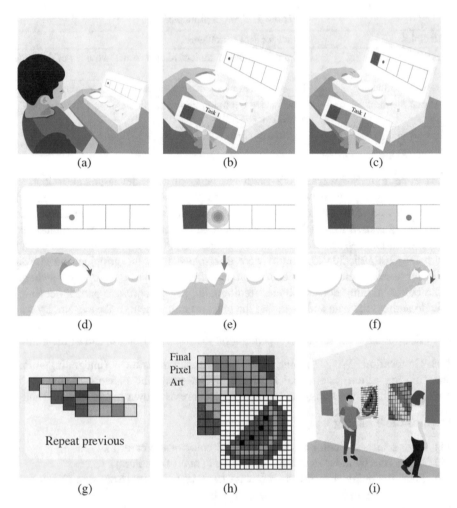

Fig. 3. Ideal scenario of design and training procedure. A tangible device with five acrylic cylinders (with D = 100, 75, 50, 25, 7 mm), and one screen for the coloring process.

Table 2. Concept details.

Concept details	
1.	A system where patients can transform their efforts in rehabilitation into something meaningful and tangible such as a self-made pixel artwork, which can then be shared with others.
2.	Five acrylic cylinders of different sizes (with D=100, 75, 50, 25, 7 mm) that followed the ADL such as opening toothpaste cap(7mm), opening bottle(25mm) and rotating door knob (75) etc. were chosen.
3.	Using their rehabilitation exercise movements to perform an input into the device for making of artwork.
4.	Finished artwork can be shared with family and friends, acting as an indication of both effort and recovery progress.
5.	Recipients can reciprocate this gift (such as fellow patients gifting one another, family members framing it at home and providing compliments). In return, patients can receive encouragements, a sense of accomplishment and emotional support, gain confidence and motivation to complete their rehabilitation.

4 Result

Prototype. A prototype of the tangible device and system are shown in Fig. 4. Through manipulating the cylinder, user can change the color by rotating, and fill the grid by clicking the button. There are six colors in the system, and users can change the colors by rotating the cylinders through every three clicks. To test this research direction and idea, we made a rough prototype with one difficulty level included to discuss with the therapist whom we did inquiry contextual with before (Table 3).

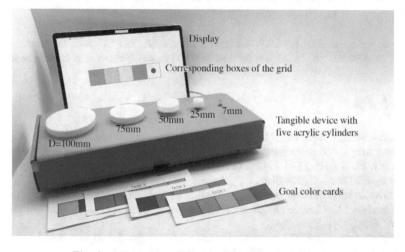

Fig. 4. A prototype of the tangible device and system.

Table 3. Therapist feedback of the prototype.

Advice	Description
• Handle design	Try more shapes of handle with reference to ADL, such as key pinching, cube shape, door handle etc.
• Task level	To fit different needs of training stages
• Colors	Provide 3/6/8 colors choices to train precision
• Grids	Provide more/less grids for different difficulty
• Sharing part	Add sharing element for family members

5   Conclusions and Future

According to the advice from therapist, the design of the handles can be varied in terms of grip style and shape, providing patient with more training options. While accommodating to different patient's needs, these designs can also form a relation to movements related to their daily activities. For the different task level design, there can be different number of colors for the patient to adjust the precision level of ULFM skill training. More colors require a more precise manipulation to rotate the handle. Moreover, number of grids and complexity of the overall artwork also determine the difficulty of the task. For the complex patterns, while it requires more grids to be constructed, the visually more pleasing outcome may give them a higher sense of accomplishment. Last and the most important point of all, the therapist recommends family members to be the party to share with, as they exhibit the most influence for motivation and are heavily relied upon for emotional support. For example, sending the pixel art from a patient's daily training and their history records may provide an instantaneous training outcome and clear progress of the rehabilitation. Peer influence is also another way to provide social support by reconnecting relations around patient. Therefore, both family and peers are considered the ideal group for sharing of their work.

Acknowledgments. Thanks to AWWA rehabilitation center and CUTE center for their kindly support on this research. Thank you for continuing to make efforts to help those in need.

References

1. Santisteban, L., Térémetz, M., Bleton, J.-P., Baron, J.-C., Maier, M.A., Lindberg, P.G.: Upper limb outcome measures used in stroke rehabilitation studies: a systematic literature review. PLoS ONE **11**, e0154792 (2016)
2. Flores, E., Tobon, G., Cavallaro, E., Cavallaro, F.I., Perry, J.C., Keller, T.: Improving patient motivation in game development for motor deficit rehabilitation. In: Proceedings of the 2008 International Conference on Advances in Computer Entertainment Technology, pp. 381–384. ACM, Yokohama (2008)
3. Glass, T.A., Maddox, G.L.: The quality and quantity of social support: stroke recovery as psycho-social transition. Soc. Sci. Med. **34**, 1249–1261 (1992)

4. Aström, M., Adolfsson, R., Asplund, K.: Major depression in stroke patients. a 3-year longitudinal study. Stroke **24**, 976–982 (1993)
5. Nicholson, S., Sniehotta, F.F., Wijck, F., Greig, C.A., Johnston, M., McMurdo, M.E., Dennis, M., Mead, G.E.: A systematic review of perceived barriers and motivators to physical activity after stroke. Int. J. Stroke **8**, 357–364 (2013)
6. Mozaffarian, D., Benjamin, E.J., Go, A.S., Arnett, D.K., Blaha, M.J., Cushman, M., Das, S. R., de Ferranti, S., Després, J.-P., Fullerton, H.J., Howard, V.J., Huffman, M.D., Isasi, C.R., Jiménez, M.C., Judd, S.E., Kissela, B.M., Lichtman, J.H., Lisabeth, L.D., Liu, S., Mackey, R.H., Magid, D.J., McGuire, D.K., Mohler, E.R., Moy, C.S., Muntner, P., Mussolino, M.E., Nasir, K., Neumar, R.W., Nichol, G., Palaniappan, L., Pandey, D.K., Reeves, M.J., Rodriguez, C.J., Rosamond, W., Sorlie, P.D., Stein, J., Towfighi, A., Turan, T.N., Virani, S. S., Woo, D., Yeh, R.W., Turner, M.B.: Heart Disease and Stroke Statistics—2016 Update. A Report From the American Heart Association (2015)
7. World Heart Federation. Stroke. https://www.world-heart-federation.org/resources/stroke/
8. National Heart Lung and Blood Institute. What Is a Stroke? https://www.nhlbi.nih.gov/health/health-topics/topics/stroke#
9. Khanacademy. Treatment of stroke with interventions. https://www.khanacademy.org/science/health-and-medicine/circulatory-system-diseases/stroke/v/treatment-of-stroke-with-interventions
10. National Yang-Ming University Hospital. Utilising Optimal Recovery Period After Stroke. https://goo.gl/UDrZSs. (把握腦中風團隊復健黃金治療期)
11. Gaete, J.M., Bogousslavsky, J.: Post-stroke depression. Expert Rev. Neurother. **8**, 75–92 (2008)
12. Antal, A.: Using social gaming to improve stroke patients motivation and engagement in rehabilitation therapy (2013)
13. Deci, E.L., Ryan, R.M.: Motivation, personality, and development within embedded social contexts: an overview of self-determination theory. In: The Oxford Handbook of Human Motivation, pp. 85–107 (2012)
14. Deci, E.L., Ryan, R.M.: Self-determination theory. Handb. Theor. Soc. Psychol. **1**, 416–433 (2011)
15. Chiang, J.-C.: A conceptual aspect of self-determination theory in practice of online learning education. J. Humanit. Soc. Sci. **7**(2), 67–75 (2011). (江瑞菁: 自我決定數位學習環境的環境要素之初探. 人文暨社會科學期刊 **7** (2011))
16. Damush, T.M., Plue, L., Bakas, T., Schmid, A., Williams, L.S.: Barriers and facilitators to exercise among stroke survivors. Rehabil. Nurs. **32**, 253–262 (2007)
17. de Jong Gierveld, J., Van Tilburg, T., Dykstra, P.: Loneliness and social isolation (2016)
18. Holt-Lunstad, J., Smith, T.B., Baker, M., Harris, T., Stephenson, D.: Loneliness and social isolation as risk factors for mortality. Perspect. Psychol. Sci. **10**, 227–237 (2015)
19. Valtorta, N.K., Kanaan, M., Gilbody, S., Ronzi, S., Hanratty, B.: Loneliness and social isolation as risk factors for coronary heart disease and stroke: systematic review and meta-analysis of longitudinal observational studies. Heart **102**, 1009–1016 (2016)
20. Boden-Albala, B., Litwak, E., Elkind, M., Rundek, T., Sacco, R.: Social isolation and outcomes post stroke. Neurology **64**, 1888–1892 (2005)
21. Rimé, B.: Interpersonal emotion regulation. Handb. Emot. Regul. **1**, 466–468 (2007)
22. Lazarus, R., Folkman, S.: Stress, Appraisal and Coping. Springer, New York (1984)
23. Uchino, B.N., Cacioppo, J.T., Kiecolt-Glaser, J.K.: The relationship between social support and physiological processes: a review with emphasis on underlying mechanisms and implications for health. Psychol. Bull. **119**, 488 (1996)

24. Gable, S.L., Reis, H.T.: Good news! capitalizing on positive events in an interpersonal context. In: Advances in Experimental Social Psychology, vol. 42, pp. 195–257. Elsevier (2010)
25. Schachter, S.: The psychology of affiliation: experimental studies of the sources of gregariousness. (1959)
26. Basteris, A., Nijenhuis, S.M., Stienen, A.H., Buurke, J.H., Prange, G.B., Amirabdollahian, F.: Training modalities in robot-mediated upper limb rehabilitation in stroke: a framework for classification based on a systematic review. J. NeuroEng. Rehabil. 11, 111 (2014)
27. Standen, P.J., Threapleton, K., Connell, L., Richardson, A., Brown, D.J., Battersby, S., Sutton, C.J., Platts, F.: Patients' use of a home-based virtual reality system to provide rehabilitation of the upper limb following stroke. Phy. Ther. 95, 350–359 (2015)
28. Holtzblatt, K., Wendell, J.B., Wood, S.: Rapid Contextual Design: A How-to Guide to Key Techniques for User-Centered Design. Elsevier, San Francisco (2004)

Sensory Engineering and Emotional Design

Open Tool for Collecting Physiological Data: Collection of Emotional Data During Gameplay

Victor Moreira[1(✉)], Rodrigo Carvalho[2(✉)],
and Maria Lúcia Okimoto[1(✉)]

[1] Graduate Department of Design, Federal University of Paraná,
Curitiba, PR, Brazil
victoremmoreira@gmail.com, lucia.demec@ufpr.br
[2] Graduate Program in Electrical Engineering, Federal University of Pará,
Belém, PA, Brazil
rodrigo.carvalho@itec.ufpa.br

Abstract. Research in the emotional design area and games try to understand emotions that users have to relate to the artifact. To do so, the researchers use many tools like: questionnaires, interviews, self-report, eye tracking, facial expressions and physiological responses. However, researchers with low purchasing power suffer with the high costs of data collection instruments. The aim of this work consist in the development of a low cost physiological data collection tool with a reasonable and open level of precision, allowing the improvement of the tool. We describe the software and hardware development process as well as the process of data collection and analysis. As a preliminary test of the tool, we collected the data of 4 people playing a game called "Limbo". The data shows the limitations of the tool and the possibilities of use.

Keywords: Physiological data · Research tool · Open software and hardware

1 Introduction

Emotional design tries to understand emotions that users have to relate to the artifact [1], therefore, many researchers are looking for tools with the objective of cataloging such emotions [2]. Some tools like questionnaires, interviews, and self-reporting provide data that depends on the user's ability to interpret the questions and express themselves. On the other hand, we have the eye tracking, facial expressions, heart beat (HR) and galvanic skin response (GSR) that exposes measures of user physiology. Both of them are very important for composing psychophysiological results and thus measure the emotional state, dimensionally [3].

Games User Research (GUR) is a research field that combines knowledge of human-computer interaction, game design and psychology, with the aim of improving the player's experience [4]. Mandryk and Nacke [5] explain that some of the most common methods for GUR are:

© Springer International Publishing AG, part of Springer Nature 2019
S. Fukuda (Ed.): AHFE 2018, AISC 774, pp. 123–131, 2019.
https://doi.org/10.1007/978-3-319-94944-4_14

- Psychophysiological player testing. Controlled measures of gameplay experience with the use of physical sensors to assess user reactions.
- Eye tracking. Measurement of eye -fixation and attention focus to infer details of cognitive and attentional processes.
- Persona modeling. Constructed player models.
- Game metrics behavior assessment. Logging of every action the player takes while playing, for later analysis.
- Player modeling. AI-based models that react to player behavior and adapt the player experience accordingly.
- Qualitative interviews and questionnaires. Surveys to assess the player's perception of various gameplay experience dimensions [5].

These methods are able to detect emotions regarding the actions of real-time players allow to understand a set of information. However, these techniques require more work and experience, and have a higher cost [5]. So, in the GUR, they combine methods that mix the data from interviews, think-aloud, psychophysiological data or games analysis. The multiple approaches help the researcher to balance the advantages and disadvantages of each method to achieve a sustainable empirical result. Therefore, in this research we will focus on the collection of psychophysiological data of video game players, in order to measure the emotional experience through sensors.

The physiological data collection tools require a financial cost that cannot be afforded. Companies like iMotions and Noldus offer hardware and software to collecting physiological data, however, the prices of these devices are not accessible for low budget research. There is little research with open platform devices (hardware and software) that allow replication and improvement. For this reason, the objective of this work is to develop tools for collecting physiological data with the following characteristics: Low cost, to make the device accessible to low budget research; Accuracy level reasonable, so that the data collected will follow the specifications necessary for the proposed study examples; Open-Source hardware, allowing greater accessibility to components and replication and Open-Source software, turning the code available for application in other programming languages.

2   Method

The structuring of the research method was based on Design Science Research [6]. This method lists 5 types of artifacts that this process can generate: Constructs, Models, Methods, Instantiations, and Design Propositions. March and Smith [7] describe the creation of instantiations more deeply through Design Science and Research. For the authors, instantiations are the artifacts that operationalize other artifacts (constructs, models and methods). Instantiations should tell us how to implement or use an artifact and its possible results.

Using the method proposed by Manson [8] (Fig. 2), it is possible to notice that the problem awareness is its first stage and was described in the introduction of this research. The suggestions (the second step of the method) are made based on project requirements as: low cost, open platform capable of replication and updates.

Subsequently, the development consists in the programming period of software and hardware. The evaluation stage, understood as a game session, was to test the instrument in order to verify the operation and format of the files for later analysis. The conclusion is that the results will be presented through the synthesis of the data collected, which should contribute to the solution of the problem presented in the first step (Problem Awareness) (Fig. 1).

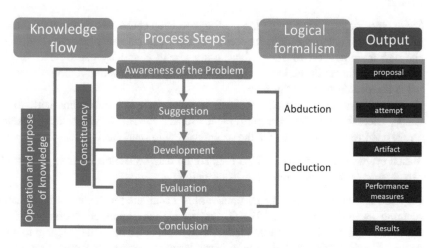

Fig. 1. Design science research method by Niel Manson [8].

Based on researchers Nacke and Mandrik [5], Nogueira et al. [9], and Picard [10], we discovered that the most recurrent instruments are: Heart rate and Galvanic skin response, the latter being more frequently cited in emotional stimuli. The facial electromyogram also appears in several studies, but because it is more invasive than the others, we chose not to use this sensor in the evaluation phase. Another requirement of the project is the recording of the gameplay so that the researcher can later analyze the actions of the player. In order to promote triangulation of the data, the device must also record the video of the player's face, so that it is possible to verify the facial expressions.

3   The Instrument of Data Collection

The process of developing the collection device began with the research and purchase of the sensors. The Grove GSR sensor met the requirements for a low price, easily accessible and platform open-source sensor. The heart rate sensor has a greater number of possibilities and will vary according to the purpose, for example if it is necessary to measure the heart rate of an individual in movements one can use a sensor with a chest strap, while for individuals in you can use a finger sensor or ear clip. In this case, the researchers chose the ear clip because it is a less intrusive sensor, easier to hold and showing more constant data during use. The development of the software took into consideration the following aspects: communication with Arduino, ability to record computer screen and webcam and possibility to export the program to other platforms. In this case, we opted to develop the software in the Java language, and to comply with the requirement of open software, we made the code available on GitHub platform.

During the development phase, the researchers raise a number of factors and some caution regarding the use of the instrument. Among the most relevant, we can cite the following:

- Static electricity: this can affect the reading and even cause electric discharge in the individual. We recommend the inspection of the electrical structure of the collection site and a simple test to verify this failure. Static electricity can be checked by placing the sensors in the individual and checking if there is a variation of reading when the person touches the floor.
- Reading errors can happen for several reasons such as: participant with rough hands, excessive heat in the room or poor sensor contact. In these cases, reading errors need to be interpreted by the researcher, who must have.

Based on the requirements we also defined the data collection form as well as the step by step to perform the collection listed below:

1. Preparation. In this process, the researcher must install the sensors in the interface of Arduino as in the scheme of Fig. 7, and verify the supported operation of the Arduino program using the serial Plotter. You should also open the program that we named Physiometrics and do preliminary tests to confirm the operation of all sensors.
2. Before starting to collect data, the researcher must open the Physiometrics program and fill in the fields with the name of the collection, participant's name and others. Place the sensors in the participant. At this moment the researcher must be ready to collect the data, for example: the data collection program (Physiometrics) and the game that the participant will play must be already running. Before starting to collect data, the researcher must open the Physiometrics program and fill in the fields with the name of the collection, participant's name and others.
3. Data recording. With all the sensors connected properly the researcher must click on "start" and, when the session is finished, can click on "stop" to finish the data collection.
4. Files generated. The data collection phase generates 3 files, one with the gameplay video, one with the webcam video, and one Excel .xls file where the sensor data is located.

4   Data Analysis

For the criteria of a preliminary data collection, we performed a game session with 4 participants (3 men and 1 woman) with the mean age of 26 years. We select the game Limbo because it is known for cause fear without being terrifying and because it causes relief after overcoming an obstacle. At first, there was a turn of 30 min game for each participant. However, in all sections, participants played for about 45 min in order to know what would come in the game. A computer with good gaming performance was used, aided by a GTX 980 TI video card, using a Wi-Fi joystick as input for the game (Fig. 2).

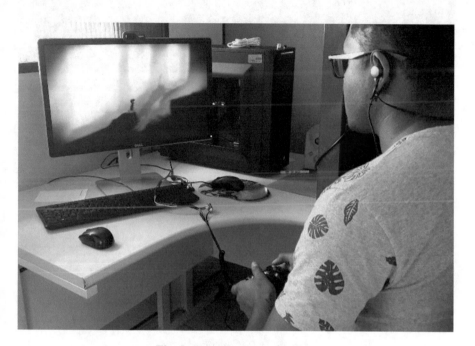

Fig. 2. Setup for data collection.

Before starting the data analysis, it is necessary to verify the consistency of the data, as these may contain some errors of reading that can be serious, causing the loss of data or light that can be adjusted by the researcher himself. For example, in Fig. 3 the Microsoft Excel program was used to create a graphic with time information (on the X-axis) and Heart Rate (in Blue) and GSR (in Orange). It is possible to notice that there are some errors in the data of the HR, however these do not invalidate the collection. The Fig. 4 shows data with a critical error, causing the loss of the information that was collected. At the time of the collection represented in Fig. 4, the researcher noticed that the participant's hands were sweating and the environment was hot.

After the verification of the data, it is possible to start the video analysis of the gameplay. In Fig. 3 we plot the graphic with the total time of the data collection and, at the top, we adjust the video of the gameplay so the timelines can match. At 08:39, a particular event happens in the game, the character of the game gets stuck in the spider's web and the player cannot do anything to escape. At this point, it is possible to realize the tension increasing through the graphic of the GSR, but the apex happens soon after. This shift occurs because these physiological responses are not instantaneous. Stern and Ray et al. [11] explain that the responses can occur between 1 and 3 s

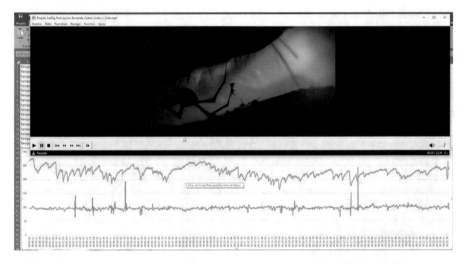

Fig. 3. The analysis of physiological responses and gameplay video.

Fig. 4. Data with critical error.

after the stimulus, and both the magnitude of the wave and its time interval in relation to the stimulus depend on the level of the individual's reactivity.

Heart rate variation was not significant among participants who played Limbo. However, in the development phase the researchers did several tests with other games like: Dota 2, GTA V and Overwatch; and in all collections that had validated HR data, there was an increase in heart rate at times of great action. We understand that the fact that the game Limbo does not contain moments with frantic action like the games mentioned above, the values of heart rate do not show great variation.

To analyze facial expressions you can use the Affedex platform. It is possible to use the software development kit (SDK) that allows integration with other applications. The SDK demo application code and the SDK are available to download at: http:// www.affectiva.com/sdk. For parameters of this preliminary study, we analyzed the video and webcam of the player when it was in high or low activation. In Fig. 5 you can see a moment of joy as you overcome a challenge. In Fig. 6, the participant demonstrates anger and sadness during his various attempts to solve the puzzle. In most of the webcam videos players remain serious and focused, at hush times (up or down) the GSR is the main moment where facial expressions happened.

Fig. 5. Participant demonstrating joy.

Fig. 6. Participant showing anger and sadness.

5 Analysis of the Collection Instrument

The results obtained in the preliminary data collection, detailed in the previous session, were obtained with the use of Anduino Nano v3 as interface with the system. The sensors were connected to the Arduino (Fig. 7) which is responsible for collecting and sending data to the Physiometrics program. The C++ code used in Arduino is also available in the repository in Github. In this version of the device there is also an electromyogram sensor that was not used for preliminary data collection.

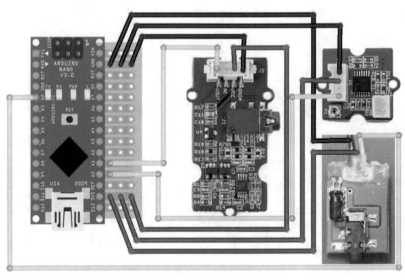

Fig. 7. Connection of the sensors in the Arduino.

An important limitation of this device is that the GSR sensors are placed on the fingers, so we use a joystick to control the game. During the development and testing phase the researchers noticed the difficulty about moving the fingers with sensors. This limitation restricts the use of certain games and players who have the ability to handle a joystick.

6 Conclusion

In this research, we proposed a tool to help the collection of physiological data. These data, along with interviews, self-reports, or think aloud, help to formulate psychophysiological data. This type of data collection can help researchers with limited financial resources to conduct the study in the emotional design area or GUR. The way this tool is designed allows other researchers to contribute to the improvement of both the physiology program and Arduino hardware. In this article, we tried to describe the development process, step by step, how to use, possible data analysis strategies and limitations.

By analyzing preliminary survey data, we verify the accuracy of the information matching the actions in the gameplay with the facial responses. We compared the graphics and values obtained in the data collection with the works of Lobel et al. [12], Mendoza-Denton et al. [13] and Mandruk and Nacke [5].

In a future work we propose a system that can learn from the data and, for example, report reading errors. The collection device (hardware) needs a stable container that can hold all the sensors. While the software can be more intuitive to other researchers and plotting the graphics in real time, it can help to verify certain errors in the collection.

References

1. Demir, E., Desmet, P.M.A., Hekkert, P.: Appraisal patterns of emotions in human-product interaction. Int. J. Des. **3**, 41–51 (2009)
2. Thoring, K., Bellermann, F., Mueller, R.M., Badke-Schaub, P., Desmet, P.: A framework of technology-supported emotion measurement. In: Design & Emotion Conference 2016, Amsterdam, NL (2016)
3. Desmet, P., Hekkert, P.: Framework of product experience. Int. J. Des. **1**, 57–66 (2007)
4. Isbister, K., Schaffer, N.: Game Usability (2008)
5. Mandryk, R.L., Nacke, L.E.: Biometrics in gaming and entertainment technologies. In: Biometrics in a Data Driven World Trends, Technologies, and Challenges, pp. 191–224 (2016). https://doi.org/10.1201/9781315317083-7
6. Dresch, A., Lacerda, D.P., Júnior, J.A.V.A.: Design science research: método de pesquisa para avanço da ciência e tecnologia (2015)
7. March, S.T., Smith, G.F.: Design and natural science research on information technology. Decis. Support Syst. **15**, 251–266 (1995). https://doi.org/10.1016/0167-9236(94)00041-2
8. Manson, N.: Is operations research really research? ORiON **22**, 155–180 (2006). https://doi.org/10.5784/22-2-40
9. Nogueira, P.A., Aguiar, R., Rodrigues, R., Oliveira, E.: Computational models of players' physiological-based emotional reactions: a digital games case study. In: Proceedings - 2014 IEEE/WIC/ACM International Joint Conference on Web Intelligence and Intelligent Agent Technology - Workshops, WI-IAT 2014, vol. 3, pp. 641–661 (2014). https://doi.org/10.1109/wi-iat.2014.178
10. Picard, R.W.: Affective Computing. MIT press, Cambridge, pp. 1–16 (1995). https://doi.org/10.1007/bf01238028
11. Stern, R.M., Ray, W.J., Quigley, K.S.: Psychophysiological Recording. Oxford University Press, New York (2001)
12. Lobel, A., Gotsis, M., Reynolds, E., Annetta, M., Engels, R.C.M.E., Granic, I.: Designing and utilizing biofeedback games for emotion regulation. In: Proceedings of the 2016 CHI Conference Extended Abstracts on Human Factors in Computing Systems - CHI EA 2016, pp. 1945–1951 (2016). https://doi.org/10.1145/2851581.2892521
13. Mendoza-Denton, N., Eisenhauer, S., Wilson, W., Flores, C.: Gender, electrodermal activity, and videogames: adding a psychophysiological dimension to sociolinguistic methods. J. Sociolinguistics **21**, 547–575 (2017). https://doi.org/10.1111/josl.12248

Development of a Speech-Driven Embodied Entrainment Character System with a Back-Channel Feedback

Yutaka Ishii[1(✉)], Makiko Nishida[2], and Tomio Watanabe[1]

[1] Faculty of Computer Science and Systems Engineering, Okayama Prefectural University, 111 Kuboki, Soja, Okayama, Japan
{ishii,watanabe}@cse.oka-pu.ac.jp
[2] Graduate School of Computer Science and Systems Engineering, Okayama Prefectural University, 111 Kuboki, Soja, Okayama, Japan
macky@dgn.oka-pu.ac.jp

Abstract. We have already developed a speech-driven embodied entrainment CG character system called "InterActor" which generates communicative motions and actions such as nods for entrained interaction from speech rhythm based on only voice input. Conventional InterActor performed the entrained communicative body movements and actions based on speech input without an audio back-channel feedback. In this study, we develop an embodied character system with the back-channel feedback based on the embodied interaction model. Moreover, the effectiveness of the system is demonstrated by the sensory evaluation in the experiment.

Keywords: Human-agent interaction · Embodied entrainment
Back-channel feedback · Nodding

1 Introduction

Recently, many communication robots have been proposed to communicate with human such as "Robovie" and "Pepper" [1, 2]. Further, conversational robots with natural expression and human-like spoken dialog agents have been developed by face image composition and dialog control [3]. In addition, attractive voice interaction systems have been proposed and developed by combining speech synthesis, speech recognition, and learning function [4–6].

On the other hand, humans communicate smoothly by synchronization of not only verbal messages but also nonverbal information such as nodding or body motions. Embodied interaction sharing causes talkers mutual familiarity and preference in the same communication space. We already analyzed the entrainment between a speaker's speech and a listener's nodding and blinking in face-to-face communication, made an interaction model called InterRobot Technology (iRT) and developed a speech-reactive system in which computer graphics representing human facial expressions simulated nodding and blinking in response to speech input [7]. Various studies using robots with iRT have demonstrated the effectiveness of the model [8]. Moreover, we developed a

© Springer International Publishing AG, part of Springer Nature 2019
S. Fukuda (Ed.): AHFE 2018, AISC 774, pp. 132–139, 2019.
https://doi.org/10.1007/978-3-319-94944-4_15

speech-driven embodied entrainment character called InterActor which automatically generates communicative motions on speech input in order to activate embodied interactions in the cyberspace [9].

However, conventional InterActor did not have the audio back-channel response function, because InterActor was developed for the interface of human-human communication based on only users' voice. Therefore, in this study, we develop an embodied character system with the voice back-channel feedback based on iRT. Moreover, the effectiveness of communication support is evaluated by two experiments.

2 InterActor

2.1 Overview of InterActor

InterActor is a Computer Generated (CG) character that has functions of both a listener and a speaker. The listener performs embodied entrainment behaviors, such as nodding and other body motions, to a user's voice [7]. The speaker performs rhythmical sympathetic motions to a user's voice. InterActor responds to utterances with an appropriate timing by means of its entire body motions and actions in the manner of a listener and a talker. In addition, InterActor can transmit the talker's message to a partner by generating a body motion similar to a speaker on the basis of a time series of speech, presenting both the speech and the entrained body motions simultaneously.

The information transmitted and received by this system is only through speech. Thus, the InterActor generates the entrained communicative movements and actions based on speech input and supports the sharing of mutual embodiment in communication.

2.2 Interaction Model of InterActor

A listener's interaction model of the InterActor includes a nodding reaction model that estimates the nodding timing from a speech ON-OFF pattern and a body reaction model linked to the nodding reaction model. A hierarchy model consisting of two stages, macro and micro (Fig. 1), predicts the timing of nodding. The macro stage estimates whether a nodding response exists or not in a duration unit that consists of a talkspurt episode $T(i)$ and the following silence episode $S(i)$ with a hangover value of 4/30 s. The estimator $M_u(i)$ is a moving-average (MA) model, expressed as the weighted sum of unit speech activity $R(i)$ in (1) and (2). When $M_u(i)$ exceeds the threshold value, the nodding $M(i)$ is also an MA model, estimated as the weighted sum of the binary speech signal $V(i)$ in (3). The body movements are related to the speech input at a timing over the body threshold. The body threshold is set lower than that of the nodding prediction of the MA model, that is expressed as the weighted sum of the binary speech signal to nodding. The mouth motion is realized by a switching operation synchronized with the burst-pause of speech. In other words, when the InterActor works as a listener for generating body movements, the relationship between nodding and other movements is dependent on the threshold values of the nodding estimation.

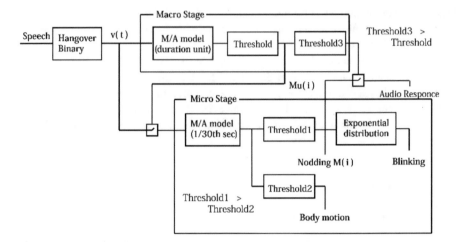

Fig. 1. Interaction Model for InterActor.

$$M_u(i) = \sum_{j=1}^{J} a(j)R(i-j) + u(i) \tag{1}$$

$$R(i) = \frac{T(i)}{T(i) + S(i)} \tag{2}$$

$a(j)$: linear prediction coefficient
$T(i)$: talkspurt duration in the i-th duration unit
$S(i)$: silence duration in the i-th duration unit
$u(i)$: noise

$$M(i) = \sum_{k=1}^{K} b(j)V(i-j) + w(i) \tag{3}$$

$b(j)$: linear prediction coefficient
$V(i)$: voice
$w(i)$: noise

The body movements of the speaker are also related to the speech input by operating both the neck and one of the other body actions at a timing over the threshold, that is the speaker's interaction model estimates as its own MA model of the burst-pause of speech to the entire body motion. Because speech and arm movements are related at a relatively high threshold value, one of the arm actions in the preset multiple patterns is selected for operation when the power of speech is over the threshold.

3   InterActor with a Back-Channel Feedback

3.1   Concept

Concept of the system is shown in Fig. 2. Conventional InterActor did not have the audio back-channel feedback function, because InterActor was developed for the interface of human-human communication based on only users' voice. It is necessary to examine the audio response to perform interaction with conversation robot more smoothly. This system could realize more interactive response of a virtual agent which returns audio feedback as well as visual feedback with auto-generated embodied movements.

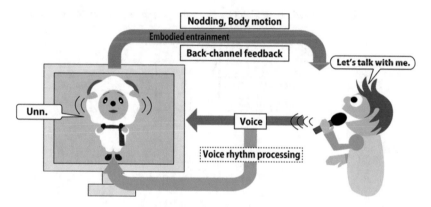

Fig. 2. Concept of InterActor with a Back-channel Feedback.

3.2   System Configuration

The interaction model of InterActor also can see in Fig. 2. First, the iRT model predicts the timing of nodding based on the threshold of the above model. Then, audio response is added based on the higher threshold for nodding (Threshold3). Moreover, audio back-channel feedback is started after 300 ms for the auto-generated nodding motion. Because, in the previous work about greetings in face-to-face communication, a lag of about 300 ms was desirable for a familiar greeting. The audio response is fixed to "Unn" in Japanese casual audio response in face-to-face communication. The length of utterance is 220 ms, and the motion of nodding is 500 ms.

3.3   Preliminary Experiment for the Frequency of Back-Channel Feedback

Preliminary experiment was performed for the frequency of back-channel feedback using the system under four modes. The four modes are the condition of 0% audio feedback (No utterance), 42% (Average of Americans), 70% back-channel feedback for all nodding (Average of Japanese), and 100% (All with nodding) by sensory evaluation

[10]. The subjects are 11 Japanese male and female students aged between 20 and 23 years. We confirmed that they were familiar with its operation before starting the experiment. They were then introduced to the four operational modes and the differences between them while using the system. Next, the subjects were instructed to perform a pairwise comparison of each mode for an overall evaluation. Four comparisons were required; therefore, the experiment was conducted six (= $_4C_2$) times.

The results of paired comparison for the three modes are shown in Table 1. Figure 3 shows the calculated results of the evaluation provided in Table 1, based on the Bradley–Terry model given in Eq. (4). Mode C, "70% back-channel feedback for all nodding" was evaluated most affirmatively, with Mode D, B and A following in that order.

Table 1. Paired comparison result in the preliminary experiment.

	0%	42%	70%	100%	Total
0%	–	2	1	1	4
42%	9	–	2	6	17
70%	10	9	–	7	26
100%	10	5	4	–	19

$$P_{ij} = \frac{\pi_i}{\left(\pi_i + \pi_j\right)}$$

$$\sum_i \pi_i = const.(= 100) \tag{4}$$

(π_i: intensity of i, P_{ij}: probability of judgement that i is better than j.)

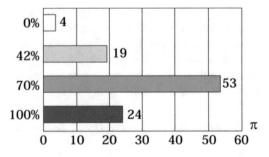

Fig. 3. Preference based on the Bradley–Terry model.

4 Evaluation Experiment

4.1 Experimental Setup

After the preliminary experiment, we performed an experiment to evaluate the proposed system using InterActor with a back-channel feedback. We compared three operational modes for Japanese students. In Mode A, InterActor moved as conventional InterActor without audio feedback. In Mode B, Character moved as InterActor with 100% back-channel feedback for all nodding. In Mode C, Character moved as Inter-Actor with 70% back-channel feedback for all nodding (Average of Japanese).

The subjects (24 Japanese male and female students aged between 19 and 23 years) used the system, and we confirmed that they were familiar with its operation. They were then introduced to the three operational modes and the differences between them while using the system. Next, the subjects were instructed to perform a pairwise comparison of each mode for an overall evaluation. Three comparisons were required; therefore, the experiment was conducted three (= $_3C_2$) times. The questionnaire was examined using a seven-point bipolar rating scale from –3 (not at all) to 3 (extremely); a score of zero denotes "moderately." Subjects evaluated three modes from the viewpoint of five items; Preference, Enjoyment, Interaction, Affability and Usability. Each subject was presented with the three modes in a random order to eliminate any ordering effect. A video editor recorded the communication experiment using a video camera, as shown in Fig. 4.

Fig. 4. Example of the experiment scene.

4.2 Results

Figure 5 shows the result of the sensory evaluation in the evaluation experiment. Significant differences between each of the three modes were obtained by administering Friedman's test. In Mode B and C, the results from sensory evaluation were rated positively compared with Mode A for all five factors. Significant differences were also obtained by administering the Wilcoxon rank sum test for multiple comparisons. A significance level of 1% was obtained for the "Enjoyment," "Sense of unity," "Relief," "Excitement," and "Preference" factors between Modes A and B and for the "Enjoyment," "Sense of unity" and "Excitement" factors when comparing Modes A and C.

The results of paired comparison for the three modes are shown in Table 2. Figure 6 shows the calculated results of the evaluation provided in Table 2, based on the Bradley–Terry model given in Eq. (4). Mode C, "InterActor with 70% back-channel feedback for all nodding" was evaluated most affirmatively, with Mode B and Mode A following in that order.

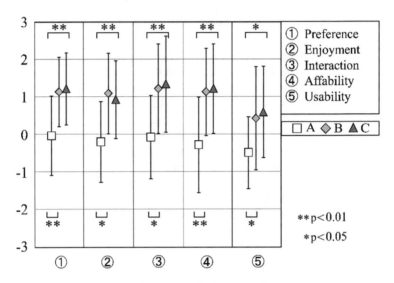

Fig. 5. Seven-point bipolar rating.

Table 2. Paired comparison result in the preliminary experiment.

	A	B	C	Total
A	–	7	2	9
B	17	–	8	25
C	22	16	–	38

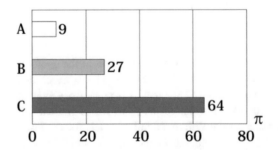

Fig. 6. Preference based on the Bradley–Terry model for the result provided in Table 2.

5   Conclusions

In this study, we develop an embodied character system with the voice response based on InterRobot Technology (iRT). Moreover, the effectiveness of communication support to InterActor is evaluated by two experiments. As a result of the first preliminary experiment, the condition of 70% back-channel feedback for all nodding (Average of Japanese) was evaluated affirmatively in comparison with 0% (No audio feedback), 42% (Average of Americans), and 100% (All) by sensory evaluation. The second experiment was for 24 Japanese students under three conditions. As a result, the condition of 70% was evaluated affirmatively in comparison with 0% (No audio feedback), and 100% (All) based on the results of seven point bipolar rating for five items; Preference Enjoyment, Interaction, Familiarity, Usability.

Acknowledgments. This work was supported by JSPS KAKENHI Grant Number 16K00278.

References

1. Kanda, T., Ishiguro, H.: Human-Robot Interaction in Social Robotics. CRC Press, Boca Raton (2017)
2. Pepper: https://www.softbank.jp/robot/consumer/products/. Accessed 28 Mar 2018
3. Kawamoto, S., Shimodaira, H., Nitta, T., Nishimoto, T., Nakamura, S., Itou, K., Morishima, S., Yotsukura, T., Kai, A., Lee, A., Yamashita, Y., Kobayashi, T., Tokuda, K., Hirose, K., Minematsu, N., Yamada, A., Den, Y., Utsuro T., Sagayama, S.: Design of software toolkit for anthropomorphic spoken dialog agent software with customization-oriented features. IPSJ J. **43**(7), 2249–2263 (2002). (in Japanese)
4. Lee, A., Oura, K., Tokuda, K.: An Open-Source Toolkit Realizing Attractive Voice Interaction Systems: MMDAgent. IEICE Technical Report, NLC2011-51, SP2011-96, vol. 111, No. 364, pp. 159–164 (2011). (in Japanese)
5. Watanabe, T.: Human-entrained embodied interaction and communication technology. In: Emotional Engineering. Springer, London, pp. 161–177 (2011)
6. Yamamoto, M., Watanabe, T.: Timing control effects of utterance to communicative actions on embodied interaction with a robot and CG character. Int. J. Hum.-Comput. Interact. **24**(1), 87–107 (2008)
7. Giannopulu, I., Terada, K., Watanabe, T.: Communication using robots: a perception-action scenario in moderate ASD. J. Exp. Theor. Artif. Intell. (2018). https://doi.org/10.1080/0952813x.2018.1430865
8. Watanabe, T., Okubo, M., Nakashige, M., Danbara, R.: InterActor: speech-driven embodied interactive actor. Int. J. Hum.-Comput. Interact. **17**(1), 43–60 (2004)
9. Maynard, S.K.: Conversation Analysis. Kurosio Publishers, Tokyo (2002). (in Japanese)
10. Ohshima, N., Ohyama, Y., Odahara, Y., De Silva, P.R.S., Okada, M.: Talking-Ally: the influence of robot utterance generation mechanism on hearer behaviors. Int. J. Soc. Robot. **7**(1), 51–62 (2014)
11. Krogsager, A., Segato, N., Rehm, M.: Backchannel head nods in danish first meeting encounters with a humanoid robot: the role of physical embodiment. In: Kurosu, M. (ed.) Human-Computer Interaction. Advanced Interaction Modalities and Techniques. HCI 2014. Lecture Notes in Computer Science, vol. 8511. Springer, Cham (2014)

Epoxy Resin Lamps Base Design Based on Users Emotional Needs

Xiong Wei and Wang Yalun[✉]

South China University of Technology,
381, Wushan Road, Guangzhou, Guangdong, China
6788036@qq.com, 877594129@qq.com

Abstract. Since the light bulb invention more than a hundred years ago, light bulbs and lighting in general have appeared in a variety of functions and morphological differentiation. Consumers have long been satisfied with the functional needs of lighting and spending tends to saturation. As consumer electronics becomes a more personalized technology, emphasis on emotional expression, richer aesthetics, and technology designs will broaden the boundaries of lighting beyond design. Design is based on the understanding of the user's experience of design and reflection, The LED source, which combines the highly designed space of the light source with stable, high transparency, low temperature-curing epoxy resin materials, has created manufacturing and general lighting design differences, which tend toward innovative user experience. The purpose of this paper is to investigate user emotional needs to better design better resin lamps. This paper focuses on the user-centered design and development process, the paper explores the potential aesthetic orientation and propensity to spend by the target user groups.

Keywords: Emotional design · Modular · Epoxy resin · LED lamps

1 Research Background

With the continuous improvement of modern lighting technology, new materials, new technology and consumer electronics are widely used. Lamps and lanterns broke through the simple lighting function in the past. In recent years, the LED light source is a good substitute for traditional incandescent lamp and fluorescent light sources. This broadens the space for modern lamp design. At the same time, consumers pay more attention to the pursuit of personalized expression when choosing household electronics. Products are not only functional carriers but also labels of their tastes. However, there are still a few lamps and lantern products with batch customized production methods on the market. To the present, lighting design has tended to follow the emotional design trend of industrial design toward the mass customization of original production-centered, product-centered design. rather than the production model, user-centered model of development progress, combined with consumer trends, technical achievements, innovation and a service concept development process.

© Springer International Publishing AG, part of Springer Nature 2019
S. Fukuda (Ed.): AHFE 2018, AISC 774, pp. 140–145, 2019.
https://doi.org/10.1007/978-3-319-94944-4_16

2 Theoretical Basis

The higher value of the product is to satisfy the emotional needs of the people, the most important of which is the need to establish its own image and establish and consolidate its position in society. Good industrial design not only easy to use, standardized production, respect for the environment, but also give full consideration to the user's psychological and emotional needs, in order to create higher additional value beyond the practical value. The more intuitive a design is, the more likely it is to be accepted and loved. Only by establishing the emotional bond between products, services and users, we can form the cognition of the brand, cultivate the user's loyalty to the brand and make the brand become emotional through the interaction of self-image, satisfaction, memory and other conditions Representative or carrier [1] (Fig. 1).

Fig. 1. Relationship between Consumer, Technology and Business

3 Potential User Needs Analysis

Through investigating the shape and usage of a large number of existing consumer lamps, looking at the entire process of establishing relationships between consumer and consumer lamp products, the user has little or no decision in shaping the lamp and ultimately the effect, and remains in a passive role. Luminaries are designed, manufactured, and sold until they reach the user's hands, and the user's opinion is missing from the design process, except for the final consumer choice.

According to "21st Century Business Herald" 2017 e-commerce consumer trends analysis: 90 consumer forces are rising. 90 after the consumer pay more attention to personalized consumer, pay attention to the intrinsic feelings and values of the brand, brand awareness after the formation of high brand loyalty; the other strong spending power, ahead of spending habits.

Throughout the user and the lighting products to establish the relationship between the entire process, user engagement is relatively low, the final shape of lamps and the decision-making, has been in a passive role. Lighting from the design, manufacture, sales reached the use of the process, the user's opinion is missing. (As shown in Fig. 2 below) Consumer demand for lighting a basic functional requirements, and has reached saturation, lighting design more spaces for improvement is to meet the user's advanced emotional needs to consider the individual differences in consumers.

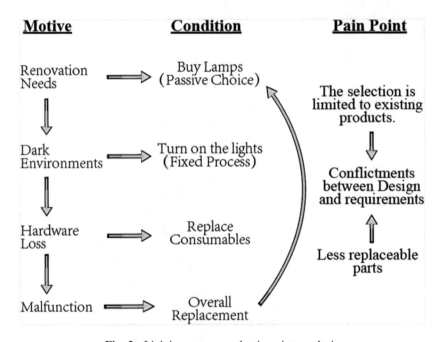

Fig. 2. Lighting process and pain points analysis

All decorative designs are created for people's aesthetic pursuit. As De Mul put forward the concept: human pursuit of aesthetics is divided into three phases: artistic value, display value and manipulation value [2]. The function of this lamp is accompanied by its artistic value. Throughout the history of arts and crafts, the lighting has enjoyed a rich and mature development at the two levels of "artistic value" and "display value," and consumers have also become accustomed to the supply and demand formed for thousands of years.

The third stage: manipulative value is still waiting to fill the blank phase. With the rapid popularization of contemporary Internet culture, young consumers are more likely to express themselves, preferring to display their iconic symbols or approach their own identity groups. This desire to express more often involved in shopping decisions, customized, personalized consumer goods has become a group of young people to establish a personal image, the expression of aesthetic carrier. Lighting design from this point to find a breakthrough point to explore the possibility of user participation in the design of the lamp (Fig. 3).

Fig. 3. Artistic value, display value and manipulation value of Mona Lisa (picture via Internet)

In the era when consumers are more and more required to express their individuality, the development mode of design and production should shift from product-centered to user-centered, and the emergence of lamps with personality differences appears. (As shown in Fig. 2, the process of using the lamp and the analysis of pain points) The demand for lamps by modern consumers can be roughly divided into three categories: lighting up the environment (basic requirements), home decoration (aesthetic requirements), and manufacturing atmosphere (emotional needs). Consumers with different backgrounds, qualifications, levels of consumption and even social circles often have different consumer electronics products. Common product designs usually can not satisfy different needs from different consumer groups at the same time.

4   Design Description

Concerned about the user needs for home atmosphere lamps, taking full account of people interested in new things or old things with memories, the use of test tube module which is easy to make and install standard parts and epoxy stable chemical and physical properties will be The combination of its micro-LED cold light source, combined with the high degree of freedom of choice of epoxy filler, making a standardized light guide module.

Design named "memory of the light", intended to personal feelings of the often fragile items sealed in a stable epoxy resin gel, the use of epoxy resin material characteristics, the only chance to create the design symbol, It can be a display of meaningless texture itself, or it can be a permanent preservation of the user's private memory. Within this transparent light-guiding medium, it is illuminated by a cold LED light source at the bottom as if it were a museum artifact, creating a retrospective or intimate, or ritualistic representation on behalf of family members, a simple combination of actions to convey a Free and open to use, reducing the user's use of the threshold, to narrow the contact between users and products to break the simple use of users and products - are used in the form of inherent interaction.

From the user's point of view, to respect the user's emotional needs, the user becomes the final light effect of the decision makers, you can replace the light guide module instead of the entire base, the replacement of the module itself can become a home decoration, to minimize waste. As the replacement module is the normal process of using the product, the light guide module and the LED light-emitting original are connected with the thread rotation method that accords with the daily usage habit, the process of replacing and replacing the module can be completed easily by one person's both hands while the LED light source uses the low-, To avoid the occurrence of electric shock accident.

Design focus on:

a. Interactivity: replaceable modular light guide module, innovative user experience;
b. Customization: resin light guide module filler can be made by the user requirements, is a customer relationship innovation;
c. The contrast of different textures, uniform in the transparent resin standard container;
d. Lightweight, with a sense of transparency, easy integration with the environment;

As a highlight of the entire lamp design, the light guide module is also a manifestation of the modular design of the product. The light guide module is connected with the LED light-emitting base by a standard connection thread, and can be disassembled and replaced at will during use. The content of resin modules can be pre-batched by the choice of combination, the content can also be provided by the user or make custom requirements, within a certain period of time by the manufacturer sealed resin irrigation.

The difference between this design and the similar products in the market is reflected in the following. Through a variety of combinations, users can freely choose modules and pedestals, innovate the user experience, and the final effect of the product

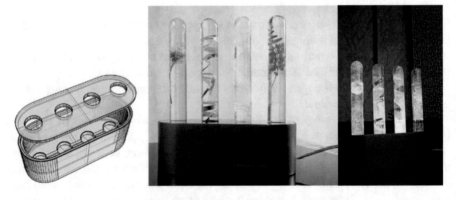

Fig. 4. Base design and display effects of Model 1

is determined by the user. Meanwhile, the epoxy resin modules are customized components, user custom made to the needs of manufacturers to produce unique lamps (Figs. 4 and 5).

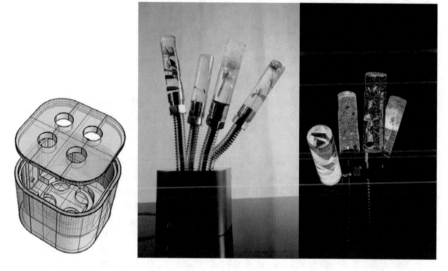

Fig. 5. Base design and display effects of Model 2

5 The Conclusion

Based on the user-centered design and development process, the design explores the potential aesthetic orientation and propensity to spend by the target user groups. This results in design of the lightweight and modularized, replaceable light guide-module lamps combined with the principle of batch customization and the new LED light source. Simple shapes, suitable for a variety of indoor use scenarios, but with the symbolic uniqueness and emotional connection with user to form emotional product design. The standard lamp base combined with the customized light module is an exploration of the production mode of Industry 4.0 and also suggests a new direction for lamp product marketing.

References

1. Newman, D.A.: Design Psychology 3 emotional design, p. 16. CITIC Publishing Group (2015)
2. de Mul, J.: Cyberspace odyssey, pp. 16–20. Guangxi Normal University Press (2007)

Construction of an Evaluation System for Automotive Interior Rendering Based on Visual Perception

Tian-tian Li[1], Dan-hua Zhao[2], and Jiang-hong Zhao[2(✉)]

[1] State Key Laboratory of Advanced Design and Manufacture for Vehicle Body,
Hunan University, Changsha, China
tintimfrank@gmail.com
[2] School of Design, Hunan University, Changsha, China
bear8213@126.com, zhaomak@126.com

Abstract. In view of the visual cognition theory, this paper aims to construct a universal evaluation system for automotive interior rendering. In terms of visual cognition and feature integration theory, this paper propose a visual cognitive evaluation framework for interior styling. Corresponding evaluation indexes are extracted and the weights of each index assigned. Application is conducted to test the effectiveness and feasibility of the evaluation system. The results show that a hierarchical relationship existed in the cognition of automotive interior rendering, and evaluation indexes can be divided into two attributes: basic index and additive index. The effectiveness and feasibility of the evaluation system is satisfied, which is effective to help designers to improve the scheme rendering specifically, and assist enterprises to make relevant decisions.

Keywords: Automotive interior · Styling rendering · Evaluation system
Visual perception · Analytic hierarchy process

1 Introduction

Unlike the automotive exterior styling, the interior is a combination composed of multiple functional systems, that the interior can be divided into about seven systems [1], each of which has multiple elements. Automotive interior is made up of a variety of parts, including steering wheel, dashboard, rear-view mirror, change gear, multi-media equipment, air-conditioning and wind gap [2]. The description of the physical set of automobile interior contains a degree of complexity. For this reason, the evaluation index based on the classification of interior elements is complicated and miscellaneous. Detailed information about each parts' design elements could not be gained because it was impossible to carry out actual experiments using so many samples in order to analyze detail design elements of each part [3], posing difficulties on the research of interior styling evaluation.

Automotive interior styling consists of two parts: one is layout design, and the other is styling design. The former refers to the spatial relationship among styling components, and the latter refers to the styling of the components. Under the background of

© Springer International Publishing AG, part of Springer Nature 2019
S. Fukuda (Ed.): AHFE 2018, AISC 774, pp. 146–157, 2019.
https://doi.org/10.1007/978-3-319-94944-4_17

the application and popularization of electric vehicle technology and HMI, the layout of the traditional interior and the existence of certain components are facing problems, posing new challenges to the evaluation of interior styling schemes.

This paper aims to construct a visual cognitive evaluation framework for interior styling in terms of visual cognition theory and the effectiveness and feasibility of the system are under verified by case application.

2 The Perception and Evaluation Framework of Automotive Interior Scheme Rendering

2.1 Rendering

2.1.1 The Role of Rendering in the Process of Design Communication

Shannon's communication model has been highly influential in design theory [4], either directly, or through its more general influence on how communication is conceived [5]. He represented the process of information transmission as five elements of information source, transmitter, channel, receiver and destination [6]. Mono has applied this basic model of communication to the study of product design [7]. Suri and Buchenau coined the term "experience prototyping", which have been shown to be powerful design communication. An experience prototype can be anything from a model to a tangible object, a visual representation or any other process/performance that is able to envision the "experience" the design seek to achieve [8]. According to Nima Norouzia, in terms of architecture, the architect is the sender and the client is the receiver, and the proposed design is the message [9]. Hence, we regard the evaluation process of the automotive interior renderings as a communication between the two main subjects of the designer (D) and the manager (M), that the individual or the design team transmit the design message through a tangible object (rendering) while the client or manager perceive the information with visual sense, forming opinions and responses to the design presented by the rendering. Design ideas need to be communicated and made "accessible" to different people involved in the design and decision making process [10]. It can be seen that, on the one hand, the rendering is a carrier of design intentions for the designer, acting as the transmitter of design information. On the other hand, the rendering is a medium that communicates the designer and the manager (Fig. 1).

2.1.2 Scheme Rendering

Relating to the different kinds of interactions that designers have with their sketches, Remko sums up three types of sketches - thinking sketch, talking sketch, and storing sketch [14]. According to Gary R. Bertonline & Eric N. Wiebe, two types of drawing exist during the preliminary ideas statement, ideation drawings and refinement drawings [11]. In architectural presentations, the term "presentation drawing" are similar to "design drawing", describing a design proposal in a graphic manner intended to persuade an audience of its value [12]. The formalized expression of graphic thinking is {quick sketch; subject sketch; comparison sketch; scheme sketch; scheme rendering} [13]. A sketch are incomplete and can be interpreted in different ways [14]. This referred to as 'ambiguity' or 'indeterminacy' [15]. The scheme rendering, or its

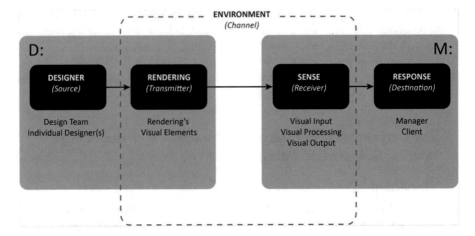

Fig. 1. Communication model of the automotive interior styling rendering evaluation.

synonym, presentation drawing, is more formalized, integrated and detailed among others. The rendering referred to this paper is scheme rendering of the automotive interior in the design review meeting.

2.1.3   Visual Elements of the Rendering

In the development of a freehand drawing, Paul Laseau presents a three step process of structure, tones, and details [9]. Edwards notes that there are five basic component skills and charts a sequential process of developing perceptual skills: edges, spaces, relationships, light and shadow and the whole [16]. Tovey and porter decompose drawings of automotive design into form lines, components, form shading and non-form shading [17]. Sixteen explicit visual communication rules for car exterior rendering are proposed by transportation design, Art center college of design during the teaching stage, including perspective, line quality, composition, proportion, contrast, 1-2-3 cube, focal point, 1st/2nd/3rd read, light source, application of media/craftsmanship, side view for every perspective sketch, interior/glass indication, door shut lines, rendered wheels, warm and cool color shift, story/drama. The above mentioned visual elements are used as a reference for the following work of the evaluation system construction.

2.2   The Basic Cognition on Automotive Interior Styling

Automotive interior styling is a combination composed of different modules to realize the established function. The carrier of the automotive interior styling is the room and space. Space is the basis of the interior design that determines the range and the scale of the object. The overall ambiance of automotive interior styling expresses the tone within the interior space, which is related to the user's psychological and emotional perception. Under the specified space scale and tones and ambiance, the form and function of the automotive interior are realized through the organization of a series of components, thus facilitating the unity of space, atmosphere, and components.

2.3   Process of Visual Perception on Automotive Interior Scheme Rendering

According to the theory of feature integration [18], visual processing is a process that is characterized by bottom-up processing and has a local interaction. The process of visual processing is input by the combination of various features. Then the objects are divided into two pathways, parallel processing the coding of each feature and integrating them after visual attention. Mishkin, Ungerleider and Macko claimed that The two main pathways for visual processing in the biological vision system are the ventral "what" pathway and the dorsal "where" pathway [19]. The suggested function of the "what" pathway is to recognize objects based on their visual appearance. The "where" pathway, on the other hand, computes spatial information about objects [20]. The feature extraction process is completed in multiple processing channels, and the output of various simple image features (color, texture, shape and space) are independent. For different features, the corresponding neural processing pathways are different, and the processing speed and processing methods are also different. In addition, the functions of different features in image information processing are not the same too. All these reflect the characteristics of diversity and hierarchy in the process of processing visual features.

This paper decomposes the cognitive object of automobile interior styling scheme rendering into three levels of {spatial information, atmosphere information, component information}, corresponding to the simple image features in visual cognition. The spatial information corresponds to the spatial relationship, which is mainly encoded via the 'where' pathway. The atmosphere information corresponds to the color and texture and is mainly encoded via the 'what' pathway. The component information corresponds to the shape and is encoded by the 'what' pathway. In the pre-attention stage of visual perception, three layers of information coexist independently, and only at a later stage, three layers of information are integrated together to achieve the overall cognition of the object of automobile interior scheme rendering. Figure 2 shows the visual perception process summarized in this article.

2.4   The Perception and Evaluation Framework

The perception and evaluation framework of automotive interior scheme rendering are constructed based on index category, index definition and index variable (Table 1). We initially define the spatial information, the atmosphere information, and the component information as follows. The spatial information refers to the information that determines the spatial scale and scope of the interior styling, depicting the three-dimensional spatial environment on a two-dimensional sheet of paper. The atmosphere information refers to the environmental tone and the influences within the surroundings. While the component information refers to the component styling and configuration.

By combining the definition, 13 visual elements of the evaluation are proposed and will be later used for the weight assignment experiments and extraction. Details of the framework are shown in Table 1.

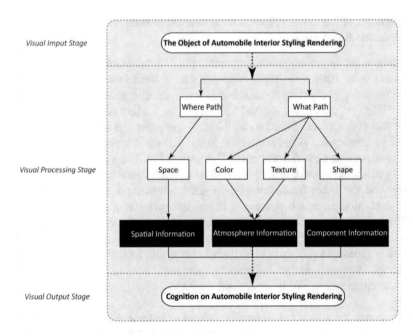

Fig. 2. Process of visual perception on automotive interior scheme rendering

Table 1. The perception and evaluation framework of automotive interior scheme rendering

Index category	Index definition	Index variable	Notation
Spatial information	Information determining the spatial scale & scope of the interior styling	Perspective accuracy	(C_1)
		Spatial depth	(C_2)
		Reasonable composition	(C_3)
Atmosphere information	The environmental tone and the influences within the surroundings	Harmonious tone	(C_4)
		Primary & secondary light shadow	(C_5)
		Texture expression	(C_6)
		The overall contrast	(C_7)
		Cold and warm color shift	(C_8)
Component information	The components' styling and configuration	Surface & structure expression	(C_9)
		Details expression	(C_{10})
		Primary & secondary focus	(C_{11})
		Good quality of line	(C_{12})
		Correct size of components	(C_{13})

3 Value Assignment and Extraction of the Index of the Perception and Evaluation Framework

3.1 Value Assignment

The analytic hierarchy process (AHP) is used to obtain the weight of the variables. Analytic Hierarchy Process (AHP), since its invention, has been a tool at the hands of decision makers and researchers, and it is one of the most widely used multiple criteria decision-making tools [21]. The AHP consists of three main operations, including hierarchy construction, priority analysis, and consistency verification [22] Yaahp is a software supporting diverse functions of hierarchical model construction, judgmental matrix data entry, sorting weight calculation and computational data export [23]. In this paper, the YAAHP software is used to assist the above works.

3.1.1 Hierarchy Construction

According to the AHP, a hierarchical model is established for the evaluation framework of the rendering. A simple AHP model has three levels of goal, criteria and alternatives though more complex models with more levels could be formulated [24]. The goal level refers to the predetermined goal of the problem. In this paper, it refers to the general requirement for the quality improvement of a rendering (A). The criteria level refers to the criterion of goal achievement. In this paper, it refers to the three categories of spatial information (B1), atmosphere information (B2) and component information (B3). The alternatives level refers to the measures to achieve the goal, which is the corresponding variables under the category. Then, 4 judgmental matrices are constructed for 13 variables of 3 categories: A-B, B1- (C1-C3), B2- (C4-C8), and B3- (C9-C13).

3.1.2 Subjects

The weight acquisition is carried out by expert valuation. The participants are 13 car styling design domain experts who are designers from main domestic automotive companies in China and PhD. candidates from Hunan universities.

3.1.3 Measurement

The fundamental scale (1-3-5-7-9) without intermediate values is used as the criterion. Experts are asked to make pair-wise comparisons to judge on comparative attractiveness of elements in the 4 judgmental matrices using the scale. The corresponding questionnaire data were obtained.

3.1.4 Consistency Check of the Questionnaires

As priorities make sense only if derived from consistent or near consistent matrices, a consistency check must be applied [25]. The consistency check are carried out to identify the acceptance of the questionnaire data. If C.R. (the consistency ratio) is less than 10% (C.R. < 0.10), then the matrix can be considered as having an acceptable consistency [17]. The inconsistent matrix can be automatically corrected by the Yaahp software. In this paper, if the judgemental matrix meets the requirement of consistency (C.R. < 0.10) by the minimal change algorithm or the maximal improved direction algorithm (changing one element) in the software, it is considered that the revised result is acceptable, and the questionnaire is valid. On the contrary, If the data required to be amended account for 20% or more of the total input data when the judgmental matrix is modified with the maximal improvement direction algorithm in the software, it is judged to be unqualified and the whole questionnaire abandoned. Among the 13 expert questionnaires, there were 10 qualified questionnaires that met the conditions of consistency detection and automatic revision. The unqualified questionnaires are not included in the results of group decision calculation.

3.1.5 Results of the Value Assignment

The qualified questionnaires are calculated with the software Yaahp. The power method is used as the calculation method. The maximal eigenvalues (λ_{max}) of the judgemental matrix and the corresponding maximal eigenvector (W) are calculated respectively. The weighted arithmetic average of the sort vector of each expert is used as an expert data aggregation method. Through the software calculation, the weight of the index is obtained, as shown in Table 2.

Table 2. Weight table of evaluation index on automotive interior scheme rendering

Goal	Weight	Criteria	Weight	Alternatives	Weight
The general requirement for the quality improvement of a rendering (A)	1.0000	Spatial information (B_1)	0.4519	Perspective accuracy (C_1)	0.1877
				Spatial depth (C_2)	0.1391
				Reasonable composition (C_3)	0.1250
		Atmosphere information (B_2)	0.2868	Harmonious tone (C_4)	0.0479
				Primary & secondary light shadow (C_5)	0.1001
				Texture expression (C_6)	0.0395
				The overall contrast (C_7)	0.0607
				Cold and warm color shift (C_8)	0.0386
		Component information (B_3)	0.2612	Surface & structure expression (C_9)	0.1258
				Details expression (C_{10})	0.0441
				Primary & secondary focus (C_{11})	0.0689
				Good quality of line (C_{12})	0.0267
				Correct size of components (C_{13})	0.0587

3.2   Index Extraction

According to the weights ranking result, 3 lowest indexes are removed and the final variable space is controlled to 10 (Table 3). The top three indicators are "Texture expression (C6)", "Cold and warm color shift (C8)" and "Good quality of line (C12)". Through the steps of evaluation and extraction, the construction of the evaluation system of automobile interior styling rendering is realized.

Table 3. Weights order and contribution rates order of automotive interior scheme rendering

Elements	Weight	Rank
Perspective accuracy (C1)	0.1877	1
Spatial depth (C2)	0.1391	2
Surface & structure expression (C9)	0.1258	3
Reasonable composition (C3)	0.1250	4
Primary & secondary light shadow (C5)	0.1001	5
Primary & secondary focus (C11)	0.0689	6
The overall contrast (C7)	0.0607	7
Correct size of components (C13)	0.0587	8
Harmonious tone (C4)	0.0479	9
Details expression (C10)	0.0441	10
Texture expression (C6)	0.0395	11
Cold and warm color shift (C8)	0.0386	12
Good quality of line (C12)	0.0267	13

3.3   Discussion

The result of weight assignment shows that the importance of spatial information is greater than the atmosphere information, and the importance of the atmosphere information is greater than that of the component information (Table 2). The result is in accordance with the initial definition of the category, indicating that there is a hierarchical relationship among the categories of the automobile interior styling rendering. At the same time, the weight of spatial information is obviously greater than that of atmosphere information and component information, and the weight of atmosphere information is close to component information. This indicates that the importance of atmosphere information and component information is more consistent for improving the quality of the rendering.

3.3.1   The Attributes of the Index

According to the result of the weight ranking (Table 3), it can be considered that the variables have two attributes of basic index and additive index. In this paper, the top ranking variables are mainly the basic indexes, the variables at the end of the ranking are the additive indexes. The basic index express the basic and indispensable needs of developing a rendering. (perspective, spaciousness, composition, component structure, contrast, etc.). Some additive indexes have a beneficial effect on the improvement of the overall quality of a rendering. For example, The effect reflected by the index "Good quality of line (C12)". Some additive indexes have a beneficial effect on the basic index. For example, index "Cold and warm color shift (C8)" creates a contrast atmosphere between intrinsic color and light source color, which further express the trend of the surface, and better distinguish the distance of the space in a rendering. which has a beneficial effect on the index "Surface & structure expression (C9)" and "Spatial depth (C2)". The index "texture expression (C6)" is established on the basis of the overall atmosphere by brightness and contrast creation. The improvement of texture information is mainly reflected in the follow-up work of CMF department, which may be another reason for the lower weight of the index at this stage.

It is possible to think that a rendering meets the requirements of the basic indexes is enough to be called a satisfactory rendering with a stable and reliable quality. However, in order to further improve the quality, it is also acceptable to take the additive indexes into account. This depends on the preferences or specific appeal of a certain company or department on the quality of the rendering.

4   Case Application

In order to prove the validity and feasibility of the evaluation system of interior styling rendering, a case application is carried out as follows: The 6 proposals provided by the state key laboratory of advanced design and manufacture for Vehicle Body (Hunan University) for a domestic automobile company is considered as the stimulus (Fig. 3).

Fig. 3. Automobile interior scheme renderings as testing samples

25 students major in transportation design were invited to score the results of the 6 car interior design using the indexes extracted.

The scoring criteria were divided into 5 consecutive scales (10-8-6-4-2), representing very good, good, general, poor and very poor. The product of the score of each index and its weight is the weighted score.

As can be seen from Table 4, the total scores of the 6 proposals are between 5.7 and 6.9. The total score of the proposal 1 is 6.9184, ranking in the first place. The total score of the proposal 4 is 6.7717, ranking in the second place. The proposal 1 got 4 first of single items, as shown in boldface in Table 4. They are "Perspective accuracy (C1)", "Reasonable composition (C3)", "Harmonious tone (C4)" and "Correct size of components (C13)". The proposal 4 got 5 first of single items, namely, "Spatial depth expression (C2)", "Primary and secondary light & shadow (C5)", "The overall contrast (C7)", "Details expression (C10)", "Primary and secondary focus (C11)". The total score and the score of the single item of the case evaluation is basically consistent with the impression of the rendering, which proves the validity and feasibility of the evaluation system of interior styling rendering.

Table 4. Results of testing samples

Sample	Spatial information			Atmosphere information			Component information				Total score	Rank
	C_1	C_2	C_3	C_4	C_5	C_7	C_9	C_{10}	C_{11}	C_{13}		
P.1	**1.531**	0.934	**0.990**	**0.360**	0.660	0.412	0.895	0.299	0.391	**0.441**	6.918	1
P.2	1.291	0.879	0.820	0.310	0.608	0.412	0.885	0.292	0.440	0.375	6.317	4
P.3	1.291	0.923	0.935	0.344	0.656	0.364	**0.915**	0.317	0.474	0.422	6.645	3
P.4	1.336	**0.990**	0.930	0.329	**0.704**	**0.446**	0.835	**0.303**	**0.496**	0.399	6.771	2
P.5	1.246	0.879	0.860	0.287	0.616	0.369	0.613	0.208	0.347	0.385	5.812	5
P.6	1.366	0.823	0.780	0.252	0.560	0.310	0.654	0.227	0.380	0.366	5.722	6

5   Conclusion

According to the theory of visual cognition, this paper puts forward the basic cognitive evaluation framework of interior styling, extracts the evaluation index of interior design scheme rendering, completes the weight assignment and extraction of evaluation index, and constructs the evaluation system of interior styling scheme rendering. The results confirm that there are hierarchical relationships in the cognition of the interior styling rendering and it is considered that the evaluation index can be divided into two attributes of the basic index and the additive index.

The application results of case evaluation show that the evaluation system of the interior styling rendering is effective and feasible.

The evaluation system of automobile interior styling rendering can be used to provide specific and targeted opinions for the improvement of the rendering, and also to provide references and support for the decision making of enterprises.

Acknowledgments. We would like to thank National Nature Science Foundation of China (51605154) for providing this research with financial support.

References

1. Stuart Macey, C.: H-Point, The fundamentals of Car Design and Packaging. Design Studio Press, Southern California (2013)
2. Yun, M.H., You, H., Geum, W., et al.: Affective evaluation of vehicle interior craftsmanship: systematic checklists for touch/feel quality of surface-covering material. In: Human Factors and Ergonomics Society Annual Meeting Proceedings, pp. 971–975 (2004)
3. Jindo, T., Hirasago, K.: Application studies to car interior of Kansei engineering. Int. J. Ind. Ergon. **19**, 105–114 (1997)
4. Beniger, J.R.: Who are the most important theorists of communication? Commun. Res. **17**, 698–715 (1990)
5. Crilly, N., Good, D., Matravers, D., Clarkson, P.J.: Design as communication: exploring the validity of relating intention to interpretation. Des. Stud. **29**, 425–457 (2008)
6. Shannon, C.E.: A mathematical theory of communication. Bell Syst. Tech. J. **27**, 379–423 (1948)
7. Mono, R.: Design for Product Understanding. Liber, Stockholm (1997)
8. Buchenau, M., Suri, J.F.: Experience prototyping. In: 'DIS' 00: Proceedings of the 3rd Conference on Designing Interactive Systems, pp. 424–433. ACM, New York (2000)
9. Laseau, P.: Graphic Thinking for Architects and Designers. Wiley, Canada (2001)
10. Leitner, M., Innella, G., Yauner, F., Northumbria: Different perceptions of the design process in the context of design art. Des. Stud. **34** (2013)
11. Bertoline, G.R., Wiebe, E.N., Hartman, N.W., Ross, W.A.: Fundamentals of Graphics Communications. McGraw-Hill Education, New York (2010)
12. Ching, F.D.K.: Architectural Graphics. Wiley, Hoboken (2009)
13. Zhao, D.: A Car Styling-based Study, the Design Intention and Interpretation. China Youth Press, Beijing (2014)
14. van der Lugt, R.: How sketching can affect the idea generation process in design group meetings. Des. Stud. **26**, 101–122 (2005)
15. Goel, V.: Sketches of Thought. MIT Press, Cambridge (1995)
16. Edwards, B.: Drawing on the Right Side of the Brain, (Rev. edn.) St. Martin's Press, New York (1989)
17. Trovey, M., Porter, S., Newman, R.: Sketching, concept development and automotive design. Des. Stud. **24**, 135–153 (2003)
18. Anne, T., Garry, G.: A feature integration theory of attention. Cogn. Psychol. **12**, 97–136 (1980)
19. Mishkin, M., Ungerleider, L.G., Macko, K.A.: Object vision and spatial vision: two cortical pathways. Trends Neurosci. **6**, 414–417 (1983)
20. Deubel, H., Schneider, W.X., Paprotta, I.: Selective dorsal and ventral processing: evidence for a common attentional mechanism in reaching and perception. Vis. Cogn. **5**(1/2), 81–107 (1998)
21. Vaidya, O.S., Kumar, S.: Analytic hierarchy process: an overview of applications. Eur. J. Oper. Res. **169**, 1–29 (2006)
22. Ho, W., Dey, P.K., Higson, H.E.: Multiple criteria decision making techniques in higher education. Int. J. Educ. Manag. **20**, 319–337 (2006)
23. Ma, X.: Build the structural model of risk assessment on the AHP method by Yaahp software: a case study on housing fund. J. Appl. Sci. Eng. Innov. **2**(6), 212–215 (2015)

24. Subramanian, N., Ramanathan, R.: A review of applications of analytic hierarchy process in operations management. Int. J. Prod. Econ. **138**, 215–241 (2012)
25. Ishizaka, A., Labib, A.: Review of the main developments in the analytic hierarchy process. Expert Syst. Appl. **38**(11), 14336–14345 (2011)

Two-Dimensional Emotion Evaluation with Multiple Physiological Signals

Jyun-Rong Zhuang[1(✉)], Ya-Jing Guan[1], Hayato Nagayoshi[1],
Louis Yuge[2], Hee-Hyol Lee[1], and Eiichiro Tanaka[1]

[1] Graduate School of Information, Production and Systems, Waseda University,
2-7 Hibikino, Wakamatsu-ku, Kita-Kyushu, Fukuoka 808-0135, Japan
gary_zhuang@akane.waseda.jp, {guanyajing1995,
h_nagayoshi}@fuji.waseda.jp, {hlee,tanakae}@waseda.jp
[2] Graduate School of Biomedical and Health Sciences, Hiroshima University,
1-2-3 Kasumi, Minami-ku, Hiroshima 734-8551, Japan
ryuge@hiroshima-u.ac.jp

Abstract. Extended roles of robots for activities of daily living (ADL) lead to researchers' increasing attention to human-robot interaction. Emotional recognition has been regarded as an important issue from the human mental aspect. We are developing an assistive walking device which considers the correlation between physical assistance and mental conditions for the user. To connect the assistive device and user mental conditions, it is necessary to evaluate emotion in real-time. This study aims to develop a new method of two-dimensional valence-arousal model emotion evaluation with multiple physiological signals. We elicit users' emotion change based on normative affective stimuli database, and further extract multiple physiological signals from the subjects. Moreover, we implement various algorithms (k-means, T method of MTS (Mahalanobis Taguchi System) and DNN (deep neural network)) for determining the emotional state from physiological data. Finally, the findings indicate that deep neural network method can precisely recognize the human emotional state.

Keywords: Emotion evaluation · Physiological data · Rehabilitation
Promotion of exercise

1 Introduction

With the advance of science and technology, the issue of the Human-Robotic Interaction (HRI) has received substantial attention. Many researchers have dedicated to improving the people's quality of life by employing the HRI technique. In human factors engineering, emotion is a significant feature to communicate with people; besides, people behavior (e.g. attitudes) may be dominated by emotion. Hence, studying emotions is crucial to the understanding of human behavior [1]. With an increase in the aging population worldwide, over the past few years, many studies have devoted to designing the assistive device for the patients, disabled person and elderly, and led them returning their normal life; thus, assistive device played a key role in the human rehabilitation field. However, most of the user intend assistive device complied

© Springer International Publishing AG, part of Springer Nature 2019
S. Fukuda (Ed.): AHFE 2018, AISC 774, pp. 158–168, 2019.
https://doi.org/10.1007/978-3-319-94944-4_18

with their mind since the assistive device is very unfamiliar to the user while wearing it. For improving the user feeling, the assistive device may enable understanding user's mental, and then follow moderately to human emotion change. It is thus necessary to investigate the emotion evaluation for human.

Several devices have been developed to identify emotional states utilizing "outer" signals, for instance, human voice, facial expressions and gestures. Nevertheless, there are limitations to these methods since it can easily control or fake these physiological signals to confuse the device occasionally. Accordingly, there are potential and actual physiological signals might be hidden or misjudged. Human emotional changes accompanied by many variations in physiological features which are difficult to be pretended. These signals thus are called "inner" signals due to their special characteristic. Therefore, analyzing the "inner" signals are great approach for researching human emotion state. There are several inner" signals, for example, through measuring human heartbeat, the instrument could compute the LF/HF ratio which was employed to assess the emotional arousal [2]. LF is an abbreviation for low frequency power which has a in connection with sympathetic component; HF is an abbreviation for high frequency power which reflect cardiac parasympathetic nerve's activity. In addition, electromyography (EMG) is an electrodiagnostic method for assessing and recording the muscle's electrical signals. Using EMG, it is able to measure the people facial muscles' activities when they located in different emotional states [3]. Furthermore, electroencephalogram (EEG) is an approach applied to record brain's electrical activity. Researching brain wave is valid for comprehending and recognizing various physiological signals. Especially, brain wave is proper signals to detect human emotion state, for instance, pleasure and arousal [1]. As indicated in previous studies, utilizing the physiological signals, it can efficiently evaluate human emotion. For evaluating the emotion, in the 1980s, Russell [4] proposed a circumplex model of affect to distinguish the human emotion via the two-dimensional plane. In this model, the horizontal dimension showed pleasure evaluation that utilized to describe the level of happy and unhappy; On the other hand, the vertical dimension exhibited arousal evaluation that illustrated the level of excited and sleepy. In our laboratory, we have developed assistive walking apparatuses which focused on assisting the ankle joint. This assistive apparatus can raise the equipped foot automatically via stretch reflex mechanism. However, currently, this assistive device only assists users using physical method [5]. In general, training and rehabilitation are not an easy and willing thing for the elderly or patients. Hence, the users' mental aspect needs to be considered when they equipped the device for training or rehabilitation [6].

In this study, we prepared the emotional stimuli experiments to inspire users' feelings and collect multiple physiological signals. The purpose of this paper is to analyze multiple physiological signals and employ several methods for precise evaluating the emotional state, and further mapping on the arousal-pleasure dimensional model.

2   Valence-Arousal Dimensional Model

Emotional models enabled to be characterized via two major features which were valence and arousal [4]. Figure 1 reveals the valence-arousal dimension model for recognizing the emotional state [4]. On the valence axis, we can judge the degree of attraction of a specific event from the human. Additionally, on arousal axis, we can respond the psychological state of the human (e.g., heart rate and blood pressure). In this study, we recognized the emotion state based on this indicated two-dimensional model.

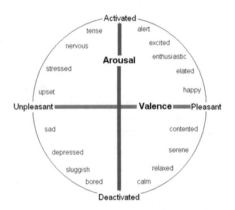

Fig. 1. Valence-arousal dimensional model

3   Experimental Setup

3.1   Normative Affective Stimuli

To elicit the real emotion, the physiological signals can effectively express emotion as the reference. Thus, we attempted to explore the correlation between physiological signals and emotion via the normative affective stimuli. Samson et al. [7] proposed a film library, a collection of stimuli, which available research the emotion variation. These films were presented to 411 subjects for a large online study to verify film clips which reliably elicited the subject emotion [7]. The rating of valence, arousal described each film. The selected film clips for the experiment were varied (with different grade of valence and arousal) to elicit various emotions from the subject. We attempted to achieve all the emotion assumptions. Therefore, we prepared 30 stimuli film through this library to elicit different emotions.

3.2   Experimental Device

In this study, we employed three physiological signals as human emotional judgments, there are EEG (Electroencephalogram), EMG (Electromyography) and HRV (Heart

Rate Variability). We used the Emotiv, Personal EMG, and MyBeat to detect the EEG, EMG, and HRV, respectively, as shown Fig. 2(a).

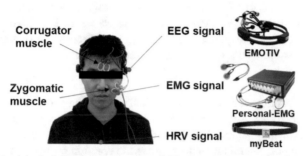

(a) Three physiological signals detector are used. The Emotiv, Personal EMG, and MyBeat to detect the EEG, EMG, and HRV, respectively.

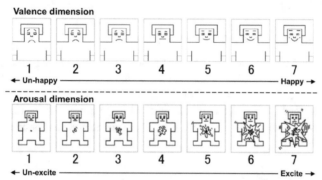

(b) Self-assessment manikins applied in this emotional elicited experiment.

Recording Baseline	Wash-out period	Emotional Stimuli	Questionnaire survey
60 sec	30 sec	30 sec	30 sec
	Rest period		Self-Assessment

Repetition

(c) The emotion experimental sequence which is as follows: 1) baseline recording; 2) rest period; 3) emotional stimuli; 4) questionnaire survey.

Fig. 2. (a) Three physiological signals detector are used; (b) Self-assessment manikins (c) Emotion experimental sequence

3.3 Experimental Protocol

We conducted experiments with 20 healthy participants (the age between 21 to 27 years old) wearing three physiological signals detector while watching the film clips. Before starting the experiment, we would record the MVC (maximum voluntary contraction) of each subject. Next, the experimental protocol was as follows: (1) Let subjects clam and rest for 1 min and record these physiological signals as the baseline that specifies the reference level. (2) Subjects watched the selected film clips (30 s for each film clip). After watching each film clips, subjects had 30 s to finish corresponding questionnaire survey (self-assessment as shown in Fig. 2(b)) and have 30 s take a rest as the wash-out period (to eliminate the effect of the previous film clip). Figure 2(c) shows the whole emotion experimental sequence.

3.3.1 Self-assessment

20 Participants would grade (from 1 to 7) their own feelings on valence and arousal dimension, respectively, after watching each film clip. Figure 2(b) demonstrates the SAM (self-assessment manikins [8]) questionnaire. SAM is visualization questionnaire that has vivid graphs, which let subjects grade the emotion on two dimensions easily.

4   Methodology

4.1   Emotional Level Identification

In this study, we processed the emotional scores (1–7) which was divided 3 levels on each valence and arousal dimension. Figure 3 illustrates the nine-emotional state. On valence axis, the 3 levels were following: 1–3, 4 and 5–7 were mapped to "unpleasure", "neutral" and "pleasure", respectively. On arousal axis, the 3 levels were considered as: 1–3 was mapped to "de-excite", 4 was mapped to "neutral" and 5–7 was mapped to "excite".

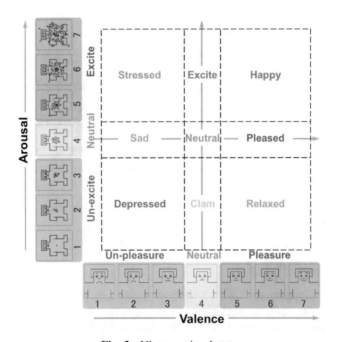

Fig. 3. Nine emotional state

4.2   Physiological Features Selection

We obtained the many raw physiological data from three physiological detectors (EEG, EMG, and HRV). Furthermore, after data processing, we extracted the 17 physiological signals as our physiological features, as shown in Table 1. To eliminate the individual difference, all selected features divided by baseline physiological value.

4.3   Classification Algorithm

4.3.1   K-Means Clustering

During our initial emotion evaluation research, we implemented the unsupervised clustering method of k-means algorithm for analyzing the physiological features and dividing the data set into several natural clusters [6]. By using k-means, it is unnecessary to give the signals with the class labels. The input signals would iteratively be calculated their centroid point (mean value of each group) until several groups could be divided evidently. Each group's feature can be thus captured [6]. We proved the emotion can be mapped on two-dimensional model into four separated areas (three groups on valence axis and two groups on arousal axis) as shown in Fig. 4. In our previous results [6], although k-means algorithm can divide the emotion to four areas, the number of subject and emotional stimuli method would significantly affect the results. To achieve more precisely results, we further expand the data set and use new emotional stimuli method (film library [7]).

Table 1. 17 extracted physiological features

	Features
1	(Max. EMG)/(MVC of corrugator muscle)
2	Max. EMG of corrugator muscle
3	(Max. EMG)/(MVC of zygomatic muscle)
4	Max. EMG of zygomatic muscle
5	Max. LF/HF
6	Max. variance of LF/HF
7	Max. heart rate
8	Min. variance of LF/HF
9	Mean LF/HF
10	Mean variance of LF/HF
11	Standardized LF/HF
12	Standardized heart rate
13	Mean Theta wave's power spectrum
14	Mean Alpha wave's power spectrum
15	Mean low Beta wave's power spectrum
16	Mean high Beta wave's power spectrum
17	Mean Gamma wave's power spectrum

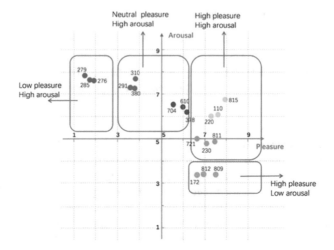

Fig. 4. Four separated groups on two-dimensional model by using k-means clustering

4.3.2 Mahalanobis Taguchi

The Mahalanobis Taguchi System (MTS) is the binary classification algorithms for diagnosis and predict multivariate data. This method utilizes the Mahalanobis distance to measure correlations between the multivariate variables, and use Taguchi method to assess accuracy of estimations via the constructed scale. To continue the emotion evaluation, we increased the number of subjects and number of physiological features, and further utilizing the T method (one of the approaches in MTS) to conduct the emotional classification. The advantages of the T method are that it does not use the Mahalanobis distance, and it can resolve mathematical constraints, and it is possible to judge the direction on emotional dimension using multivariate physiological data. We can attain overall predicted output through the weighted integration with the corresponding signal to noise (S/N ratio). In T method, we need to construct the measurement scale. Following the measured results, we can construct the measurement scale by selected physiological features together with questionnaire results, as shown in Table 2. To confirm threshold for classification, this work used a half of the difference value of two distributions' mean values.

Table 2. T method measurement scale.

	Feature 1	Feature 2	⋯	Feature k	Pleasure axis	Excite axis
Sample 1	⋯	⋯	⋯ ⋯		1	2
Sample 2	⋯	⋯	⋯ ⋯		3	5
⋮	⋮	⋮	⋮ ⋮		⋮	⋮
Sample n	⋯	⋯	⋯ ⋯		7	3

4.3.3   Deep Neural Networks

Deep neural network (DNN) is special form of Artificial Neural Network (ANN). DNN's structure comprise of the multiple hidden layers between the input and output layers. DNN used stacked autoencoders, which can process highly non-linear problems. By using this method, it can train and learn from the all data through multiple layers to calculate repeatedly, and finally obtain the great predictions. In this study, we faced the multivariate variables situations, which is better to apply the DNN to evaluate the emotional state.

We implement the emotion recognition classification of the valence and arousal state, respectively, by using DNN algorithm, and further combined two calculated results as final emotional states judgments. In our DNN system, two classifiers separately employed a stack of three autoencoder with softmax layers as shown in Fig. 5. Additionally, hidden layer 1 has 60 hidden nodes, hidden layer 2 has 30 hidden nodes, and hidden layer 3 has 20 hidden nodes. To reduce the calculated time and obtain high accuracy, it is necessary to screen the important physiological features. Thus, all physiological features would be calculated via attribute selection algorithm [9]. Finally, nine selected features used in valence classifier, and 11 selected features used in arousal classifier. The output features of the hidden layer 3 are utilized as input features for the softmax layer that can be trained as parameters at the same time. After the network completed learning of the weight and bias parameters in softmax classifier, the algorithm must fine-tune all the weights and bias parameters in the whole network simultaneously. The fine-tuning process enabled to improve all weights for all layers in the network. We employed the self-assessed emotional states (valence and arousal) as the basic fact.

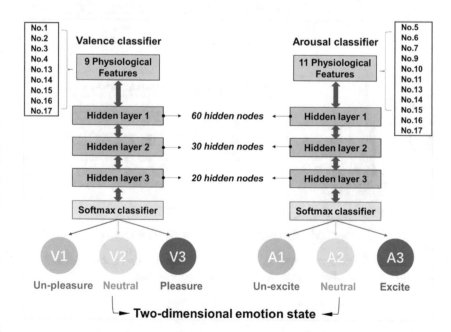

Fig. 5. Two-dimensional emotional DNN classifier structure

5   Results and Discussion

5.1   The Calculated Results of MTS

In T method, the unit space selection would significantly affect the calculated results. We respectively applied the MTS method to the valence and arousal state. On valence scale, we specified the unit space was score 7. The findings reveal that the valence state can effectively be divided into two groups (happy and unhappy); besides, we also found the distribution of each score that exhibits the single directional tendency from negative (Score 5 and 6) to positive direction (Score 1, 2 and 3), as shown in Fig. 6(a). However, we specified the unit space was score 2 on arousal scale, and the classification results demonstrate the slight lack of precision as shown in Fig. 6(b). To classify results of the arousal scale effectively, we think it may need appropriate unit space to obtain vivid two distributions (excite and un-excite). The classification accuracy of T method was 77% for valence state, and 47% for arousal state.

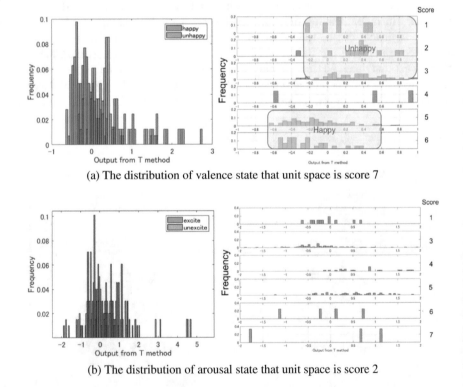

(a) The distribution of valence state that unit space is score 7

(b) The distribution of arousal state that unit space is score 2

Fig. 6. Calculated results of T method. (a) The distribution of valence state that unit space is score 7; (b) The distribution of arousal state that unit space is score 2

5.2   The Calculated Results of DNN

Deep neural network (DNN) can effectively process the multivariate variables problems for emotion recognition. In our DNN procedure, we utilized the three hidden layers to learn the relevant features from the input physiological signals and further predict the two emotional state. The results reveal that the DNN can achieve the classification accuracy of 69.8% and 77.4% for dividing three valence states and three arousal states by using 17 physiological features. Moreover, by using attribute selection algorithm, we attained the important features (9 features for valence classifier and 11 features for arousal classifier) which can be as input signals of DNN on two classifiers. Thus, the findings show the DNN with selected features can increase the classification accuracy to 79.2% and 81.1% for classifying three valence states and three arousal states. Figure 7 shows the comparison classification accuracy through using the different algorithm.

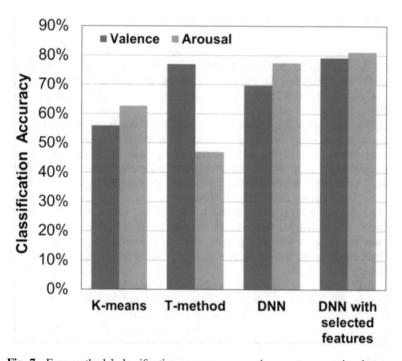

Fig. 7. Four methods' classification accuracy comparison on two emotional states

6   Conclusions

This study aims to develop a new emotion evaluation method for the two-dimensional valence-arousal model by using multiple physiological signals. We applied the k-means, T method of MTS and DNN for confirming the emotional state from physiological data. In previous results, we can employ k-means to divide three groups on valence axis and two groups on arousal axis by using less subjects' measured results. We further enlarged the number of subjects, and the findings showed it can successful divide three groups on each axis; however, the classification accuracy only can reach to 56% and 62.7% on valence and arousal state. Furthermore, we employed the T method of MTS, and we found this method can possible distinguish the valence state (the classification accuracy is 77%). However, on arousal state, this method was less significant (the classification accuracy is only 47%). Finally, we proposed our DNN procedure for processing the emotion evaluation. By using DNN with selected features, we can clearly classify three groups for each dimension; besides, the classification accuracy can achieve to 79.2% and 81.1% for valence and arousal state, respectively. To conclude, the findings indicate that DNN (deep neural network) can precisely recognize and effectively divide the human emotional state. In the future, we will progress the real-time emotional recognition for combining the assistive walking device to improve the feeling of rehabilitation and assistance.

References

1. Chanel, G., Kronegg, J., Grandjean, D., Thierry, P.: Emotion assessment: arousal evaluation using EEG's and peripheral physiological signals. In: International Workshop on Multimedia Content Representation, Classification and Security, pp. 530–537. Springer, Heidelberg (2006)
2. Tanaka, E., Muramatsu, K., Osawa, Y., Saegusa, S., Yuge, L., Watanuki, K.: A walking promotion method using the tuning of a beat sound based on a two-dimensional emotion map. In: Proceedings of the AHFE 2016 International Conference on Affective and Pleasurable Design, pp. 519–525. Walt Disney World, Florida, USA, 27–31 July 2016 (2016)
3. Turner, J.R.: For distinguished early career contribution to psychophysiology: award address 1988. Psychophysiology 26(5), 497–505 (1989)
4. Russell, J.A.: A circumplex model of affect. J. Pers. Soc. Psychol. 39(6), 1161–1178 (1980)
5. Tanaka, E., Muramatsu, K., Watanuki, K., Saegusa, S., Yuge, L.: Development of a walking assistance apparatus for gait training and promotion of exercise. In: 2016 IEEE International Conference on Robotics and Automation (ICRA 2016), Stockholm, Sweden, pp. 3711–3716, 16–21 May 2016 (2016)
6. Zhang, Z.Q., Tanaka, E.: Affective computing using clustering method for mapping human's emotion. In: 2017 IEEE International Conference on Advanced Intelligent Mechatronics (AIM2017), Munich, Germany, pp. 235–240, 3–7 July 2017 (2017)
7. Samson, A.C., Kreibig, S.D., Soderstrom, B., Wade, A.A., Gross, J.J.: Eliciting positive, negative and mixed emotional states: a film library for affective scientists. Cogn. Emot. 30(5), 827–856 (2015)
8. Bradley, M.M., Lang, P.J.: Measuring emotion: the self-assessment manikin and the semantic differential. J. Behav. Ther. Exp. Psychiatry 25(1), 49–59 (1994)
9. Karegowda, A.G., Manjunath, A.S., Jayaram, M.A.: Comparative study of attribute selection using ratio and correlation based feature selection. Int. J. Inf. Technol. Knowl. Manag. 2(2), 271–277 (2010)

"We Dream Our Life Could Be": A New Undergraduate Course Aimed at Bringing Emerging Technologies Its Breakthrough Applications

Qian Ji[1(✉)] and Min Wan[2]

[1] Industrial Design Department, School of Mechanical Science and Technology, Huazhong University of Science and Technology, Wuhan 430074, Hubei, China
jiqian@mail.hust.edu.cn
[2] Landscape Department, School of Architecture and Urban Planning, Huazhong University of Science and Technology, Wuhan 430074, Hubei, China
Wanmin01@sina.com

Abstract. This paper describes the insights that have been introduced in the course "Specific Theme Design" for the fourth year undergraduate students in Industrial Design Department, School of Mechanical Science and Engineering (MSE), Huazhong University of Science and Technology (HUST), and that is called "We dream our life could be".

Each year a specific topic will be chosen in "Specific Theme Design" according to the development of the emerging technology. In 2017 the course is focused on the new intelligent flexible materials application.

Students produce novel, unexpected solutions and define feasible product idea based on technological opportunities of flexible materials, relevant user needs, anthropological research and scenario writing.

Keywords: Specific Theme Design · Intelligent flexible materials
Product design

1 Introduction

Many new products are the result of what is often called "Technology push", the result of new techniques, new materials or new methods [1]. Within MSE, HUST a lot of research, both fundamental and applied, is carried out. Too often it happens that the results of this research remain in a theoretical phase and don't find their way to the industry because they lack a "breakthrough application". The course Specific Theme Design teaches students of Industrial Design background to design products about new

Supported by National Natural Science Foundation of China (Grant No: 51708236).
Supported by Seed Foundation of Huazhong University of Science and Technology (Grant No: 2016YXMS273).

technology that was developed by the faculties of the MSE, HUST in a two- month course.

The "We dream our life could be" project in "Specific Theme Design" was introduced for the fourth year undergraduate students in 2017 to keep up with the rapid changes of intelligent flexible materials technology, resulting into new product profiles requiring a shift in teaching scheme and methods.

Different from the traditional teaching scheme of lecture or exercise, the course consists of a mix of project work, lectures, brainstorming, exercises and presentation. The students can just turn their attention to the screen, communicate with technicians or team partners and some 20 or 30 min later they are working on their project again, while the teachers are giving detailed explanation to groups or individuals. Project will be assessed both on a group result as well on an individual basis. A product presentation, including the process for the design decisions, and user scenarios at the end of the project are part of the assessment.

This course is an exploration for new material design application [2], the intelligent flexible material often have several functions to support their intelligence, they are typically more rugged, lighter, portable, robust, lightweight compared to traditional rigid substrate counterparts[1]. While the intelligent flexible material introduced in the course developed by the research team in MSE, HUST, is as thin as "onion skin" affixed to the users' skin and can detect the data of user's body. It has aroused great interest and enthusiasm among the students.

Students are trained to analyze context, generate innovative product ideas and use their design skills to develop product concepts based on the material characteristics. Moreover, they have to be able to manage the innovation process and control all critical aspects by systematically performing interdisciplinary verifications. The development of these "dream" products requires a merging of different knowledge domains and various design skills when comes to the ability to adapt to the environment, the autonomy, the human interaction, the multi-functionality, the reactivity, the ability to react emotionally and to operate in cooperation with other products.

2   Comprehensive Objective of the Course

The course offers the students the opportunity to further investigate the materials and techniques. Students will understand applicability and constraints of intelligent flexible material, explore applicability of the specific material of existing products and applicability of it for new products, analyze problems, define product and technology requirements, communicate with interdisciplinary team of designers, researchers and experts as well as customers, finally, develop solutions [3].

They have to learn that in looking for opportunities, characteristics of a product or material can create new opportunities. During a brainstorm session a member of the

---

[1] Wong, William S., Salleo, Alberto: Flexible Electronics Materials and Applications. Springer US (2009).

innovation group suggested to consider it as breakthrough. This lead to research in products where the material could be used.

After defining the product functions and the user requirements of the expected products [4], students have to develop the interaction concept and the system design, describing all components and the software requirements. After that students define the product concept and the graphic user interface. Specific attention is given on the usability and aesthetics. The result is presented using technical 3D drawings. It will provide a platform for students to gain design knowledge through problem analysis, design research, communication, teaming work and solutions development using design techniques in teamwork, care about the society calculating a sharp sense of new technology development and its social application [5]. Table 1 gives an expecting outcome as well as a comprehensive objective of the course.

Table 1. Comprehensive objective of Specific Theme Design course.

OBJECTIVE1: DESIGN KNOWLEDGE	OBJECTIVE2: PROBLEM ANALYSIS	OBJECTIVE3: DEVELOP SOLUTIONS
It can solve the problem of product design with the system expertise of basic design theory, design method, design procedure and application design, and it can provide the right design target for a clear market positioning.	Based on the basic principles of material science, modeling technology, mechanical structure manufacturing and ergonomics, we can express and analyze product problems through preliminary research and practice research, so as to get effective conclusions.	It can design practical products to meet the needs of specific users, and reflect innovation awareness in the design process, taking into account technological feasibility, material, form, culture, social and environmental factors.
OBJECTIVE4: DESIGN RESEARCH	**OBJECTIVE5: DESIGN TECHNIQUES**	**OBJECTIVE6: DESIGN AND SOCIETY**
Based on the people-oriented and user centered design method, we can do research on product design, including questionnaires, interviews, testing and interpreting data, and get reasonable and effective conclusions through information synthesis.	We can choose the appropriate computer technology in the design, express the specific form, internal structure, material, color, interaction function and layout of the product, and understand its limitations.	It can be reasonably analyzed based on the relevant knowledge of industrial design, and evaluate the impact of product design solutions on society, health, safety, law and culture, and understand the responsibilities that should be undertaken.
OBJECTIVE7: TEAM WORK	**OBJECTIVE8: COMMUNICATION**	**OBJECTIVE9: LIFELONG LEARNING**
A good team spirit and ability to play a role in a multidisciplinary team.	To effectively communicate and communicate with industry peers and the public on the issue of industrial design, and have certain international vision, communicate and communicate under cross cultural background.	It has the consciousness of self-study and lifelong learning, and has the ability to learn and adapt to development.

3   Examples of Design Outcomes

3.1   Athletes Monitoring Clothes

The concept of special clothes is designed for professional athletes based on the intelligent flexible materials, which is rugged, lighter, portable, robust, lightweight and can be worn tightly and snug against their skin. Athletes' movements will cause physiological changes, like muscle stimulation, breathing and heart rate acceleration, sweating, body temperature elevation and other physiological phenomena related to the amount of exercise and duration of exercise. Therefore, the design of clothes aims to monitor the physiological parameters of athletes, help them maintain reasonable physical strength, determine the proper amount of exercise to adjust the athlete's physical condition to the best condition and monitor the athletes' index and performance in real time and send more precise real-time data to coaches.

As the largest organ of the human body, the skin is the protective barrier of the human body, and it is also the most important medium for obtaining the physiological signals of the human body. The sensor is placed on the surface of the skin, can collect physiological waves, respiratory frequency, body temperature, blood pressure and metabolic activities and other information, and these data not only can be used for daily health monitoring and tracking of the disease diagnosis, can also monitor the movement efficiency and muscle fatigue (Fig. 1).

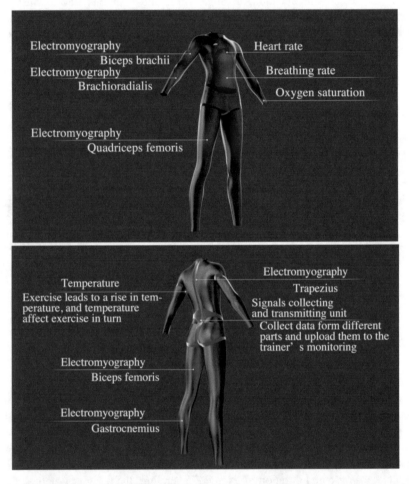

Fig. 1. Physiological and motion parameters can be directly measured by sensor material, and computer system can be analyzed and processed in real time. It can find out the relationship between the characteristics of a sportsman and the winning parameters. Through the analysis of real-time collection data in training, the training strategy can adjust timely, which can effectively prevent the athletes from overwork and injury. During the competition, the athletes' physical parameters are measured and the help participants adjust to keep the athletes' physical strength. In normal sports training, by analyzing the changes of physiological parameters, we can determine the reasonable amount of exercise and adjust the athlete's physical condition to the best condition before entering the competition. Designed by Yuqi Long and Jiawen Min.

3.2   Intelligent Nail Polish for Pregnant Women

Unlike the traditional nail polish, the nail polish can be continuously monitoring the users' emotional state 24 h a day, and the information will be presented vividly on the nails through the changing color and pattern on the nail surface, emotional information has been classified into various grades, each grade will be shown with different designed patterns and colors matched up with happiness, sadness and tension in order to realize the relationship between pattern and color semantics and different physiological information.

Customize a nail polish according to the shape of the user's fingernails. The sensor chips installed in the smart flexible material that is attached to the surface of the user. Through monitoring and analyzing the relevant factors of users, the user and caregiver will get the monitoring results of their emotions.

The smart function of emotional assessment, recording, real-time display will more add pleasure in daily life and may relieve pressure for users compared with traditional nail polish products. More detailed body healthy information feedback can be viewed by the users through mobile App at real-time, as well as by their caregivers or doctors to observe the users even from long distance.

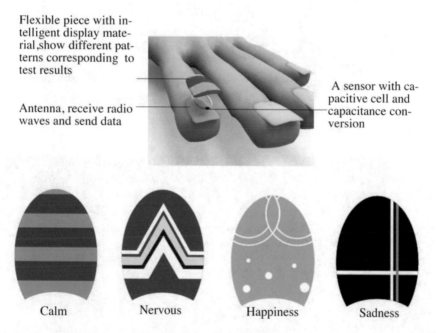

Fig. 2. Real-time monitoring of pregnant women's mood and sleep status, Real-time feedback monitoring results. Relieving pressure of pregnant women. Sustainable use of postpartum. Sometimes pregnant women have emotional fluctuations, sensitive, needing more accompany and more confidence and their desire for beauty is stronger than normal times, the design of emotional correspondence of color patterns is the innovative thinking of the product. Blue, white, green and horizontal lines represent calm. Purple, yellow, dark red, and triangle represent tension. Orange, the pink and the heart, the curve shape represent happiness. Black, gray, dark blue and square represent sadness. Designed by Hui Yan and Mengdie Jiang.

The design is stylish, intelligent, no harmful and can effectively avoid the discomfort, interval monitoring, wearing complex and other problems of traditional monitoring wearable products. It will lead a new fashion trend in Manicure industry (Fig. 2).

3.3   Intelligent Cosmetic Contact Lenses

Based on the characteristics of intelligent flexible materials, the intelligent cosmetic contact lenses can change the color according to the environmental change, check the health of the eyeball and prevent the potential disease on eyeballs. The induction system imbedded in the intelligent flexible material can acquire the information of the surrounding environment, automatically adjust the color pattern, and the user can also choose and control the color pattern freely through App, and can also remind the users to have proper rest according to the eyeball fatigue state. It will lead a new fashion trend in cosmetic industry (Fig. 3).

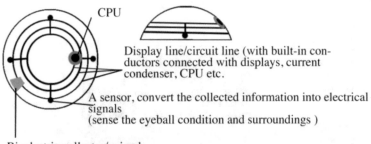

CPU

Display line/circuit line (with built-in conductors connected with displays, current condenser, CPU etc.

A sensor, convert the collected information into electrical signals
(sense the eyeball condition and surroundings)

Bioelectric collector/animal

Fig. 3. Compared with traditional Design of Cosmetic contact lenses, it can not only correct visual acuity, but also change the color, check the health of the eyeball based on smart flexible material. Designed by Yingying Feng and Fangzhen Liu.

3.4   Emotion Communication for Long-Distance Lovers

The product helps lovers from long-distance get to know each other's emotional status by wearing ring-shaped tattoo based on intelligent flexible materials. It aims to increase interest and get close quarters experience for long-distance lovers, through the smart sensor of the product, it can help them understand his or her mood change, when he or she is angry and sorrowful, the care will be sent to each other soon, which will add more pleasure, warm and sweet feeling for long-distance lovers. The sensor technology with visual interaction is on the human finger skin, and the user's personality and status can transmit through the visual language. It shows the scene of "interaction" with the feeling exchange between lovers, which embodies the blend of emotion and aesthetics. More information and interactive functions will connect to the App on smart phones (Fig. 4).

3.5   Intelligent Flexible Facial Mask

The intelligent flexible facial mask is designed to detect users' skin condition then provide intelligent scheme for users based on it.

The silver ions on the mask produce the adsorption force under the action of the biological negative electrode, and adsorb and purify the harmful substances such as lead and mercury which remain on the skin in our daily cosmetics.

The smart chip on the mask is used to read the skin state data through the silver ion electrode, and then push the water molecule hyaluronic acid on the mask to the deep skin.

Personalized mobile phone tracking technology App remind: nursing is controlled by a App, can be adjusted according to the feeling of facial intelligent biological wave, at the same time nursing, mobile phone App will get your facial skin condition data, and remind the skin care method for the surrounding environment according to the weather changes.

The host with intelligent mask can open the corresponding App and choose the proper frequency and intensity of flapping and electric shock (Fig. 5).

Fig. 4. Designed by Yiming Zhao and Yi Yu.

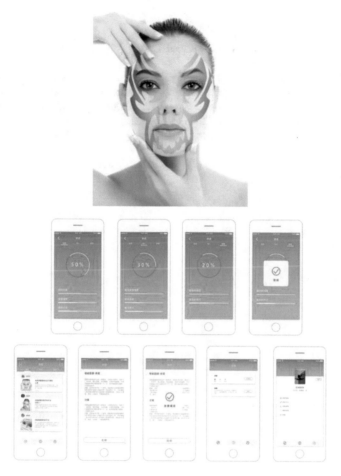

Fig. 5. The intelligent flexible facial mask is designed by Xinwan Zhang and Ming Zhang.

4 Conclusion

The course "Specific Theme Design—we dream our life could be" educates students to analyze material characteristics, generate innovative product ideas and use their design skills to develop product concepts based. This unique merging of different knowledge domains creates a mix to stimulate new ideas that will contribute to the creativity and effectiveness of the product design.

References

1. Archer, B.: Design Awareness and Planned Creativity in Industry. Thorn Press Limited, Toronto (1974)
2. Miller, D. (ed.): Material Cultures: Why Some Things Matter. UCL Press, London (1998)
3. Ji, Q., Zheng, J.: Industrial design education mode based on outcomes-based Education. In: Proceedings on Seminar of Industrial Design Education of China and International Industrial Design Forum 2017, pp. 131–136. China Light Industry Press, China (2017)
4. Hekkert, P., Schifferstein, H.: Product Experience, p. 334. Elsevier Science, Amsterdam (2008)
5. Dorst, K., Cross, N.: Creativity in the design process: co-evolution of problem-solution. Des. Stud. **22**(5), 425–437 (2001)

Intelligent Nail Film Product Design Based on the Application of Flexible Electronic Materials

Qian Ji[✉], Yu Zhang, and Jiayu Zheng

Industrial Design Department, School of Mechanical Science and Technology,
Huazhong University of Science and Technology, Wuhan 430074, Hubei, China
{jiqian,zhangyu0815,jiayuzheng}@hust.edu.cn

Abstract. This paper describes the design process of an intelligent physical health detection nail film product using flexible electronic materials. Unlike traditional nail polish, the "nail film" can continuously monitor the users' physiological state 24 h a day, and the information will be presented vividly on the nails through the nail surface. The innovation of this product is how to transform complex physiological signals into intuitive visual language based on the technological possibilities of intelligent flexible material. In the product, physiological information has been classified into various grades, each grade will be matched up with different design patterns and colors, i.e., red, orange, yellow and green are respectively matched up with superior, good, medium and poor, in order to depict the relationship between patterns and color semantics and different physiological information. Additional body health information feedback can be viewed through mobile APP in real-time by the users, as well as by their caregivers and doctors to observe the users from long distances.

Keywords: Intelligent nail · Physiological monitoring
Flexible electronic materials

1 Introduction

Manicure is a fashion that has been sought after by women, like make-up, hairdressing, body beautification and other activities. Although nail polish is popular, there are some disadvantages to the traditional nail polishing products. For example, unsanitary, of short duration, single function, tedious replacement and inability to be changed at any time.

The design of intelligent nail film is a combination of nail polishing products and a wearable device. It's stylish, intelligent, does no harm and can effectively avoid the discomfort, interval monitoring, wearing complex and other problems of traditional nail polishing products, as well as intelligently monitoring wearable products. Compared to

Supported by National Natural Science Foundation of China (Grant No: 51708236).
Supported by Seed Foundation of Huazhong University of Science and Technology (Grant No: 2016YXMS273).

the traditional product, nail film is soft, slight, produces no radiation or side-effects on the human body, and the function of physiological detection is based on flexible electronic material. The appearance, color, graphics, and flexible electronic materials characteristics will be the innovative part of the product [1].

2   Product Development

2.1   The Initial Idea of New Product

The intelligent flexible electronic materials and Microfluidic technology applied to the nail products.

On the one hand, it is imminent to design an intelligent product that can detect users' body condition in time and operate simply, easily.

The product aims to provide a brand new mode of using experience, users can update exclusive and personalized nail patterns, colors and style at any time, and also can receive physiological detection and data analysis and visualization, the function of detecting physiological situation will be the most distinct difference from the existing smart nail machines and traditional nail polishing products.

On the other hand, it also highlights the personality and the status of identity in order to attract more users and expand market demand for women, fashion chaser, except for its aesthetic functions.

The design points are initially summarized into four parts, the design of nail film surface, the connection of color language and body state, the visual presentation of more detailed body index, and the whole software interface design.

2.2   Brief Introduction of Material and Technology

Microfluidic technology is an emerging technology of controlling, operating and detecting complex fluid under microscale. With the combination of flexible electronic technology, a new flexible microfluidic electronic technology is developed, which is expected to play an important role in the fields of deformable electrodes, wearable electronics, flexible antennas and others[1].

According to the design assumption for the intelligent nail film, we take advantages of microfluidic technology and intelligent flexible electronic material, which is portable, transparent, lightweight, stretching, extensible and no harm to the human body, at the same time providing multiple functions such as heart rate detection and monitoring and so on [2].

---

[1] Yin, Z.P., Wu, Z.G., Huang, Y.A.: Flexible Microfluidic Electronics: Materials, Processes and Devices. J. Material China, Vol. 35, pp. 108–117 (2016)

2.3   Target User

In the context of accelerating the pace of life, women are under more pressure and burden due to changes in traditional concepts and family responsibilities. They aspire to fashion, the pursuit of individuality, like to pay more energy and money for a product with the pleasing and showing personality function. They are more in pursuit of a healthy life, sensual and have a certain cultural accomplishment, they not only have a strong demand for external image, but also have a strong desire for physical health [3].

Meanwhile, female's fashion needs have also been given consideration during the product design and development, city white-collar women, elderly women, pregnant women and the school students, all fashion leading women have been accommodated as the target users. In the early stage of users' research, it has been revealed that beauty and health were most important topic of women's concerns, especially those during pregnancy or elderly women. They do not want to give up the opportunity to maintain beautiful because they are pregnant or older. This design will not only meet the functional requirements, but also satisfy the emotional needs of self-seeking and self-centeredness.

In consumer demand research, it is found that they are happy when they are told that they can change their beauty and form to form and material, such as physical health examination function, but at the same time do not lose their manicure function. What they need is a product that is harmless, low in consumption and beautiful in fashion.

Design requirements analysis:

(1) Comfortable: flexible electronic materials with portability, transparency, light weight, stretch, easy to bend and so on, not only light and soft, but also make people work without binding, no obstruction.
(2) No harm: The use of new flexible electronic materials, which not only no radiation, but also with the traditional liquid nail polish also no chemical hazards. [4].
(3) Physiological monitoring: Flexible electronic materials are widely used in biomedical, wearable devices, medical sensors, flexible optoelectronic devices and other fields. The design will utilize its sensing capabilities to analyze and apply the detected physiological information.
(4) Beautiful and fashion: There are specific pattern, color and style data base, catering to the individualized aesthetic needs and tastes of different users.

2.4   Functional Positioning

Functional positioning is suitable for the real-time monitoring of the information of the user's temperature, pulse, ECG, heart sound, force, attitude and so on.

It gives real-time feedback for monitoring the results through pattern transformation. It can help the users or the people around you realize your health status in time, and also provide detailed physical data for the users and their care givers. In addition to the physiological condition of the body itself be detected, it can also be used to detect the external environment, like ultraviolet radiation intensity, air temperature and others in order to remind users to pay attention to sunscreen, sunstroke prevention measures on time.

Interest: On the basis of satisfying the user's aesthetics, the user can select the pattern type according to his/her preference, realize the interaction, and add real-time monitoring functions to allow the user to feel loved and respected.

Sustainability: Compared with the traditional manicure, it has the advantages of long endurance and long storage time, which is both environmental and cost saving.

3 Product Display

The whole system aims at monitoring and analyzing the relevant factors of users, it gets the monitoring results of its physiological data, determine the color and pattern display of nail plate, and helps users understand their physical condition. In the next section we explain the processes of pattern and color design. The focus of pattern design is a personalized customization and selection, while color design focuses on the use of color semantics to display the user's body state.

3.1 Pattern Design

All the patterns are stored in the pattern library in the APP. The pattern library has three sub-items: user patterns, alarm patterns, and user's contribution boxes.

In the user's pattern, the user can select a favorite style on the mobile APP according to mood, makeup, dress, and occasion. Various styles include themes, colors, emotions, creative designs, art, food styles, military, Chinese style, etc., and various topics are divided into various sub-items. For example, the theme includes a girl theme, a mom theme, an office worker theme, a cartoon theme, a text theme, a cool theme, a tile theme, a Chinese style theme, etc. Emotional topics include sentimental, lonely, funny, happy, inspirational, and healing, the color sub-items include orange, yellow, green, mint green, blue, purple, gray, luxury gold color and gray as a main color for users to choose (Fig. 1 shows an example).

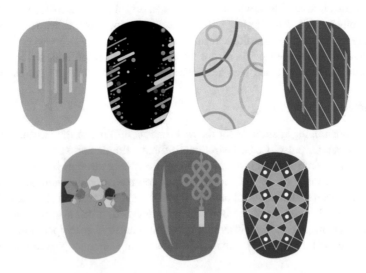

Fig. 1. Pattern design

The alarm pattern is designed as a fixed pattern, because all kinds of alarm charts in life have formed a fixed cognition in people's mind, and the design of colors and patterns will also be close to people's cognition. In the design, the characteristics of the warning light will be used to gradually deepen the color from the inside to the outside, forming a gradient, with a sense of hierarchy, giving people a feeling of intense heart (Fig. 2 shows an example).

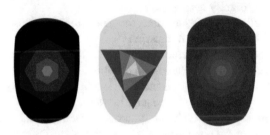

Fig. 2. Alarm pattern design

The purpose of the user's paper box option is to see what pictures and landscapes they like, or what they feel in life and what they think about the product. All of them can be submitted to developers in the form of pictures and texts. The designers will adopt and design, update to the pattern library in real time, and follow the trend to meet the needs of users.

3.2   Color Design

We design the pattern and the choice of colors so that it is covered in an area of only 1 cm^2 on the fingernail. Unlike the photos we saw in our lives, we could clearly see the above content and easily express the content. The small area of the fingernail makes it impossible for us to put a vivid picture on it, but we need to express the meaning of the picture which requires the collocation pattern and color design on more effort, to express what we want all kinds of language style. For example, what kind of picture is used to express the four seasons. Pictures can be expressed by photos of natural scenes taken in different seasons, but as a piece of film, it is not so easy. So the use of color is particularly important here.

The red, yellow and blue in the color are three primary colors, and they match each other between the three and can be matched with thousands of different colors. Just because of its color and its complex phenomena and colorful appearance, it has created a magnificent and colorful material world. For example, the characteristics of the four seasons in China's climate are very obvious. Each season has its unique color appearance: the buds in the early spring are gradually exposed, and the treetops are green. Afterwards, all kinds of flowers are also competing openly, colorful and enchanting:

Midsummer Green - The trees are shaded; the fruits are golden and dark red in autumn, and they are mature and noisy. In winter, it is a period of recuperation. The white shows peace and tranquility.

Green - Green is a mixture of yellow and blue, symbolizing life, soothing and peaceful, giving people a vibrant atmosphere stimulating their emotions. Create an enthusiastic, warm, elegant, stylish, lively, cute and other different impressions.

Blue - Blue gives a deep and peaceful feeling, fresh and refined. Symbolizes quietness, reflection, dignity and wisdom.

Yellow - The yellow light is moderate, the most luminous color in all hue, giving a light, brilliant and hopeful color impression.

Orange - orange is a mixed color of red and yellow, symbolizing happiness, and representing excitement, lively, cheerful and bright

Purple - Purple is a mixture of red and blue, symbolizing nobility and devotion. It is characterized by elegance, modesty, beauty, mystery, luxury and beauty.

Different colors represent different symbolic meanings, and users have different preferences. Because of the popularity of the season and the different skin qualities, it is possible to spread the range of colors as far as possible, echoing patterns and cater to users' personalized needs.

The level of brightness echoes the state of physical health. On the basis of hue, four levels with different lightness from high to low were used to respond to four levels of excellent, good, medium, and poor physical conditions. The higher the brightness, the better the physical condition and the lower the lightness, the worse the physical condition (Fig. 3 shows an example). When the status of the body is poor and lasts for more than 24 h, the APP terminal will provide the relevant physical health information and content according to the physical condition. And through the intelligent data evaluation, real-time feedback to the mobile phone APP to arouse the attention of the user. The mobile application gives corresponding prompts and countermeasures by analyzing the data.

The physical condition of the display color is humanized, is not mandatory, the user has the right to change. Especially if the user is on a special occasion or does not want to disclose his or her health and privacy, he needs to set the background running mode on the APP. You can change your own health display pattern by selecting your favorite pattern again. But at the mobile terminal, we can still understand the user's current physical state, that is, the transformation on the nail surface will run on the phone, including the alarm bell and the alarm pattern when the body is in bad condition. There are two sections in the background running mode, one is a pattern discoloration map, and the other is a more detailed physiological data map.

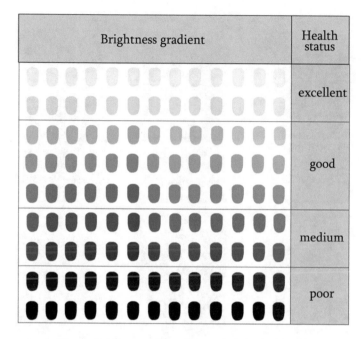

Fig. 3. Brightness echoes the state of physical health

3.3   APP Interface Design

The display of the nail condition on the physical condition is only an approximate figure, and a more detailed physiological indicator is displayed on the mobile APP.

The enhancement and weakening of heart sound is the diagnostic basis of human poisoning, anemia, inflammation, emphysema and other diseases [5]. The enhancement and weakening of the pulse is the basis for the diagnosis of diseases such as human respiratory system disease, pain disease, thermo STD and so on [6]. The change of blood pressure also often reflects the diseases of the human body [7]. If the symptoms mentioned above are detected, the users will be reminding to be hospitalized in time. If the symptoms are light, it will remind the user to pay attention to the related matters, such as the law of diet, the guarantee of sleep, and more exercise (Fig. 4).

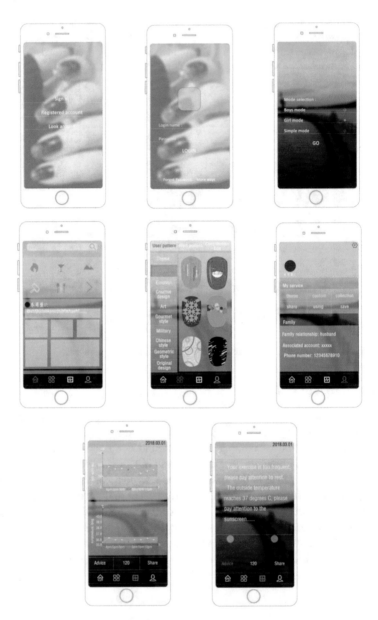

Fig. 4. App design interface

4   Conclusion

This paper refers to market demand research on nail polish film, analysis of the feasibility and flexible smart technology, product use situation simulations, visual design of physiological monitoring, information feedback reflection, and mobile APP information interface design.

References

1. Yang, L.L., Zhang, X.Z., Hu, J.Q., Shi, H.Q., Wen, X.J., Deng, H.Y.: Research and application of flexible wearable physiological monitoring equipment. J. Gen. Rev. (2017)
2. Wong, W.S., Salleo, A.: Flexible Electronics: Materials and Applications. Springer, New York (2009)
3. Zhang, N.R.: Analysis of the behavior of durable consumer goods in the female market. J. Bus. Res. (2017)
4. Cao, W.C., Guo, J., Liu, Z.P., Huang, X.Q.: Design of a portable human physiological index detector. J. Autom. Instrum. (2015)
5. Wang, W.H., Chen, D.R., Chang, Y., Tian, Z.F., Shi, M.: The development of portable phonocardiogram analysis instrument. J. China Med. Equip. J. (1994)
6. Qiao, Z.S., Shen, Y.H., Gao, Z.: Design of wearable pulse monitoring system based on Internet. J. Sens. Microsyst. **36**(12), 102–103 + 107 (2017)
7. Wang, X.L., Xu, Y.L.: Progress in the combined application of three blood pressure measurement methods. J. Nurse Study Mag. (2016)

Deeper User Experience - Emotional Design

Shan Wang[⊠] and Huaxiang Yuan

Huazhong University of Science and Technology, Room 401, E-1 Building,
Wuhan 430074, Hubei, China
601824331@qq.com, 1941288718@qq.com

Abstract. In the perspective of psychology, emotion is the heart of personality. In this view, products are expected to satisfy people's emotional needs providing users with pleasant aesthetic experience. With the coming of "experiencing economy" era, "emotional" has become one of design trends for current products. At present, designers should pay more attention to make products be closer to people and trigger changes in human's emotion, rather than being addicted to cool visual effects, thus enhancing user viscosity. This paper intends to provide designers with a full understanding of the concept of "people-oriented" products, showing how they can deeply explore user needs and eliminate cognitive misconception, so that they can produce products able to better satisfy user's physical and mental needs.

Keywords: Emotional design · User experience · Emotional experience

1 Introduction

As today's material resources have become extremely rich, we cannot satisfy people with only functional products. There is a shift from "forms follow functions" to "forms follow emotion", with the spiritual experience becoming gradually a core factor of competitiveness in products. Even though a lot of products have the same function, some of them are preferred to others, or uninstalled after several minutes of usage. What could make this happen? This is related to products' usage experience, which is well knew as user experience. User experience refers to users' subjective feeling while using products. Pleasing users is the most important part of product design. Moreover, products should be easy-to-use. With the increasing fierce competition on the market, people's psychological needs have received unprecedented attention. It will be the inevitable trend of the market to design more products to meet the psychological needs of consumers.

2 Emotion and Emotional Design

2.1 Emotion

According to "The Comprehensive Dictionary of Psychology", emotion is defined as attitudinal experience, which depends on whether the objects can meet our needs. It means that emotion is an intuitive response to interactions, when people interact with

© Springer International Publishing AG, part of Springer Nature 2019
S. Fukuda (Ed.): AHFE 2018, AISC 774, pp. 188–195, 2019.
https://doi.org/10.1007/978-3-319-94944-4_21

the world. It depends on the need and expectation. When such needs and expectations are satisfied, pleasure emotions will be generated. Otherwise distress and disgust emotions will be produced. Due to its obscurity and abstraction, emotion focuses on the feeling and experience, reflects the relation between objects and individual needs, and is a stable psychological reaction with social meanings [1].

2.2 Emotional Design

Emotional design was proposed by Donald Norman in his book, "Emotional design". The common definition of emotional design is experience, which is based on the emotional and spiritual needs, create congenial products, and finally make people happy. In this book, Norman reveals three levels of human nature from the perspective of perceptual psychology: instinctive, behavioral and reflective [2]. Norman considers the relation between three levels of design and Characteristics of products as: instinctive level of design-appearance; Behavioral level design-fun and efficiency; Reflective level of design-self-image, personal satisfaction and memory [3] (Fig. 1).

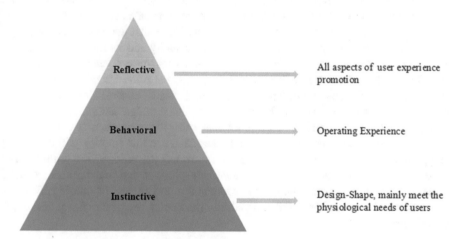

Fig. 1. The corresponding relationship between three levels of design and product characteristics

Instinctive design: Users get design information through sensory experience. This layer mainly meets the physiological needs of users. Human beings are visual animals and the more instinctive the vision is, the more likely it is to be accepted.

Behavioral design: Users receive design information, which pursues practicability, based on operation experience. The design of behavior level is probably drawn our most attention. The four aspects of design in good behavior level are as following: function, comprehensibility, usability and physical sense. The usage of products is a series of operations. Whether the first impression brought by beautiful interface critically can continue depends on whether the task can be accomplished effectively and whether it is fun experience.

Reflective Design: get design information mainly from reaction experience and value experience It focuses on significances of the information, cultural heritage products or product utility. The reflective design is related to the meaning of the object, which is influenced by the environment, culture, identity and so on. This level, in fact, is related to the customer's long-term feelings and is required to establish a brand or a long-term value of the products. Only if the bonds between products, services and users were built, and the interaction with products can affect users' self-image, satisfaction, memory, users can have brand awareness and cultivation of brand loyalty.

The instinctive design, which pays attention to the visualization, bring users with the intuitive feelings. The behavior design, which focuses on the operation, presents the user experience from operations. The Reflective design, which focuses on the emotion, which is equivalent to the improvement of user experience.

2.3   Why Should Emotional Design Be Carried Out

On the one hand, with the increasingly competition in the market, people's psychological needs have received unprecedented attention. It will be the inevitable trend of the market to design more products to meet the psychological needs of consumers. On the other hand, according to the above analysis, we know that the emotional design is to establish a relationship based on the user's personality and its main purpose is to draw users' attention and elicit the user's emotional response (Conscious or unconscious) to improve the design of performing a specific behavior. Users' thinking could be substituted by the emotion design, thus making the users' unconscious behavior and bringing more intuitive operation. Finally, the emotional design can make the product have personality charm and increase emotional care of the product. If the user has positive emotions in the process of product interaction and this emotion will make the user produce happy memory, thus being willing to use the product; the influence of positive emotions makes users be in a relatively pleasant and relaxed state, which increasing their tolerant on difficulties encountered in the process of use. If you want the product to be invincible in the competition, the importance of emotional design is self-evident.

3   The Relationship Between Emotional Design and User Experience

User experience is a subjective feeling that could be built during the interaction with objects, the users experience is used to make products and services match user's operating behaviors, thus decreasing users' cost on learning and improving usage. The user experience is divided into three categories: 1. sensual experience: presents the audio-visual experience to the user and emphasizes the comfortableness. 2. Interactive user experience: the interface gives the user the experience, when they use or communicate with objects, it emphasizes the interactive and interactive characteristics.3. Emotional user experience: give users psychological experience and emphasize psychological recognition. This classification corresponds entirely to the three-level design proposed by Norman.

In the book "Designing for Emotion" published by Walter, he combines emotional design and Maslow's Hierarchy of Needs. Being similar to human's physiology, security, love and belonging, self-esteem and self-realization [4], characteristics of products can also be divided into four levels: functional, dependable, usability and pleasure. The emotional design is the top "pleasure" level [5].

In terms of the application of interfaces in symbolic linguistics or linguistics, user interface is the medium between user and designer. Therefore, users should first understand what the designers express and communicate. At the bottom level: the user understands the meaning of the interface. Next, the user is satisfied with sense level and appreciate this interface, namely aesthetic feeling. Some studies have proved that aesthetic sense greatly affects the user interaction experience, but interaction is always more important than art workers. And next is emotion. If a product is based on emotional design, it can certainly receive the greatest degree of user adhesion and loyalty. Combined with the theories of Norman and Walter, it means that emotional design is a deep, top-level user experience.

4   How to Do Emotional Design in User Experience

The most important part in emotional design is to find a way to "empathize" with users, that is, to be able to resonate with the user. Resonance requires the sender to transmit the message to the receiver, who simultaneously produces the same thought or idea, causing telepathy. Because of both sides share common values, ways of thinking and codes of conduct, they can identify with and recognize each other.

4.1   Product Resonance

Product resonance is that integrating resonance thinking into the product and making the product become the sender and the user become the receiver and bring the user resonate via the content transmission. The factors that produce resonance are affective consensus, value consensus, knowledge consensus, thinking consensus and so on. There are four methods that could produce resonance: 1. emotional commonality: share emotion with people who have similar experience. 2. scene resonance: brings the user to the scene by designing the scene. This resonance takes us into the mind resonance via simulating the scene. Although there is not similar experience, simulating a set of scenarios to take us into plots and generate resonance via the development of the plot. 3. The collision of ideas resonance: there is a saying that "no discord, no concord". From strangers to know well, we can have a shared thought after the ideological collision, thus producing recognition or even the feelings of worship and causing resonance. 4. Aspiration resonance: People always hope others can realize his unfulfilled wish. When they realize the aspiration, they will have a great sense of achievement and, then producing resonance. Without these methods, a simple picture or some text is very difficult to reach the user's feelings. With the assistance of the method of dramatization, it can be very good to resonate with the user. For example, an advertisement of Nongfu Spring mineral water expresses the high quality of mineral water by telling the story of the source warden, which is more than a simple slogan "we don't produce water and we are just the porters of nature ". The advertising story can be recognized by more consumers."

4.2   How to Design Product Resonance

Empathy is generated via the expression of emotions and thoughts. According to methods of resonance, the methods could be considered as either a similar experience, people and things around you, thoughts capture you, you are brought into a situation, or you have a desire. To generate resonance between product and you, it need to relate to you in the story, scene, mind and the character and then project them into the product and create a sense of resonance.

Because everyone's living environment, emotion and mood are different, it is difficult to make a product, which can satisfy everyone. The first step is to target the consumer group. Resonance is also designed along with the user group. During making story productization, we should grasp the commonness and characteristic, character and emotion of the user group, thus accurately finding the users' resonance and arousing them. The general steps for designing resonance are: 1. find out the resonance 2. Awaken the user resonance 3. Resonate. The following examples are used to illustrate these viewpoints.

First step is to look for resonance points. Take tourism products as an example. As today's living qualities had been improved, many people want to have a travel. The increasingly popular short vacations show people's desire on travelling. After finding this resonance point, we need to target the group. The two necessary conditions for travel are time and money, which can divide users into four categories: I: have both money and time, II: only have money, III: only have time, IV: have neither money nor time. People in the first category, which is a small group, pay more attention to quality, it is difficult to design for this group. For people who are in the second category, the possibility that the platform wants to make a profit is too low, so is the possibility of implementation. Because people who are in the third group do not have strong execution, it is difficult to implement. Even though people who have time and no money have an incentive to travel, they are usually limited by economic conditions. If certain benefits are offered, it is highly possible that they will be encouraged to travel. So, people who have time and no money are the most likely candidates to have resonance.

Step two: awaken the user. According to the above contents, we can start from the same experience, ideological collision, and story introduction desire. Firstly, for people who have the same experiences and desires, we can encourage people who have been the same sightseeing, to share their travel experience, thus attracting user's attention. The platform can share the travel experiences of travel experts, analyze the users' travel purpose and desire by collecting users' visiting trace, then pushing experience of similar group to users and triggering their emotional resonance. This is similar to the tracking function of search software. We can design the function of the product as following: travel Notes, intelligent recommendation, travel evaluation. Secondly, let us focus on the story introduction, which can also be created by travel talent, who can use the story of talent to create culture, food, local customs and other channels, thus introducing users into the product scene, and triggering the resonance. The product functions that can be established include: travel talent recommendation and sharing such as food, beauty, local customs and culture. From the perspective of identity, the most important resonating point for target users who have no money and time is how to make them spend less money to realize travel. The platform can provide some strategies and guidance on economical solution, and adopt the rational analysis to interest the users. For example, where the hotel is the most affordable, how to save money on transportation, where we can buy cheap and fine commodities and so on. The product functions that can be designed include: money saving strategy, talent guidance, consumption evaluation, cost calculation. In addition, we can also design a function, which can make friend with others, to enable sympathetic users to share experiences, and convey positive emotions, thus allowing products to be quickly promoted.

Step three: resonance production. Through the function guidance, the user will be continuously introduced into the product scene. Based on a series of process, including setting up a scene for the user, analyzing user behavior, recommending intelligently, we can evoke user resonance. To reduce the strangeness between users, the relationship among people is simplified, so that the relationship bridge will be built between unrelated users. Stories narrated by acquaintances are more compelling than that of strangers. Resonance cannot be achieved via a simple function or a scene. It is the most advanced application of products and requires a composite function, scenes and group to build a platform to produce resonance.

Resonance is a communication between hearts. To analyze the resonance of a product, we should firstly start from the target group, understanding the user's habits, behavior, hobbies, culture, personality and other aspects. The more you understand the user, the easier it is to get close to the user, find the resonance point and awaken the user's sense of resonance.

4.3   The Integration of Elements of Emotional Pleasure

The emotional cognitive model (Fig. 2) agrees that: emotion is generated by individual evaluation on physiology arousal and perception of environment. The core of emotional design is to induce users' cognitive pleasure and bring positive emotional experience to users.

The pleasure elements include:

1. Sense of control: Most users close the window within 2 s when the interface is not prompted. However, when there is prompted text, it greatly reduces the user's departure rate. So it is important to make the user clearly confirm the current state and produce a sense of controllability thus forming an operational expectation and guiding the user to continue the operation.
2. Social interaction: The basic attribute of human beings is sociality. People's sense of belonging and respect can be satisfied with social interaction. In the design, if being good at using social interaction factors, we can greatly enhance the user's participation and resonate with each other.
3. Social reference: people always have group psychology. Most people want to pay less cognitive resources to gain the maximum return. For example, when we do not know which brand of cosmetics is better, we will choose brands that most people have used. We can obtain effective information resources by referring to other people's information.

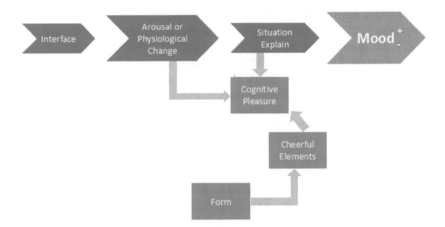

Fig. 2. The emotional cognitive model

4. Availability: in 1995, Jacob Nielsen proposed the concept of "usability". Based on the characteristics of users' perception, cognitive patterns and personality, product usability is used to make the design respect the user's psychological model and make the product to be "good to use, easy to use". The interface with strong usability is not only convenient for users to operate, but also brings positive emotional experience to users. At the same time, it also brings immeasurable commercial value to enterprises.

In addition, emotional design elements also include challenges, discovery, colloquial, achievement, fluid experience. They can bring users with pleasant cognition. Designers can integrate more these elements into designs. Let the user feel the heart and intimate of the product.

5 Epilogue

Good user experience and design reflects the use rational methods and techniques to finish a design, which make users generate positive emotions and accept it emotionally via the interaction with the product. To achieve this, it is necessary to focus on "user-centered" concept and excavate and mobilize the user's most hidden emotion. An effective emotional design strategy mainly includes two aspects: 1. create a unique and excellent concept of style so that users have a positive response; 2. continue to use the concept to create a whole set of designs with a personality level. The real emotional design should be based on "profound knowledge and strive to achieve certain kind of psychological interaction with consumers". In addition, good design should make the people who use this design feel that the product connected to their hearts. In general, emotional design is to pursue from a rational point of view to meet the emotional needs of people and be demonstrated with a reasonable and harmonious means. Hope this article play a role in inspiration and causing more thinking and concern and better use of emotional design to solve practical problems.

References

1. Li, X., Men, D.: A study on the Expression of Emotional Concept in Visual Communication Design. South China University of Technology, Guangzhou, pp. 10–15 (2012)
2. Norman, D.: Emotional Design. Electronic Industry Press, Beijing, pp. 47–65 (2006)
3. Liu, Y.: Emotional design based on user experience. J. Sci. Technol. Commun. 119–128 (2016)
4. Walter, A.: Designing for Emotion. Happy Cog (2011)
5. Jin, Y., He, Y.: Emotional design in interactive design. J. Mod. Decoration Theory, 16–17 (2017)

Empathy and Design. Affective Participations for Clothing Design in Colombia

Claudia Fernández-Silva, Ángela María Echeverri-Jaramillo^(✉),
and Sandra Marcela Vélez-Granda

Universidad Pontificia Bolivariana, Medellín, Colombia
{claudia.fernandez,angela.echeverri,
sandra.velez}@upb.edu.co

Abstract. Both practice and education in clothing design in Latin American countries such as Colombia, have been focused on the stylistic production of forms directed towards accelerated consumption, under the premises of fashion trends. Consequently, different social problems centered in the relation body-artifact are left behind, especially those from minority groups, like the people who live in poverty, disability, rural contexts, and forced displacement, amongst others. These conditions require an approach from clothing design that addresses, not only the operational and functional aspects, but also those linked to an understanding from the intuitive, affective and emotional factors of their surrounding world.

This paper shows the outcomes of several projects carried out by the Faculty of Clothing Design at Universidad Pontificia Bolivariana. They have focused on the improvement of various human conditions like literacy, civics and communal living of children, teenagers and adults who coexist in violent contexts. It also shows the outcomes of projects where the creation and use of wardrobe devices work as collective memory reconstruction processes with victims of the armed conflict, and self-management processes with communities to identify and let them participate in their own needs resolution.

These projects use different methodologies that appeal to literature reviews and interviews with experts, while the class is taken to the place where the communities coexist and carry out their everyday routine. It is here where the design students and teachers join the daily activities of the different groups in participant and non-participant observations, audio-visual record, statistical study of the population and characterization of the communities, as well as an analysis of the devices and environments where their daily life takes place. We find, then, outcomes that are artefacts working as narratives which suggest a different way of telling the story of our Country.

These projects' development and outcomes are the result of a co-creative and empathic job amongst the different communities, the teachers and students from the Faculty of Clothing Design. The aforementioned has fostered feelings of affective participation among the different participants of the projects, thus, linking subjects and practices.

The development of this project and its results also raise different considerations about the teaching of design in Colombia, strongly related to the more contemporary positions of human centered design, critical design, autonomous design, and design for behavioral change, all which appeal to the commitment design professionals have with the people, their contexts and their particular situations.

© Springer International Publishing AG, part of Springer Nature 2019
S. Fukuda (Ed.): AHFE 2018, AISC 774, pp. 196–204, 2019.
https://doi.org/10.1007/978-3-319-94944-4_22

Keywords: Clothing design · Emotional design · Latin-American design

1 Introduction

Colombia is located north in South America and it has the particularity of being both in the Caribbean and the Andes. This location has granted it a number of thermic lands, life zones, ethnicities, languages and socio-economic conditions that are different in its whole territory. Homogeneity is not a defining characteristic of the country, which conditions Human Science disciplines to reach out to different and less orthodox ways to study and understand its reality, while applying them to the teaching and learning practices of design.

To get an idea of the heterogeneity characteristics in Colombia, one can turn into demographic data that portray said diversity. The projected population in Colombia as of June 30, 2018, is approximately 49,834,000 people, of which 23% live in rural areas, 6.3% have some type of disability according to DANE's[1] census in 2005 and the same prevalence rate as in 2015. Population over 60 years old exceeds 11%, and 17% of the overall country's population is in poverty. Colombia's Gini[2] coefficient in 2016 was 0.517, there are 87 indigenous ethnicities, three different afro-Colombian peoples, the ROM or gypsy people, and 67 different languages (Spanish, indigenous languages and creoles). Taking these conditions into account, clothing design in Colombia and Latin-American must focus on different issues for stylistic production, accelerated consumption and world trends.

It is important to understand that design in Colombia was promoted –mid XX century– by industrialization policies that sought –and still seek– competitiveness through products' export. However, at the end of the XX century, it is established that Colombia and other Latin-American countries would concentrate their efforts in the export of raw material, which would later create an unstable field for design actions. This way, the vast majority of design practices in Colombia have been focused on external issues, making it difficult for those processes that involve inner and contextualized problems.

Escobar, in his book Autonomy and Design, shares some reflections about design in the past few years. He states that "the importance of the social context to succeed in design, well beyond the functional or commercial applications of the products or the effectiveness of services" is one of the focus design must have today. Besides, he points out how "we all design, which gives room to ethnographic, participative and collaborative design while rethinking the whole design system" [1].

---

[1] National Administrative Statistics Department in Colombia.

[2] Gini coefficient is an inequality measurement created by Conrado Gini. It is used to measure the inequality in a country's income. 0 indicates total equality and 1 the maximum inequality. So, having a value of 0.517 –close to 1– indicates that there is inequality in the country. Colombia is in the 7th place in the world based on inequality, according to the World Bank.

The call from design today to Colombia and Latin-America is to be relational, participative, diverse and plural, re-establishing the connection body-artifact in a holistic understanding of what is theirs. Thus, the diverse and plural can be addressed and diversity becomes an opportunity for local design. This stance differs from the ones that have been the base of emotional, affective and pleasurable design, which are established from generalized and standardizes methodologies.

Among the founding guidelines of this design line are found –in respect of emotional design- the postulates of Donald Norman, who suggests it consists of the design of services and products that are, not only functional and understandable, but also pleasurable –as evidenced at the moment of consumption/use. As stated by Norman, said perspective requires an analysis of all the factors related to the Object-Man-Environment triad, as well as an expansion of its users and its own questioning range concerning the perception, emotion, sensation and experience fields, which answers are subjective [2].

Relative to affective design is also the *Kansei* engineering. It was created in the 1970's at Hiroshima University and it is presented as a methodology towards the development of new projects which objective is to facilitate the emotional expectations translation. According to this particular approach, these emotional expectations can be impressions -or stimuli- understood to be the consumer's judgement about a product and which take form, mentally, as luxurious, natural, and remarkable, to characteristics and technical design specifications. Thus, *Kansei* engineering joins feelings and emotions with the engineering discipline aiming at letting the designer gather the emotional needs of the user which can be transferred to the developed product while satisfying the consumer's expectations [3].

In addition to the aforementioned, Jordan's affective design proposal suggests that products and services should satisfy three needs in a particular order, as follows: Level 1, functionality: it solves a specific problem; level 2, usability: the product is comfortable, easy and safe to use; level 3, pleasure, besides fulfilling its functional and usability needs, it provides emotional benefits [4].

This classification refers to Abraham Maslow's postulate, which suggests that human needs are set in a pyramid shaped hierarchical organization and are targeted in an ascending way. This, meaning that once the lower level –the base– needs are met (physiological needs such as homeostasis and nutrition), the other level's needs will also begin to be met, reaching the top of the pyramid where needs related to self-actualization (morality and creativity) are located. Furthermore, Jordan claims that if a product is nor functional, it will not be easy to use. Hence being unappealing and unable to reach the consumer's emotional aspects.

From the previous stances it is clear that affective, pleasurable and emotional design is located on top of the pyramid, as the highest of human needs, as long as basic needs such as nutrition, health and work are resolved. However, positions cannot be transferred to the Latin-American reality for their generalization ignores the particularities of its people and its context.

As long as the emotional needs are not clearly defined, it is difficult to respond to them, for there can be a big difference among German, Vietnamese and Colombian people regarding their socio-economic, political or academic situation. Hence, the question arises whether the emotional aspects can only be addressed when the basic needs have been satisfied or, on the contrary, addressing emotional needs could even prompt the solution of other needs located somewhere else in the pyramid.

The previous idea appears as a reflection from some of the projects from the Faculty of Clothing Design, which will be presented further on.

2 Findings and Discussion

2.1 Affective, Pleasurable and Emotional Design Devices: Human Emotions Styling to Seduce and Promote Provisional Identity Bonds

In addition to what the presented authors have built about affective, pleasurable and emotional design, -as a design form that addresses human emotions generally disintegrated from the solution of problems considered linked to the efficiency and usability of the designed device-, it is also possible to observe a constant in the formal characteristics of the products which are assigned to this type of design. Said features are related to what Van Rompay and Ludden [7] define as the different types of embodiment in design[3]: (1) Anthropomorphism, (2) Relational properties: image schemas and symbolic meaning, (3) Meaningful sensory experiences, (4) Movement and product action embodiment. The first type is the most evident in emotional design products. Here, the consideration of the emotiveness, given to humans by devices and artifacts, happens through iconic body representations –in the way of styling[4]–, appealing to the expressiveness of artifacts and not specifically to the experience of use.

As stated by the authors, embodiment is not the only explanatory framework for said expression for semiotics has also played an important role in this understanding. Nonetheless, from those four ways, the first three are still subject to visibility and meaning and not so much to a located perception and a contextually embodied practice. Being the later what we think to be the ways to evaluate an empathy experience from the use that takes us to consider whether the interaction between people and devices effectively leads to emotion, affectivity and pleasure within a specific time frame.

---

[3] The notion of embodiment, defined by the anthropologist Csordas [5] has as its premise a relation between perception and practice from the concepts of pre-objective (Merleau-Ponty) and habitus (Bordieau). These authors are also analyzed by the sociologist Entwistle [6] to suggest an understanding of clothing that takes into account that while it is being a social and individual experience, it is also a phenomenon that evolves around discourse and practice.

[4] Styling is a concept used to refer to one of design's philosophies that emphasizes on making products that are attractive to the consumer in order to be able to sell it. This philosophy is opposed to functionalism and its main representatives was the the American industrial designer, Raymond Loewy.

Devices with human or animal form, that hug each other, that seem to smile or imitate any human emotion or –as in the second type of embodiment– rational properties: images schemes and symbolic meaning, appeal to the way in which we perceive the world from visual and spatial relation patterns, measured by image schemes. For instance, the verticality scheme has connotations in language as pride, dominance and success that, according to Van Rompay and Ludden [7], can be interesting to design at the moment of supplying for this type of expression. This type of characteristics perceived in objects, such as closeness and warmth or coldness and distance, can also affect one's proximity to them.

If we ask ourselves why we create artifacts with these semiotic expressive premises, we can see that the majority seek to reach empathy with the consumer, appealing to identification feelings. This is why, when it is time to choose what product buy among a wide range of similar products, I will choose the one with which my imaginary record finds more connection points. This is neither good nor bad, but it is important to highlight that there may be other ways to understand emotional affective and pleasurable design, that is not only focused on seducing at the time of purchase, or on the satisfaction of having and using objects that we think talk about ourselves, our likes and world views, and that said ways can become very strong ways to think about how we act, how we relate, and even how we use and consume in the contexts we live in.

Regarding the aforementioned, other approaches to design –as critical design and behavioral change design– also appeal to our daily emotional bonds with the devices we use to make us think about concepts, ways of acting, using and consuming that are already natural to us. The products of these kinds of designs seek empathy through interrogation and set a place for possibility to change in those context where the product is needed.

In the Faculty of Clothing Design at Universidad Pontificia Bolivariana in Medellín, Colombia, we inquiry about, not only about what is implied in the relation body-clothing-context but also, the different ways clothing design products can question us about the society we live in and suggest parallel alternatives to the commonly accepted. These search has resulted in valuable experiences, such as the ones we will see in the following paragraphs.

2.2   Affective and Emotional Clothing Design in Colombia: Emotions to Thing, Question and Trigger Actions and Transformation

From its name, the program of Clothing Design of Universidad Pontificia Bolivariana in Medellín, Colombia, aims at a conception of the profession and discipline that goes beyond the assumptions and archetypal linked to clothing design. The reason to tae the name *Clothing Design* and not the more traditional on *Fashion Design*, is based on the desire to address questions regarding clothes and clothing in a broader sense, non-circumscribed to the regular, constant change, called fashion.

In this context, clothing design projects have been developed. They resort to methodologies that vary between literature review, interviews with experts and expansion of the classroom by taking the classes to the place where the involved group carry out with their daily lives. Here, designers in training and teachers join the everyday activities of the people in non-participant and participant observation exercises, audiovisual register, population statistical study, community characterization and analysis of the devices and spaces found in these contexts. In this co-design exercise, the Participatory Action Research (PAR[5]) methodology is adopted, for it invites the community to have an active role, and enhance horizontal relations among designers and users as well as constantly validating the process.

As a result, clothing devices and services are created parting from an understanding of the different agents involved in the design process. At this stage, through their personal and contextual history, their approaches and their values, agreements can be made among the parts –designers and people– about possible needs and what the true design opportunities will be found in said context. Additionally, the results of these projects work as narratives, which suggest another way to tell our country's history. An example of this is the project *Soy Estigmatizada (I'm stigmatized)*, which depicts the stereotypes around the way women dress and points out some of the gender issues in our country (Photograph 1).

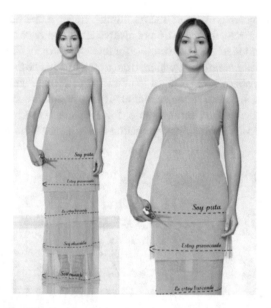

Photograph 1. Project *Soy Estigmatizada (I'm stigmatized)*. Clothing Design students, 2017.

It is also possible to find projects that, as explained in the previous paragraph, from the critical design appeal to emotions, affection and, in general, to the empathy between humans and devices. This, aiming at questioning and social and political provoking, but also cultivating values and raising awareness. For instance, the project *70% agua (70% water)*; this project has two parts: The first one, clothes 100% cotton were dyed

---

[5] This kind of research was created by Orlando Fals Borda, in the 1970s, for educational environments. In the Faculty of Clothing Design, we have transferred it to the co-design process.

in five different ravines in Medellin; the second one, samples from the water of said ravines and information of their high pollution percentage; by asking '*would you wear these clothes?*' awareness is raised regarding this situation (Photograph 2).

Photograph 2. Project *70% agua (70% water).* Clothing Design students, 2017.

This kind of projects go beyond the traditional precepts that design has set for the creation of design objects, for these do not always aim at body comfort or efficiency in different physical activities. A good example for this is the project *Dos caras de la misma moneda (To sides of the same coin),* which, from a clothing device that houses two bodies at the same time, forces those wearing it to be facing each other and, after a while, start talking. This project is placed in the school environment and is seen as an invitation to strangers to know each other, setting aside, for a moment, the barriers promoted by stereotypes, which foster exclusion, segregation and violence (Photograph 3).

Photograph 3. Project *Dos caras de la misma moneda (Two sides of the same coin).* Clothing Design students, 2017.

Another project is *Unidos por la paz (United for peace)*, which seeks to improve the cohabitation among children in a school in the 8 commune[6] in Medellin, through didactic and dynamic activities. In this project, the clothing device encourages team play from joining all the children's bodies to generate group dynamics that allow healthy socialization (Photograph 4).

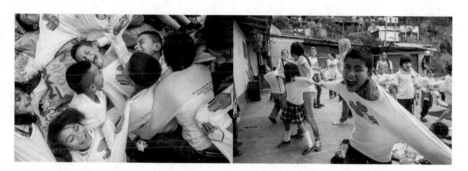

Photograph 4. *Unidos por la paz (United for peace)*. Clothing Design students, 2014.

3 Conclusions

Design study in Colombia requires that the design discourses and theories generated in other latitudes are critically reviewed through a reconceptualization of them. This, because their homologation can lead to, not only decontextualized proposals but also, a design incapable of operating under specific situations of a place and people, making design overall a useless discipline.

Rethinking the concept of affective, pleasurable and emotional design for contexts similar to the Colombian one –and in general, non-Anglo-European centered–, implies reevaluating the considerations that said designs are only related to the needs of the top of the pyramid (Maslow). On the contrary, it is the most essential realities of humans, those that, in context like ours, require the designer to focus on the core of emotions and pleasure principles in the everyday relation to tangible culture, as ways to reach dignity in every area of life, which is entirely measure by devices.

Emotional, affective and pleasurable design as we have known it since the XX century, requires to be updated for the XXI century, for the main call for design is to rebuild a world that suffers the results of bad practices in which design itself has been taking part. It is the most recent approaches to design –critical design, behavioral change design, transition design– the ones that recognize nowadays how the empathy relations with design products should go beyond just convincing the user to buy. They should, above all, make the consumer more critical about their time, raise awareness, and promote values also using irony, humor, surprise, pleasure and emotion.

---

[6] 8 commune in Medellin is located east in the city and has extreme poverty conditions. That is the reason why its population are in a high vulnerability position.

References

1. Escobar, A.: Autonomía y Diseño: la Realización de lo Comunal. Universidad del Cauca, Colombia (2016)
2. Norman, D.: La Psicología de los Objetos Cotidianos. Nerea, Madrid (1990)
3. Fundación Prodintec: Diseño Afectivo e Ingeniería Kansei: Guía Metodológica, Asturias (2011)
4. Jordan, P.: Designing Pleasurable Products: an Introduction to the New Human Factors. Taylors & Francis, London (2000)
5. Csordas, T.: Introduction: the Body as Representation and Being-in the World. Thomas Csordas, Cambridge (1994)
6. Entwistle, J.: Fashion and The Fleshy Body: Dress as Embodied Practice. Fashion Theory, London (2000)
7. Van Rompay, T., Ludden, G.: Types of Embodiment in Design: the Embodied Foundations of Meaning and affect in Product Design. Department of Product Design, Netherlands (2015)

Principles of Developing an Effective Environment for Affective and Pleasurable Design Team

Amic G. Ho[✉]

School of Arts and Social Sciences Creative Arts,
The Open University of Hong Kong, Ho Man Tin, Hong Kong
amicgh@gmail.com

Abstract. Referring to previous studies in past decades, the nature of design has changed as a tool for social interaction with attached with affective concerns. Some design scholars proposed that the environment for designing (such as the working space and working schedule) should be changed accordingly. Some design scholars propose better ways to design a working schedule and facilities for the design process. However, there are still obstacles that hinder the creation of an effective environment for designing affective and pleasurable design. Some design scholars believe that social interactions among designers would be helpful to optimise their working performance. These interactions need an appropriate setting and are influenced by the physical working environment; environments that differ to the traditional working spaces which are oppressively symbolic of hierarchy and discipline. Traditional working spaces are giving way to more effective environments to facilitate collaborative and flexible design process for affective and pleasurable design outcomes. So, this is a study to investigate how can designers make the working space more effective. Some principles would be proposed to seniors for creating the appropriate environment which would allow time and space for the design team to de-stress, open their minds, and inspire each other.

Keywords: Emotion · Working environment · Design process
Interaction

1 Introduction

According to one discussion about effective and pleasurable designs, the nature of design has changed, becoming a tool for enabling social interaction that focuses more on the interactions between users and designers. Some scholars proposed that the environment for creative should be modified accordingly. Some scholars proposed methods to design effective design process schedule and facilities for creatives. However, there are still obstacles that hinder the creation of an effective environment for creative. Some scholars believed that designing affective and pleasurable design needs an appropriate setting and is influenced by the physical creative environment; environments that differ to the traditional working space which are oppressively symbolic of hierarchy and discipline. Traditional working spaces seem unable to

© Springer International Publishing AG, part of Springer Nature 2019
S. Fukuda (Ed.): AHFE 2018, AISC 774, pp. 205–214, 2019.
https://doi.org/10.1007/978-3-319-94944-4_23

provide effective environments to facilitate collaborative and flexible creative work. So, how can designers make more effective creative spaces? How should design team leaders create an appropriate environment which would allow time and space for designers to de-stress, open their minds, and inspire each other?

2 The Influence of Social Constructivism on Design

Before further investigating these questions, it is necessary to understand what the current development of design is. Influenced by the concepts of constructivism, the active role of creative has shifted. The users take the main role. Which means, designers no longer provide information and functions into the design outcomes only [1]. Sustainable Solutions to complex problem is regarded as the key to effective community engagement and partnerships. And there is another important point: they have to investigate how to communicate with users. When the process of understanding and constructing knowledge becomes an active and continuous process, communication will be interactive and ever changing. Therefore, training the designers will mean they required the ability to investigate new knowledge on finding the potential affective linkages [2] instead of considering functions logically only. For example, the design process for constructivism focuses on problem solving. Designers manipulate the design process from conducting an inquiry, investigating topics, and allocating various resources to explore solutions. When designers are investigating, they propose solutions and conclusions; those are the foundation for continuous exploration [3]. In most cases, a design team leader takes the role of guiding the designers in the design process. With this trendy concept of design fundamental, the design process next faced the influence of social constructivism. Since social constructivism values cultural background and humanistic factors, the design team leaders have to consider more from cultural aspect and human factors, and lead their design team on these perspectives.

3 Collaborative Design Process

In social constructivism, it is proposed that collaborative design is applied in the workplace. It is a process of interactions among designers which is directed by their design team leader. During the process, designers use the context of collaboration to better understand their thinking process. Designers acquire new knowledge and relate them to prior knowledge through creative [4]. Discussion (among the design team leader and designers interactively) [5] can be encouraged by the introduction of specific concepts, situation or inquiries. And it is guided by different approaches to effective questions, the presentation and clarification of information, and knowledge from previous experience [6, 7]. Therefore, designers are active roles to create the communication framework and messages for communication [8]. Design activities under collaborative design approach include group projects, debates, study teams, etc. [9]. The design process enables designers to set up their working network [10]. Designers share and construct new concept or knowledge in the collaborative design process.

In e-creative field, the construction of knowledge in collaborative design can be connected together through internet.

4 Proposed Collaborative Design Path

According to the investigated theories, social constructivism influenced creative and generated the development of collaborative design approach. The detail of collaborative design was reorganised, and then a new collaborative path and model is proposed (Fig. 1).

(1) The first step is total collaboration. At this stage, design team leaders and designers process the creative tasks together as partners. The creative experiences were authentically developed during co-planning and co-assessing session. Also, creative experiences with rich technology support are prepared. Designers obtain benefits from the support of design team leader throughout the process from creative experience planning, the process of training and assessing the performance. Total collaboration provides the experience to designers the framework of their creative and sufficient supports those are the important foundation for their further creative stages.

(2) The second step is limited collaboration. At this stage, designers are able to make use of the known foundation for total collaboration and further manipulate their creative actively in confident. They would request assistance from design team leader when necessary. Design team leader and designers mainly co-planning the design experience but designers would take the initiative in investigation and construct the design experience. Limited collaboration is the key stage to help designers building confidence for understanding and judging in the process of investigation. It still able to be modelled by the support of technology. Technology tools would provide specific help in certain areas such as delivering instruction to designers, obtaining comment from the design team leader and other designers.

(3) The third step is coaching collaboration. At this stage, design team leader work as support in backstage more. They support the designers to execute the co-planned design schedule and brainstormed ideas for projects. After experiencing planning and training, designers still need some guidance on the way to manipulate their investigation effectively. Therefore, coaching collaboration is the stage that allows designers to obtain feedback on their manipulation of the design process.

(4) The fourth step is monitoring collaboration. At this stage, designers are independent to manipulate their design process. Based on the feedback obtained in coaching collaboration, designers would develop the information they investigated and constructed their design outcomes and experience. At the same time, they would be expert among their peers. A design team leader is not the solo sources that provided advise. This step also creates a sustainable foundation for further design tasks.

The first and second steps (total and limited collaboration) are design team leader-leading collaboration stages, the third and fourth step is (coaching and monitoring collaboration) are designer-leading collaboration stages.

Fig. 1. The proposed collaborative design process.

5 The Development of Design Community

The application of collaborative design creates a new communication approach that promotes close attachments amongst design team leader and designers as part of a design community. 'Design community' refers to a group of people who have a common concern about their creative in a regular pattern. Some study explored the relationships between the creative components include community, practices, meanings, and identity. Creative is the vehicle for a person to evaluating practices and development, and transforming identities as a kind of social participation [8]. Therefore, five crucial activities for founding a design community were proposed: (1) events, (2) connectivity (3) membership, (4) creative projects and (5) artefacts. All these activities would be supported by creative beliefs and behaviours [9]. These creative beliefs involve the psychological concerns of people in the community (including design team leader and designers). They support the team creative behaviour such as construction, constructive conflict, and co-construction. Therefore, psychological concerns are important factors to support the creative in collaborative design. There are limited studies focusing on this aspect and it is important to understand how they relate if an effective creative environment is to be developed.

6 The Application of Studio-Based Design

Referring to the development of collaborative design in couple of years, its relationship social constructivism and its influence on encouraging interactions among learners is recognised. The emphasis on collaborative design and the concept of design community lead some courses to apply studio-based design. The objective of this approach is for providing designers to engage creative experience in an authentic situation [10]. They would obtain feedback about their work from their peers and design team leader immediately. However, even adopted studio-based design in the classroom, there were still problematic points found in the general studio [11, 12] such as:

- All designers worked separately within the group.
- There were varied creative responses of designers to individual and shared projects.
- Whether the work reflected the intentions of the designers, the tutor, and the project brief.
- The individual's work, sharing their perspective with others, and developing critical thinking by analysing their own and others' work.

Therefore, it has been identified as the greatest source of designer's dissatisfaction.

7 Identified Difficulties of Creative and Design

It was found that there are still some difficulties identified regarding creative. Several typical points would be observed:

(1) Reliance on mental shortcuts, which are usually habitual patterns of thinking and doing. Typically, these operate from a position of 'this is how it is' rather than a 'this is how it could be'.
(2) Verbal criticism is given before new ideas have a chance to develop.
(3) The design schedule is too tight. There is not enough time to discuss other people's ideas.
(4) Stress among designers.
(5) Visual clutter and confusion.
(6) Lack of investigation on the design of physical environment (such as space planning, interior design and facilities) that would promote interactional activities among design team leader and designers.

8 Experiment for Investigating the Effectiveness on Design Abilities in Various Environment

8.1 Research Process

This study sought to investigate whether design attributes are influenced by the environment. The experiment was conducted in two stages. The first consisted of an examination into the extent to which design abilities such as communication skills and creativity are influenced by the external environment. After the confirmation that the external environment can affect designers' abilities, the second stage consisted of an explanation as to how different environments influence designers' performance. In the first stage, participants were divided into four groups, all of which were asked to manipulate a design process for a user interface design. The participants took turns to conduct meetings to discuss the design tasks in two rooms, which were decorated with themes – pleasurable and negative atmospheres. The room with the pleasurable atmosphere featured positive images and pleasant music, while negative images and sad music were present in the other room. The presentations that took place in the rooms were secretly recorded. In the second stage of the experiment, the influence of various types of external environment on designers was investigated. To this end, four rooms were set up. In the first, images about lifestyles collected from magazines were used for decoration; in the second, relaxing music was piped out as background music; in the third, the working tables were grouped together; and the fourth room served as a control sample setting, featuring a traditional working space. Video recording equipment was also placed in these four rooms, and each group was randomly assigned one room as their working space. Observation of the design process was conducted, and feedback from designers collected through questionnaires.

8.2 Participants

A total twelve participants took part in this study, all of whom are professional designers who have been working in creative industry for three to six years. In addition, all had graduated from creative-related bachelor's degrees courses, and were aged between 25 and 30.

8.3 Research Result

In the first stage, it was found that emotional reactions, performance in communication, and feelings about the design were all influenced by the external decoration of the environment in the discussion rooms. The different influences of the various settings of the external environment were examined in the second stage, and it was found that the third room, which grouped working tables together, encouraged the participants to interact with each other. The feedback from the designers also indicated that positive music and images helped them to work more creatively.

9 Suggested Principles of Developing an Effective Collaborative Design Environment

Based on the understanding of the theories and the training experience in design studies under studio-based design in the past decade, the importance of an effective environment of creative were understood. Some scholars investigated how psychological concerns after adopting the psycho-evolutionary concept. They proposed that psychological concerns are involved in evaluative patterns which enable designers to react the situations around them in an effective way [13]. A response of designers to the psychological concerns could be regarded as the preceding process of decision-making and reactions [14]. Such process could encourage designers to aware on and identify critical information from the issues and situations around them. This is the concept of "judgement process." Also, engagement was found as one of the energy to lead designer interact with design team leaders as well as take part in peer-to-peer collaboration It is positively influence the creative performance of designers [15]. Furthermore, this process is reflected the psychological concerns is based on thinking. The process of understanding, interaction with the external situations, and shaped concepts were featured [16]. Scholars further explored how the psychological concerns would influence participants' responses on through visual and audio material. During the study, the scholars directed designers, those had various cultural backgrounds, developed an online platform [17]. The scholars invited some participants to present their preferences according to their perceptions and psychological concerns. They investigated that the preconceptions and psychological concerns of the participants affected different kinds of complex judgment processes and feedback from the participants. Their study investigated how psychological concerns was attached with the judgment of participants and images those displayed. Thus, some principles of developing an effective environment were suggested.

- Develop an "everything are possible" atmosphere, which encourages creative for problem-solving, with praise and material rewards that will motivate designers.
- Humour and pleasurable playfulness to be introduced to reduce stress.
- Always create interactions, challenges and conflicts.
- Collect verbal, audio, images and physical signals for new behaviours for problem-solving design.
- Spare some time slot for generating 'useless' ideas even it is in a tight design schedule.
- Examine concepts through prototyping.

10 Examples of Designing Physical Environment to Enhance Creative

The proposed principles above are not for creating an attention-grabbing environment. They are principles required to develop a tailor-made physical environment to facilitate the development of new knowledge within the design community. Based on these principles, some examples of design in the physical environment to enhance creative were proposed:

(1) Provide appropriated spaces for encouraging interactions and teamwork. For example, a space for effective creative in which designers can confidently carry out ideas and to break rules freely.
(2) A setting of classroom for collaborative design is different to the usual classroom to enhance creative ability and problem-solving.
(3) Interior design and music, which will help to relax the mind and encourage interactions for inspiration among designers.

11   Comparison Between Collaborative Design and the Suggested Principles of Preparing an Effective Collaborative Design Environment

The suggested principles of preparing an effective collaborative design environment are based on collaborative design, but with some differences [18]. Collaborative design is mainly concerned with how to encourage designers to go through the design process together [19]. It leads designers to maximise their creative resources and knowledge: for example, questioning each other to gain understanding and evaluating each other's ideas. While the suggested principles of developing an effective collaborative design environment are based on collaborative design, the focus is more on environmental design. Utilising spaces (interior, classroom setting, etc.) leads designers to be more motivated in participating in the interactions of collaborative design.

12   Challenges in Implementation

The suggested principles of developing an effective collaborative design environment would help designers to enhance interactions in their design process and encourage the creative experience in real situations [21]. However, there are still some challenges in implementation. For example, the suggested principles prepare the basic setting for the formation of collaboration, but there are some factors which would hinder team building and the interactions of the design community [22]. These factors may include designers' personal character. Some team building activities could help to solve these limitations.

It takes longer to transfer creative experience from collaborative design to reflection and knowledge. Hence, it is better to conduct the suggested principles of developing an effective collaborative design environment over the long-term rather than the short-term, which would not provide any obvious improvement.

13 Conclusion

Some scholars proposed better environment for creative. However, there are still some difficulties were identified. More investigation are still needed to encourage social interactions and promotes collaboration and flexibility. In this paper, theories about social construction, collaboration creative, and design community have been referred to. It has been found that there are some psychological factors involved in the design process of problem solving. It seems to indicate that psychological factors are possible aspects that should be considered during the environment design process. In this paper, some principles of developing an effective environment were suggested. It is a preliminary step to further investigating the approach which can create an appropriated environment and allow time for designers to de-stress, open their minds, and inspire each other.

References

1. Richards, C., Tanquilu, N.: Sustainable solutions to 'Complex Problem-Solving': a key to effective community and industry engagement and partnerships by University researchers. Indonesian J. Plann. Dev. **1**(1), 1–10 (2014)
2. Richards, C.: Using a design research approach to investigate the knowledge-building implications of online social networking and other Web 2.0 technologies in higher education contexts. In: Alias, N., Hashim, S. (eds.) Instructional Technology Research, Design and Development: Lessons from the field, pp. 117–140. IGI Global Press (2012)
3. Tarman, B., Kuran, B.: Examination of the cognitive level of questions in social studies textbooks and the views of teachers based on bloom taxonomy. Educ. Sci. Theory Pract. **15** (1), 213–222 (2015)
4. Sigogne, A., Mornas, O., Tonnellier, E., Garnier, J.L.: Co-engineering: a key-lever of efficiency for complex and adaptive systems, throughout their life cycle. In: Complex Systems Design and Management, pp. 19–37. Springer, Cham (2016)
5. Rahmawati, Y., Utomo, C., Anwar, N., Setijanti, P., Nurcahyo, C.B.: An empirical model for successful collaborative design towards sustainable project development. J. Sustain. Dev. **7** (2), 1 (2014)
6. Eckert, C.M., Clarkson, P.J.: Planning development processes for complex products. Res. Eng. Design **21**(3), 153–171 (2010)
7. Melo, A.F., Clarkson, P.J.: Planning and scheduling based on an explicit representation of the state of the design. In: ASME 2002 International Design Engineering Technical Conferences and Computers and Information in Engineering Conference, pp. 89–98. American Society of Mechanical Engineers (2002)
8. Wenger, E.: Communities of Practice: Learning, Meaning, and Identity Learning in Doing: Social Cognitive and Computational Perspectives. Cambridge University Press, New York (2000)
9. Cox, M., Arnold, G., Tomás, S.V.: A review of design principles for community-based natural resource management. Ecol. Soc. **15**(4), 38 (2010)
10. Chase, J.D., Uppuluri, P., Lewis, T., Barland, I., Pittges, J.: Integrating live projects into computing curriculum. In: Proceedings of the 46th ACM Technical Symposium on Computer Science Education, pp. 82–83. ACM (2015)
11. Burke, A.: Group work: how to use groups effectively. J. Effect. Teach. **11**(2), 87–95 (2011)

12. Brown, S.E., Karle, S.T., Kelly, B.: An evaluation of applying blended practices to employ studio-based learning in a large-enrollment design thinking course. Contemp. Educ. Technol. **6**(4), 260–280 (2015)

13. Buede, D.M., Miller, W.D.: The Engineering Design of Systems: Models and Methods (2016)

14. Gratch, J., Marsella, S.: Psychology for the design of lifelike characters. J. Appl. Artif. Intell. Some Lessons Emot. **19**, 215–233 (2005)

15. Carmeli, A., Dutton, J.E., Hardin, A.E.: Respect as an engine for new ideas: linking respectful engagement, relational information processing and creativity among employees and teams. Hum. Relat. **68**(6), 1021–1047 (2015)

16. Zhang, X., Zhou, J.: Empowering leadership, uncertainty avoidance, trust, and employee creativity: Interaction effects and a mediating mechanism. Organ. Behav. Hum. Decis. Process. **124**(2), 150–164 (2014)

17. Fox, K.C., Girn, M., Parro, C.C., Christoff, K.: Functional neuroimaging of psychedelic experience: an overview of psychological and neural effects and their relevance to research on creativity, daydreaming, and dreaming (2016)

18. Voogt, J., Laferrière, T., Breuleux, A., Itow, R.C., Hickey, D.T., McKenney, S.: Collaborative design as a form of professional development. Instr. Sci. **43**(2), 259–282 (2015)

19. Kangas, K., Seitamaa-Hakkarainen, P.: Collaborative design work in technology education. In: Handbook of Technology Education, pp. 1–13 (2017)

20. Wang, X., Love, P.E., Kim, M.J., Wang, W.: Mutual awareness in collaborative design: an Augmented Reality integrated telepresence system. Comput. Ind. **65**(2), 314–324 (2014)

21. Wang, P., Chung, T.S.: Recent advances in membrane distillation processes: membrane development, configuration design and application exploring. J. Membr. Sci. **474**, 39–56 (2015)

22. Dubowitz, T., Ncube, C., Leuschner, K., Tharp-Gilliam, S.: A natural experiment opportunity in two low-income urban food desert communities: research design, community engagement methods, and baseline results. Health Educ. Behav. **42**(1), 878–968 (2015)

Typing Attitudes Toward Exercise and Investigating Motivation Suitable for TPO

Masanari Toriba[1(✉)], Toshikazu Kato[1], Fumitake Sakaori[1],
and Etsuko Ogasawara[2]

[1] Chuo University, 1-13-27 Kasuga, Bunkyo-ku, Tokyo 112-8551, Japan
a13.nt6w@g.chuo-u.ac.jp, t-kato@kc.chuo-u.ac.jp,
sakaori@math.chuo-u.ac.jp
[2] Juntendo University, 2-1-1 Hongo, Bunkyo-ku, Tokyo 113-0033, Japan
eogasawa@juntendo.ac.jp

Abstract. In recent years, lifestyle diseases have become a serious problem in Japan. According to a survey by the Ministry of Health, Labour and Welfare, more than half the causes of death in FY 2005 were attributed to lifestyle diseases. To prevent lifestyle diseases, diet improvements and regular exercise are necessary. However, many people cannot continue exercise. In this research, we classified the sports consciousness of participants, measured their motivation during exercise in various TPOs with questionnaires, and clarified these relationships. We also considered the method of motivation for exercise.

Keywords: Health care · Motivation · Exercise · TPO (Time, Place, Occasion)

1 Introduction

In recent years, lifestyle diseases have become a serious problem in Japan. According to a survey by the Ministry of Health, Labour and Welfare, more than half the causes of death in FY 2005 were attributed to lifestyle diseases. To prevent lifestyle diseases, diet improvements and regular exercise are necessary. According to a survey conducted by the Ministry of Health, Labour and Welfare, few people are able to exercise regularly. To solve the problem of people discontinuing regular exercise, using "Motivation" attracts attention. Motivation is a method for improving enthusiasm to work on things such as exercise and study. Many studies on motivation have shown that the same approach can have different effects on different people. Therefore, motivation requires an individualized approach. Regarding the "motivation of sports," Sport England managed an effort called "Under the Skin." They conducted a questionnaire on consciousness about sports, classifying the participants into six groups by their attitudes toward sports. However, when actually motivating people, their situation always changes; therefore, it is necessary to consider not only the consciousness of exercise but also people's situation at that time.

In this research, we classified the sports consciousness of participants, measured their motivation during exercise in various TPOs (Time, Place, Occasion) with questionnaires, and clarified these relationships. We also considered the method of motivation for exercise.

© Springer International Publishing AG, part of Springer Nature 2019
S. Fukuda (Ed.): AHFE 2018, AISC 774, pp. 215–219, 2019.
https://doi.org/10.1007/978-3-319-94944-4_24

2 Related Research

As a survey by Sport England, a campaign called This Girl Can was conducted [3]. In this survey, social media were used to send one promotional video to more women, resulting in 2.8 million views. It has become apparent that women started to behave more actively after watching the campaign. This survey has also been able to motivate many people naturally when seeing a promotion. It is also obvious that there were people who did not experience the effect as a motivation because the impression and way of feeling depends on each individual. Therefore, I expect that classifying people according to similar tendencies will lead to more motivation of people.

In the investigation in Sport England, Under the Skin was conducted [2]. In this survey, we conducted a questionnaire on consciousness of sports and daily exercise habits and categorized the participants based on the results. It became clear that the participants could be classified into six groups, but when actually exercising, not only the consciousness of exercising but also the motivation effect differed depending on the situation on the spot Conceivable.

3 Approach

In this study, we obtained data about exercise by questionnaire and classified the participants to clarify their grouping and motivation to exercise in various TPOs. Likewise, we acquired questionnaires using questionnaires to clarify the relationship between "consciousness to exercise" and "TPO with increased motivation to exercise." Using these results, it is possible to examine motivation by each individual while also considering TPO.

4 Determination of TPO

4.1 Definition of TPO

People's situation can be described by the abbreviation TPO, meaning "Time, Place, Occasion." "Time" and "Place" refer to the time and place of exercise. "Occasion" can be defined two ways: the first is "external state," regarding one's personal environment and state; the second is "internal state," regarding one's physical and mental state.

4.2 Investigation on Motivation for Exercise in TPO

Participants were 11 college students (8 men, 3 women) asked to take the opportunity to exercise in four situations: Time, Place, Occasion (external), and Occasion (internal). Each answered with free explanation as to whether there was a lot or if they could do it again.

We classified the items that were similar in subjective evaluation of the results of their free descriptions. The classified results are summarized in the table above (Table 1).

Table 1. Classification of TPO during exercise.

Time	Place	Occasion(External)	Occasion(Internal)
Day Morning, Noon, Evening, Night	House Room, Living, Bath	Human Alone, With friends No peoples, acquaintance	Body Sleep, Hunger, Fatigue (Injury, Cold)
Week Weekday, Holiday	Near the House Park, Around the House	Sound Silence, Noisy Favorite music	Spirit Task, Stress
Season Before summer	Moving Train, Street	Light Bright	Awareness Feel fat, Impressed watching sports
Working Moving, Brushing teeth, Bath, Watching Working, Studying	Destination University, Company	Climate Comfortable	
After Work Overeating, Watching sports, Tedious work	Sports Facilities Gym, Gymnasium	Environmental movement Ground condition, Exercise equipment	

In the shaded part of the above table, there were differences in opinion by participants, so, when clarifying the motivation for exercise in various TPOs this time, the following items of TPO were changed and organized: "Time (1 day)," "Daily (1 week)," "Occasion (external)," "person," "Occasion (internal)," and "spiritual." This is summarized in the following table (Table 2). Additionally, as a reason why items concerning Place are not included this time, it is conceivable that many people fix the place depending on the time of day and, to some extent, the weekday. It is assumed that the place when getting up in the morning and around noon is "home," "school or company" when it is a weekday, and "home" again in the evening, so Place depends on Time. As it is not dependable to some extent, Place is not considered in this experiment.

Table 2. Classification of TPO during exercise.

Time	Occasion(External)	Occasion(Internal)
Day After waking up, Lunch break After school, Before sleep	Human Alone(Surrounding people), Multiple people(Friends)	Spirit Task, Stress
Week Weekday, Holiday		

In addition, from the above summary, we will decide questionnaire items to be used for this experiment: five elements from "Time," three elements from "Occasion (external)," and four elements from "Occasion (internal)" for a total of 12 elements, as summarized below (Table 3).

Table 3. Classification of TPO during exercise.

Time	After waking up
	Lunch break
	After school
	Before sleep
	Holiday
Occasion (External)	Situation in which no one is around
	A situation where there are people around
	With friends
Occasion (Internal)	There is something that I should do
	There is almost not that I should do
	Having a lot of stress
	Not having stress

5   Method

5.1   Exercise Habits and Consciousness of Exercise

In addition to exercise habits and daily lifestyle, we conducted a questionnaire about exercise and the participants themselves. Items, details, and answering methods are summarized in Table 4. Classification of TPO during exercise. In addition, these questionnaires were prepared based on the questionnaire used for Sport England's Under the Skin.

Table 4. Questionnaire on exercise habits and awareness about exercise

Question item	Details	Answer Method
1. Ask about you	Name, Sex, Age	Description
2. Ask you about your current exercise and sports activities	Club activity, Exercise situation	Selection formula
3. Ask about your current daily life	Meals, Sleeping, Television	Selection formula
4. Ask you about your current living environment	How to go to school and time	Selection formula
5. Ask about yourself	Consciousness to exercise	5 grades evaluation
6. What do you take care of in your life	Things that take care in life	Select all that apply

5.2   Motivation to Exercise in Various TPO

We conducted a questionnaire on motivation to exercise in TPO. For motivation to exercise in TPO, the situation was shown by combining each item of TPO obtained in 3.2, and participants responded from 1 (do not want to exercise) to 7 (exercise).

6   Result and Discussion

The assumptions of this study are now verified, and we will discuss the results when they are announced on the day of the AHFE presentation.

7   Future Work

In this experiment, we aimed to clarify the relationship between classifications by way of thinking on exercise of participants and the motivation to exercise in TPO by conducting a questionnaire on participants' "consciousness of exercise" and "motivation to exercise at TPO." We conducted experiments.

As a future prospect, from the relationship between motivation to categorize and motivation in TPO, we examine and verify motivational approaches according to each characteristic. With these clarifications, the movement is continuing quite often. I think that it is possible to motivate individuals with varying characteristics and TPO and that more people can continuously exercise.

References

1. Health, Labour and Welfare Ministry, Heisei 27 years the National Health and Nutrition Examination Survey Results Summary. http://www.mhlw.go.jp/file/04-Houdouhappyou-10904750-Kenkoukyoku-Gantaisakukenkouzoushinka/kekkagaiyou.pdf
2. Sport England, Under the skin. https://www.sportengland.org/media/10233/youth-insight_under-the-skin.pdf
3. Sport England, This girl can. https://www.sportengland.org/our-work/women/this-girl-can/

Analysis of Intuitive Thinking in Five-Sense Design: An Example of Auditory Design

Tao Xiong, Jianxin Cheng[✉], Tengye Li, Wenjia Ding,
and Zhang Zhang

School of Art Design and Media, East China University of Science
and Technology, M.BOX 286, NO. 130, Meilong Road, Xuhui District,
Shanghai 200237, China
472353860@qq.com, 13901633292@163.com,
lty900821@qq.com, 13585758008@163.com
15618746761@qq.com

Abstract. Intuitive thinking is a creative way of thinking based on perception. Through mobilizing the knowledge and experience people already mastered, it can help human brain produce the hypothesis or image of the whole property and law of things quickly. In product design, the correct use of intuitive thinking can not only help designers to enhance the imagination and creativity, improve the level of design, but also make products more humane and associated with the user experience, thereby it will greatly reduce the learning costs and bring a better user experience.

Five-sense design is a design method with intuitive thinking. Users interact with the product through sensory organ, and their original memory and experience associate with the product, which will produce a result. Relative to visual, tactile sense in the five-sense design, listening, smell and taste are diluted a lot. But listening is also an important part to achieve information communication between users and products, which should be taken seriously.

In this paper, we will analyze the role of intuitive thinking in product design with design cases from the point of auditory design, and summarize the relationship between auditory design and intuitive thinking, so as to provide users with a more intuitive and pleasurable product experience and make a reference for later design practice.

Keywords: Intuitive thinking · Five-sense design · Auditory design
User experience

1 Introduction

With the development of technology and economy, the actual demand of people for products is also constantly increasing. For products, people are no longer satisfied with beautiful appearance and good functions, but also have higher requirements in user experience. So design should move toward a more vivid and intuitive direction. An improved user experience must be accompanied by changes in the way products used. In the process of product use, sensory use may no longer be a one-to-one relationship,

© Springer International Publishing AG, part of Springer Nature 2019
S. Fukuda (Ed.): AHFE 2018, AISC 774, pp. 220–227, 2019.
https://doi.org/10.1007/978-3-319-94944-4_25

namely no longer just with eyes to see, ears to listen, hands to adjust and so on. Accordingly, people face the challenge of adapting to redefining sensory functions. Intuitive thinking exactly summons up people's memory and experience through perception, so as to make contact and interaction with products. And the correct use of intuitive thinking can break through the constraints of one-to-one correspondence between senses and usages, allowing users to quickly adapt to changes in redefining sensory functions and gain a better user experience.

This article will take hearing as an example, and summarize the relationship between auditory design and intuitive thinking by analyzing the role of intuitive thinking in five-sense design through design examples, in order to provide users with a more intuitional and friendly user experience.

2   Intuitive Thinking in Product Design

Intuitive thinking is a rapid conjecture and identification of human brain to the objective world and its relations [1]. Different from logical thinking emphasizing analysis, intuitive thinking is usually based on knowledge structure people familiar with and perceptual experience, takes skipping as a form, and makes judgments and identifications quickly on the whole. "Epiphany" is the best word that summarizes the characteristics of intuitive thinking.

So in the state of intuitive thinking, various long-standing "latent" stored in the brain will be aroused. And they may not be combined in logical ways, but may make new connections in an unexpected form to cover the deficiency of fact and logic, which often leads to creative conclusions and results [1]. Precisely because of this, splendid achievements in mathematics, physics, biology and other areas have been made due to intuitive thinking, leaving remarkable moments in human history.

Similarly, the creativeness of intuitive thinking plays a significant role in product design. For designers, intuitive thinking is an important part of getting inspirations. In the process of design, it can help designers to develop imagination and creativity due to its irrationality, multidimensionality, burstiness and comprehensiveness, which will improve the level of design and build a better "design thinking".

For users, intuitive thinking is equally important. The formation of intuitive thinking is closely related to people's knowledge structure and life experience. Psychologically speaking, intuitive thinking is the combination of perception, imagination and understanding [1]. On the basis of perception, people use products combined with past experience and knowledge to finish the using process. If there is a lack of knowledge related to new products in user's intuitive thinking system, the cost of product using will increase, resulting in a poor user experience.

Therefore, designers should take target user's intuitive thinking system into account, which means product design should associate with user's intuitive thinking. This requires designers to have a solid, broad theoretical knowledge, keen observation and rich imagination, both with overall analytical skills and comprehensive communication skills, and also to develop the habit of keeping curiosity [2].

3 Five-Sense Design and Intuitive Thinking

3.1 Five-Sense Design

Five senses generally refer to sight, touch, smell, sound and taste. Mostly people come to understand objects through these five senses. By using five senses, five-sense design interprets information required conveying from different perspectives to achieve a full range of information delivery, and ultimately make the information reach their destinations quickly and accurately. Also, it aims comfortable and emotional sensory experience as a goal [3]. In "Designing Design", Kenya Hara explains "Five-sense design" as follows: "A man is more than just a sensory receiver, but also a sensitive memory regenerator that can reappear various images in mind based on memory. And the images appear is a grand picture composed of people's feelings and regenerative memory. And this is exactly where designers are."

Among five senses, sight and touch are the most developed senses. Sight can capture color, material, modeling, text and other product information. And touch plays a leading role when users make further contact with products, which helps people form subjective feelings about products. Therefore, sight and touch both dominate in five-sense design. Relative to the dominance of sight and touch, the role of smell, sound and taste are downplayed in product design, often easy to be ignored by designers.

3.2 Intuitive Thinking in Five-Sense Design

Five-sense design is an intuitive design method as it interacts with products through user perception.

When people interact with products, their brain responds accordingly to products' shape, color, sound, smell, etc., resulting in five senses. At the same time, sensory information previously stored in the brain will make connection with those five senses and even supplement, enrichment and development through imagination and association [3]. Just as people refer "table tennis" to "tennis" similarly, there are similarities in these two experiences on five senses. And Kenya Hara explains the meaning of "Information Construction", which is the core of Five-sense design theory, as follows: "'Impression' is the outcome that sensory organs accept external stimulus and combine them with previous memory in the brain." Design behavior, which is based on the premise of this combination, will make deliberate intervention in the combination in return. And we refer the process as "Information Construction", namely the process of intervening the formation of composite impression intentionally and systematically", which also embodies the essence of intuitive thinking.

Buddhism is the best example of taking full advantage of five-sense design, which embodies intuitive thinking at the same time. Visually, exquisite representative Buddhist buildings are numerous and Sanskrit is also highly recognizable; Touch is embodied in unique costumes and scriptures; Temple block Muyu is a distinctive musical instrument; Buddhism also emphasizes being vegetarian in taste while there are sandalwood in smell. In a word, Buddhism is just like a strong corporate brand. The unification of symbols, hairstyles, visions, slogans, apparel, and management makes people immediately think of Buddhism when they feel just one feeling of its five senses.

Therefore, the full use of intuitive thinking in five-sense design is an essential reason why Buddhism can be widely spread and become one of the three major religions. This is also an important reference for product design. In the current situation that sight and touch are widely used, trying to start from sound, taste or smell may open up a new world of intuitive thinking in five-sense design. Just as the "Five-sense theory" proposed by designer Jinsop Lee, sometimes trying to enhance one or two five-sense elements neglected before may lead to an improvement in product design (Fig. 1).

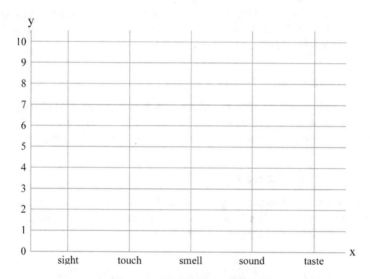

Fig. 1. The five senses graph.

4 Auditory Design and Intuitive Thinking

4.1 Auditory Design

Hearing is the most important human sense second only to sight, which is a key way for people to obtain outside information. According to statistical analysis, 83% information people get is obtained by eyes, 11% by ears, and 6% by others [4]. Therefore, as the second most important sense of information source, hearing occupies a very important position in human intuitive thinking system and should be valued in five-sense design. (1) Adding auditory elements into design will help people judge things more accurately and quickly. As eyes usually perceives and receives information selectively, some important information may be missed sometimes. As a result, intuitive thinking can't make quick and accurate judgments as some memories and experiences can't be aroused in time. But auditory information can play the role of mutual complement and conversion. At the same time, the coordination of sight and sound can achieve the product multi-mode interaction and enhance user experience. (2) Auditory design can improve people's understanding of product use. Researches show that normal sound conveys a sense of security. In product design, sound-matching is a common method, such as the click of buttons, sound of shutting car doors. These sounds have the prompt function that indicates to users that one certain step of product using has been completed. (3) The mood of users will be affected by auditory design. Through ingenious combination of different notes in time-domain, various emotional sound can be made and give people different feelings [5]. So adding moderate sound elements into products can improve the pleasure of product using. (4) Auditory design helps to improve people's memory of product using. Integrating auditory experiences into product design can deepen people's impression of using to promote understanding and memory, in order to form the corresponding impressions in the brain and enrich, complete intuitive thinking system.

4.2   Intuitive Thinking in Auditory Design

Therefore, as an important source of information, the significance of hearing to human perception is self-evident and naturally sound is a key element of intuitive thinking. Products use auditory elements appropriately often reflect the essence of intuitive thinking.

(1)   Mileage-counting Drum-cart

In ancient China, Mileage-counting drum-cart is a mechanical device that automatically broadcasts mileage. It's a typical product that intelligently uses auditory elements and fully reflects intuitive thinking (Fig. 2).

Fig. 2.   Mileage-counting Drum-cart.

The basic principle of mileage-counting drum-cart is the use of differential relationship of gear mechanism. According to records, the drum-cart looks like a normal car, which has two wooden men, one drum and one bell. And one man for drumming while the other belling. The cart is equipped with a set of reduction gear, connected with the axle. When the drum-cart runs for one mile, the gear controls the drumming man will rotate once and the man will also drum once accordingly. When the drum-cart runs for ten miles, the gear controls the bellman will rotate once and the man will bell at the same time. So when people in the car hear the drum and the bell, they will know how far the car has travelled.

In the absence of odometer, GPS and other equipment in ancient time, the drum-cart cleverly converts mileage into sound, so people can quickly know the distance. Compared to look at odometer, determining the distance by listening fully mobilizes people's intuitive thinking, which is more quickly and directly. As time passes, the concept of distance will gradually formed through the combination of listening and watching, making up for visual deficiency.

(2) Nokia Tune

Nokia's iconic ringtone "Nokia tune" is adapted from the guitar piece Gran Vals written by 19th-century Spanish musician Francisco Tárrega (Fig. 3). By 1998, the ringtone was well known and referred to as "Nokia tune". When the familiar ringtone sounds, people know that they have received a message. Although sound-matching seems to be nothing special nowadays, "Nokia tune" was the best brand logo at that time. In addition to users themselves, others will naturally think of Nokia phone when hearing the ringtone. So it plays a very good brand promotion effect. Even now, Nokia tune also has the role of affecting people's emotions. Many people are reminded of the glorious age of Nokia when hearing the ringtone, causing nostalgia.

Fig. 3. Nokia tune.

In fact, in addition to Nokia tune, other products are also trying hard to put brand label on sound, such as Apple's typical ringtones and the music of road sprinkler. So distinctive product sounds can deepen users' memory of the products and even affect their emotions.

(3)
Sound Book

Children sound books break through the limitation of traditional books with sight and touch as the main perception modes, creating a multi-mode interaction of reading

Fig. 4. Children sound book.

(Fig. 4).

Incorporating auditory experience into reading also compensates visual deficiency and helps children to form correct impression of new things quickly. For example, if children see a kitten in the book and hear a meow at the same time, they can rapidly establish the relationship between cat's image and its voice, and simultaneously set up the cognition towards cats both visually and aurally. In addition, vivid and interesting audio-visual experience can also arouse children's interest in reading. Children sound book itself reflects intuitive thinking, but also help children gradually establish and improve intuitive thinking system.

5   Conclusion

Based on the research of intuitive thinking, five-sense design and their relationship, this essay chooses sound from five senses and analyzes the role of intuitive thinking in auditory design through design examples. The conclusions are summarized as follows: Using intuitive thinking, auditory design can help people make faster and more accurate judgments; Auditory sense can make up the deficiencies of other senses in five-sense design, and even make conversions with other senses; Auditory design based on intuitive thinking is more likely to create a matching relationship between sound and product, and guides user's emotion simultaneously; Innovative fusion of sound and other senses often brings design improvements. Follow-up researches will be based on the results of this study, to make innovative attempts in auditory design.

Acknowledgments. This research was supported by master studio project of regional characteristic product research and development for "the belt and road initiatives" supported by shanghai summit discipline in design (Granted No. DC17013).

References

1. Zong, C.: Seek infinity from a limited mentality and see eternity in moment – an analysis of the embodiment of intuitive thinking rules in ancient artificial design. J. New Vis. Art. **02**, 28–33 (2013)
2. Sun, X.: Research of intuitive thinking intervention in design thinking construction. J. Art Educ. **12**, 150–151 (2014)
3. Tao, L.: Applied research of synesthesia in furniture design. Jiangnan University (2009)
4. Luo, X.: Book design research of children aged 0–6 based on auditory and tactile experience. Sichuan Normal University (2014)
5. Liu, Z.: Research of audio-visual emotion semantics and its applications. Taiyuan University of Technology (2012)
6. Chen, C.-H., Trappey, A.C., Peruzzini, M., Stjepandić, J., Wognum, N., Li, T., Cheng, J., Xiong, T., Ye, J., Zhang, Z.: The Use of Intuitive Thinking in Product Design Semantics: From Chinese Characters to Product Design. IOS Press, Amsterdam, 15 June 2017
7. Hartley, J.: Improving Intuition in Product Development Decisions. Springer, Netherlands, 15 June 2009

Applying Intuitive Thinking in Smart Home Design Based on Semantic Association

Tengye Li, Jianxin Cheng[(⊠)], Tao Xiong, Wenjia Ding,
Zhang Zhang, and Xinyu Yang

School of Art Design and Media, East China University of Science
and Technology, M.BOX 286, NO. 130, Meilong Road, Xuhui District,
Shanghai 200237, China
lty900821@qq.com, 13901633292@163.com,
472353860@qq.com, 13585758008@163.com,
15618746761@qq.com, 546467089@qq.com

Abstract. With the advance of Industry 4.0, the merging of automation and information has become an important trend of product design. Smart home products are increasingly valued and favored by consumers. However, the existing smart home product design mainly focuses on the product performance and technology sense and often ignores the users' emotional and spiritual needs, which may increase the users' cognitive difficulty and reduce users' satisfaction. Good products should not only meet the needs of usage, but also to take the user's psychological factors into consideration, accordingly to enrich the emotional experience of users. Household products are closely related to the users' life. When using the household products, users are more likely to use the visual intuitive thinking mode, thereby giving the product a special emotional meaning. This paper takes the innovation design of water equipment in kitchen as an example. After fully understanding the users' imaging and description about the future kitchen, it extracts key words such as "cleanness" that the users generally think can represent the future kitchen, to guide the user to start the semantic association and establish the users' semantic schema. Also, this paper uses the Evaluation Grid Method (EGM) to link the images produced by intuitionistic association with the design elements of products, and put forward the product design principles in line with the users' intuitive thinking to establish a product design framework based on semantic association. By testing the concept design product in practice, it proves that this framework can improve the users' satisfaction with the product. This paper proposes that the product design process should take full account of human factors, and that the users' intuitive thinking habits should be applied to product design, which provides a reference for innovative design of smart home products.

Keywords: Intuitive thinking · Smart home design
Evaluation Grid Method (EGM) · User experience
Product semantic association

© Springer International Publishing AG, part of Springer Nature 2019
S. Fukuda (Ed.): AHFE 2018, AISC 774, pp. 228–236, 2019.
https://doi.org/10.1007/978-3-319-94944-4_26

1 Introduction

With the advance of Industry 4.0, the merging of automation and informatization has become an important trend in the society and product design has already stepped into a "smart age" [1]. Household product, with a close contact with our daily life and a great variety in its category, takes a significant part of the smart design. Considering the convenience brought us by those smart designs, many families have shown their favor during the selection [2]. However, the development of smart household products has also encountered some problems recently. When paying more attention to improving the technology and simplifying the operation of products, we always ignore the emotional and spiritual needs of the users. Thus, it makes our users understand the products more difficult. Household products are generated from home culture and ideology, so the relevant selections made by our users are affected by their culture, customs, and lifestyle. Currently, most of smart home designs failed to take users habits from different cultural backgrounds into consideration. As a consequence, they hardly meet the demand of users.

Intuitive thinking has been proved to weigh the same as logic thinking in the ideology of human beings. When the user meets a normal decisional issue, he or she would more tend to solve it with the support of his or her intuitive thinking [3]. Household products closely relate to our daily life, so intuitive thinking with imagination is frequently adopted during the use of those products at home, which brings the products themselves a special meaning with our emotion. When the product function has satisfied the expectation of the users, the improvement of product symbolism and interaction would further increase user experience efficiently [4]. In future, it would be an important objective in product design to integrate users' experience and memory, and to lead the users to generate the intuitive association rapidly and precisely through product semantics. Based on the imagination and expectation on the future kitchen from the users, this paper summarizes the key elements in product design in line with users' intuitive thinking through guiding some users to conduct semantic association with certain representative emotional words. Then we continue to the innovative product design aiming for improving user experience, which would provide a reference for normal design workflow in future.

2 Literature Review

2.1 Intuitive Thinking

Intuitive thinking is an important category of innovative thinking, as well as an irrational thinking rooted in the instinct of human beings. Since Bounded Rationality proposed by Herbert Simon, many scholars have led deep research on human ideology. Eventually, Modern Cognitive Dual – Process Theory was formed. A series of research show that implicit and intuitive processing controls human behaviors in the most of cases. That means our life is greatly led by the unconscious and automatic intuitive thinking [5]. When a person analyses an issue in dual system, he or she could get a result in line with both intuition and logic, if two processes go to the same direction.

Otherwise, a conflict would emerge and final result depends on which side take more advantages in the competition. However, comparing with logic system, our intuition system possesses more speed advantage during the decision-making [6]. Meanwhile, the decision made by the users based on their intuitive thinking would be closer to their original expectation than the logic one, so they would feel more satisfaction.

2.2 Product Semantics

In 1984, the concept of Product Semantics was firstly put forward by Krippendorf and Butter and introduced in the "Product Semantics symposium" organized by Industry Designers Society of America in Cranbrook Academy of Art [7]. Product semantics breaks the traditional theory in which the factors related to human beings are simply categorized into ergonomics that only includes the physical and biological considera- tion about human beings. Instead, it extends the design to more spiritual and mental aspects. Product semantics has been applied to design practices. During this process, semantic communication lies in the key point throughout, with more aesthetic and practical elements involved as well. Recently years, more and more scholars have turned their research on product shape into decoding product meaning, which guides us to explore a new direction to further develop product semantics [8].

2.3 Evaluation Grid Method

Evaluation Grid Method is developed by Junichiro Sanui, Japanese scholar, from the Repertory Grid proposed by psychologist Kelly in 1986 [9]. Through building the Contrast System, we could understand the surroundings and happenings, and predict the results. Based on this, scholars including Sanui improves Repertory Grid into Evaluation Grid Method in which the certain factors in the Contrast System are taken out, their relationship with items in the evaluation project are clarified, and finally Laddering is involved also inside [10]. When receiving clear answers from the inter- viewees, the interviewer should continue to ask other relevant questions so that the interviewees could be guiding to further illustrate their opinions and concepts. Then all collected concepts would be organized into Evaluation Grid for showing the real ideas gained from those interviewees.

3   Research Method

Kitchen is a significant part of our home and issues about water in the kitchen are mostly concerned by the users. In our survey, we found that water in the kitchen is not only used for cooking, but also being regarded as an important supply for daily use in the house. Comparing with other electric appliance in the kitchen, the design for water system is not so smart and some shortcomings still exist. Current designs are mainly for the tap's shape or smart control. As for the way of taking and storing water, however, the research is relatively few. In this paper, the water using experience and user intuitive association are integrated into design, so that we could improve the user experience and also provide a new exploration of home design in future.

4 Method and Process

4.1 Stage 1 Key Words Selection for Semantic Association

We adopt the interview to conduct the survey for eight samples selected from Shanghai. They are females between 30 and 35 years old with long-term experience in cooking. Basically, the samples share similar cultural background and lifestyle. During the interview, each sample is required to explain the issues they met when using water in the kitchen and further describe their imagination about this water system in future, including either their own feelings or specific problems they hope to be solved. After collecting all description from eight interviewees, we transform it into emotional adjectives and sort them based on the similarity. Then the number of times that each word is mentioned is calculated (descending order) as following Table 1: Emotional Word List of Water in Future Kitchen.

Table 1. Emotional word list of water in future kitchen.

Clean 8	Convenient 7	Smart 7	Entertaining 6	Relaxing 5
Integrated 4	Communication-enhancing 4	Sharing 4	Eco-friendly 4	Safe 3
Sustainable 3	Unified 3	Invisible 2	Transparent 1	Good-smelling 1

As shown in the Table 1, more than half interviewees think that water in the kitchen should be clean, convenient, smart and entertaining, among which "clean" is featured as the most important criteria for the evaluation. During our survey, we also found that water in the kitchen is not simply used for cooking, as well as for other aspects like the house cleaning. (the reason why people do not take water from the bathroom is related to spatial design and daily habits.) Thus, "convenience" is selected as the second evaluating standard. Besides, "smart" is a crucial direction for the development of home design in future, so most of the interviewees agree that design in the new kitchen should meet the requirements of intelligence. However, there is one interviewee disagrees with it and claims that so-called "smart" hardly bring us better experience in fact. Furthermore, almost all of them mentioned that the operations in the kitchen are quite complicated and bored, especially during some waiting moments. Thus, the design for enjoyment and entertainment also count for the future kitchen.

After the summary and discussion, we finally decided to select "clean", "convenient" and "entertaining" as the key words for inspiring users' intuitive association. ("smart" is deleted at last considering the difficulties in organizing information in intuitive association, because most of interviewees hardly describe "smart" products specifically and have any further association).

4.2 Stage 2 Semantic Association from Intuition

In this stage, we invited another group of 10 females who love cooking at the age of 25–35 from Shanghai. They are required to have intuitive association from the words listed above. Then, according to Evaluation Grid Method, we try to link the answers received with the superordinate (abstract concepts) and subordinate (specific design elements) factors. Traditional method mainly focuses on seeking for a better solution from simply comparing every two pictures collected, in which the design in shape still performs as the core. Comparing with the traditional one, however, improved Evaluation Grid Method leads the exploration on the real thoughts from the users from their memory and association, and guides us to optimize the product usage and service in future.

The process of collecting semantic association is arranged as one-to-one interview. Except for the person who poses the questions, there are another two for the record and photograph respectively. The interviewee is required to search and describe any specific objects or events in their memory according to the given key words. In the following questions, we would further look for the relevant information about the superordinate and subordinate factors and build the evaluation grid. A representative interview is shown as follows:

Q1: Among your experience about using water, what kind of objects could bring you a sense of "clean"?

A: I think that is from the fountain.

Q2: Except for clean, what other feelings does the fountain bring you?

A: Fountain could also remind me pleasure and tenderness, because it stands for some good wishes.

Q3: What other elements in the fountain make you have those feelings?

A: spraying water, drawing a smooth curve and splashing drops upward make feel the same.

Through the conversation above, the intuitive images from semantic association are organized into two parts, superordinate abstract factors and subordinate specific factors. Based on this, the evaluation grid about intuitive images is built (Fig. 1).

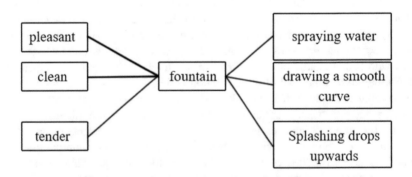

Fig. 1. An evaluation grid sample of "clean"

We repeat those questions one after another until the interviewee takes relatively longer time to response or refuse to continue (at this moment, intuitive thinking stops and recognition burden increases, so the generated imaginations have little influence in our experiment). When the intuitive association relate to all three key words has been recorded, the evaluation grid for water using in future kitchen could be fulfilled completely.

After receiving all results from 10 interviewees, we summarize and categorize all the evaluation grid. In order to avoid some obvious personal differentiations among the individuals, we only take the result with overlapping more than 3 times. The final user-oriented evaluation grid of water using in future kitchen is organized as follows (Fig. 2).

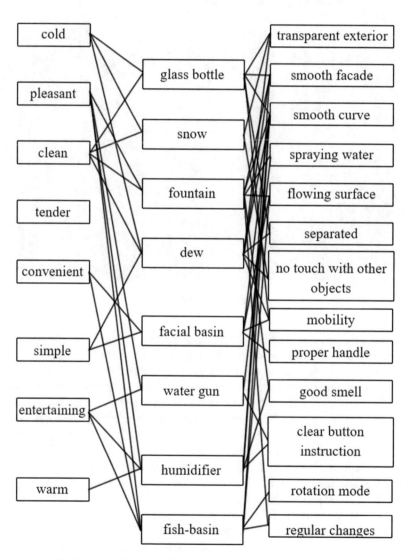

Fig. 2. Overall evaluation grid of water using in future kitchen

4.3   Analysis

In our survey, the images provided by the interviewees, including the fountain, facial basin, fish-basin (Fig. 3), are closely link to their daily habits and cultural backgrounds. Meanwhile, the results of the survey also reflect the information structure stored in their minds.

Fig. 3. Fish-basin

Through the analysis of the evaluation grid above, we found that, except for three given key words, "simple" and "entertaining" in the superordinate factors are also proved to be the critical items for evaluating the kitchen in future. During the illustration of the intuitive images for "clean", many interviewees mentioned the concept of "cold". It might enhance the impression of "clean" to a degree. On the other hand, they put forward "warm" while thinking on the images of "entertaining". Thus, it means we should adopt more neutral colors in the design when tend to emphasis the features of "clean" and "entertaining" at the same time.

As for the analysis of the subordinate specific factors, it is manifested that most of design elements relate to the actual utilization of the users which belongs to the "behavior" in the experience layers proposed by Norman. Most of interviewees suggest water using design in future should involve the features including "smooth curve and façade", "spraying water", "no touch with other objects", "proper handle", "mobility", "clear button instruction", and "regular changes". All those elements summarized provides a practical guideline for our design in the next stage.

4.4   Design Practice

After considering the integrated elements summarized above, we conducted the design practice and finally presented the related solution. In this proposal, the traditional tap is replaced by the waterspouts in the interior of the water sink. It could be detached from the platform and moved to any other places easily. When the sink is rotated, water would spray out from some separated spouts, and the users could clean their stuff and hands. In the user evaluation validation, the concept of "clean", "convenient" and "entertaining" scores relatively higher. Meanwhile, the response time from the users is shorter, so it efficiently improves user experience and satisfaction (Fig. 4).

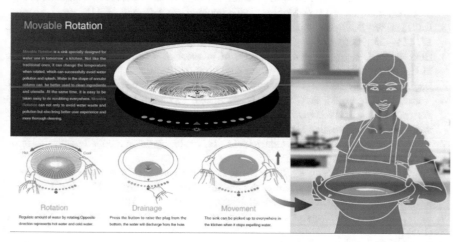

Fig. 4. Movable rotation (design based on semantic association)

5   Conclusion

This paper introduces the method to guide the users' semantic association with intuitive thinking. After collecting their expectation about kitchen in future, we select the most representative and well-accepted words, including "clean" and so on, to build the evaluation grid for water using design. We connect images from users' intuitive association and product design on the premise of Evaluation Grid Method, and propose product design principles in line with the intuitive thinking of the users. Thus, a user-sematic-association oriented design framework is established. After applying the theory to the practice, the design concept is validated and proved to be efficient in increasing users' satisfaction. The paper aims to emphasis the consideration about factors of human beings, including the involvement of users' intuitive thinking, during the product design process, which provides a reference for the innovative home design in future.

Acknowledgments. This research was supported by master studio project of regional characteristic product research and development for "the belt and road initiatives" supported by shanghai summit discipline in design (Granted No. DC17013).

References

1. Hermann, M., Pentek, T., Otto, B.: Design principles for industrie 4.0 scenarios. In: Proceedings of 2016 49th Hawaii International Conference on System Sciences, pp. 3928–3937 (2016)
2. Bian, J.L., Fan, D.L., Zhang, J.M.: The new intelligent home control system based on the dynamic and intelligent gateway. In: Proceedings of 2011 4th IEEE International Conference on Broad Band Nerwork & Multimedia Technology, pp. 526–530 (2011)
3. De Neys, W.: Dual processing in reasoning-two systems but one reasoned. J. Psychol. Sci. **17**, 428–433 (2006)
4. Hekkert, P.: Design aesthetics: principles of pleasure in design. J. Psychol. Test Assess. Model. **48**(2), 157–172 (2006)
5. Bargh, J.A., Morsella, E.: The unconscious mind. Perspect. Psychol. Sci. **3**, 73–79 (2008)
6. Amsel, E., Klaczynski, P.A., Johnston, A., Bench, S., Close, J., Sadler, E., et al.: A dual process account of the development of scientific reasoning: the nature and development of metacognitive intercession skills. Cogn. Develop. **23**, 342–471 (2008)
7. Krippendorff, K., Butter, R.: Product semantics: exploring the symbolic qualities of form. Innovation **3**(2), 4–9 (1984)
8. Zhang, X., Hu, F., Zhou, K., Sato, K.: Reflecting meaning of user experience semiotics approach to product architecture design. In: Proceedings of 2017 24th ISPE Inc. International Conference on Transdisciplinary, pp. 329–337 (2017)
9. Kelly, G.: The psychology of personal constructs. W. W. Norton & Company, New York (1955)
10. Sanui, J.: Visualization of users' requirements: introduction of the evaluation grid method. In: Proceedings of the 3rd Design & Decision Support Systems in Architecture & Urban Planning Conference, vol. 1, pp. 365–374 (1996)

Influence of Art Museum Exhibition Space Design on the Emotions of the Viewer

Binli Gu$^{(\boxtimes)}$ and Yasuyuki Hirai

Graduate School of Design, Kyushu University, Fukuoka, Japan
kohinre@hotmail.com, hirai@design.kyushu-u.ac.jp

Abstract. Design has been more and more concerned with clients' experience and continuous emotional influence. This article makes use of three typical art museums as examples to prove that there are several differences between them with respect to their emotional design elements and expressions by characteristic contrast.

Keywords: Art museums · Emotion · Space creation · Learning enthusiasm

1 Introduction

Prior related research has shown:
Pre-emptive research: Hui Ma's in Thematic Museum Entrance Designs Based on Emotion Expression conducted a preliminary study and analysis on the emotion of the Chinese Movie Museum, this type of thematic museum provides a cultural or an aesthetic education. Art museums and exhibits each have a unique culture in their fields, so as to extract their characteristics and apply their characteristics to the space environment. The overall design style of the Cinema Museum in China is based on the movie playing theme. The whole design takes the surroundings of the venue and the figurative features of the art as the starting points for the design, and expresses the theme of the entire museum by means of implication and expression respectively. The implied approach can be fully expressed by the fragmentary film materials combined with personal experiences while viewers visit the exhibition [1].

Ren Kangli's "The Representation with Abstract Painting and Deconstructionist Architecture for a Contemporary Visual Space: The Architectural and Exhibition Design of the Denver Art Museum", uses a thematic expression of art, with lines, colors, and the exhibits themselves as part of the space for soft-filled space decoration. The abstract colors in the exhibition hall highlight the intrinsic essence of the artwork under a symbolic cultural background and stimulate the spiritual resonance between the exhibits and the audience. The designer shows the entire space as a work of art in terms of painting. So that the a space produces a sense of presence, and arouses the desire of the audience to learn and imagine [2].

In Emotion Design Research in the Exhibition Design, the concept of "empathy" of designers not only plays an important role in the participation of human beings in normal social activities, but also plays an important role in the creation of display design. Designers can optimize and redesign the museum space t using empathy.

© Springer International Publishing AG, part of Springer Nature 2019
S. Fukuda (Ed.): AHFE 2018, AISC 774, pp. 237–249, 2019.
https://doi.org/10.1007/978-3-319-94944-4_27

Redesigning using empathy with exhibitors, t partners and the audiences with regional and cultural differences will enhance the audience's experience. Getting the audience's feedback is also an integral part of the designing strategy [3].

For the design of museums, the ideal condition is the attractiveness of a single exhibition hall and the smooth transition between of all sections. This creates the perfect combination of part and whole geared toward fully meeting the emotional needs of the exhibitors, thus achieving excellent display design. The whole unit and the matched layer changes are an important part of the emotional design of an excellent museum.

Linnan's research on Visual Psychology in Exhibition Space proposed that the color space, structure space, light and shadow space represent inner feelings and emotions in an extended and externalized form. In addition, the space is an important factor affecting the audience's emotion. The three museums convey this experience by using the entire space as an artwork to showcase the immersive experience the atmosphere of the entire space. Some of the museums focus on changes in the color of the space and the design of exhibitions, some of them focus on combining changes in spatial storytelling, changes in light and shade, variety of interior materials and so on, thereby arousing the audience's different emotions and desires to learn [4].

However, in Linnan's research, the affective design of these typical museums did not provide a uniform classification of the most influential elements. This paper will use a literature survey of the three museums and an analysis of the existing data, to organize and discuss reasonable conclusions.

1.1 Background

Facing a growing number of media techniques and the rapid development of new technologies, museums' exhibition space has to maintain its own traditional culture and incorporate new technologies at the time. There is also increased demand for information on spiritual culture, both new and ancient culture. This is a crucial challenge for creating new designs that provide knowledge and satisfy visitor's spiritual needs.

1.2 Purpose

Through a comparison of three museums, we will identify the aspects of special creations and exhibited designs that most affect viewers's positive emotions and arouse their desire to learn.

1.3 Research Methods

Through a large number of literature related to the three museums and from examining the experiences of the three museums' audiences, this paper will investigate and show how the research on the design of affective space can help lead to a positive impact on the viewers. We will summarize and identify the similarities and differences among them, and examine which elements have the most profound impact on the positive emotions of viewers.

2 Process

2.1 Emotional Design in Art Museums

There are many emotional design experience in art museums, such as how to create aesthetic ideas, accumulation of knowledge, design element empathy and so on. Tomiko [5] classified emotional design based on Edgar Dale's theory of "The Cone of Experience". The first stage is direct experience, the second stage is optic and auditory experience, the third stage is upstairs abstract experience.

The final goal of an audience's emotional experience in an art museum is a high degree of abstraction as shown in Fig. 1. This is also similar to the "The three dimensions" theory by Donald Arthur Norman in emotional design (Fig. 2). In this figure, the reflective level is also identified as a highly abstract emotion experience of introducing emotional content.

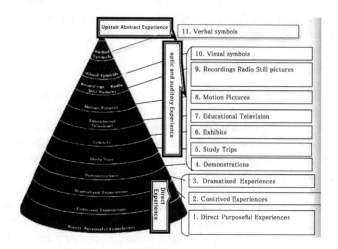

Fig. 1. The cone of experience

Fig. 2. The three dimensions

In the following, we apply these emotional experience theories to our analysis of the way the three art museums utilize principles of emotional design.

3   China National Film Museum

3.1   The Emotion Expression

The spatial emotional expression of the China National Film Museum is primarily based on creating fashion minimalism. The museum pursues this goal by using better proportions, concise structure, clean color, pure material and simple lines, within the constraints of space requirements to achieve a "less is more and better" design level. This simplicity is not an "easy simplification", but a reflection of a higher level of cultural heritage and an avant-garde way of life. Usually, this emotional expression embodies many dialectical relationships between exhibits and props, audience and space, space and lighting. It has the characteristics of sober and rational. It conveys the strong emotional and rational aesthetic in exhibition space design [6].

3.2   The Clear Theme of the Exhibition

The China National Film Museum takes the Centennial movie history as the core clue. Through use of a pentagram as the entrance, the museum incorporates images of the magical world using the development of movie history as the axis in a space over-lapped by movie and shadow. using. It shows the invention of the film to the birth and dissemination; it presents the arrangement and placement of different types of films, such as children's films, fine arts films, and educational films. It has a certain logic of story, so that the audience fulfill the need of recognition (Identity). At the same time, it conveys the history of film development in an unconscious way. The museum has a certain update cycle (Fig. 3) and regular movie updates, so it not only satisfies a single learning experience, but also satisfies visitors returning many times. Through the change and renewal of different exhibition halls, the audience's desire for learning is aroused again.

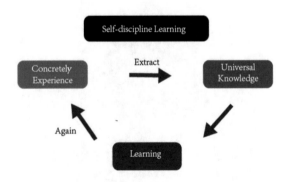

Fig. 3. Learning experience cycle

3.3   The Construction of Spatial Atmosphere

The construction of spatial atmosphere is divided into the types of form creation and atmosphere creation. The form creation includes physical form - the direction, size, gathering spaces, thresholds, form and composition - all of which have a certain psychological influence on people. These factors give the same psychological influence on virtual space. The space is classified by direction; it can be vertical space, horizontal space, or undirected space (Fig. 4).

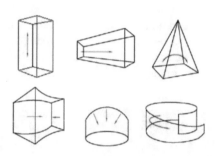

Fig. 4. Shape of space

The construction of virtual emotional space is closely related to psychological features. It may be invisible, but it is constructed by the real space and the relationship between space and audience. It seems two points can be inferred into a line; three points can be extrapolated into a surface through different entity designs, including enclosing, extrusion, folding, parcels, implants, penetration and so on. Audiences can feel the emotional hints of separation, crowding, warmth, excitement and so on.

The space scale is generally not the real size of building, but the relationship of real size and the feeling of the space. Size is an important element in the principle of space aesthetic, it is an essential factor in harmonious space creation. Aoki Yutaka believes that human perception is not only about five senses, but also about time perception and spatial perception. Spatial perception is the perception and consciousness of space size. The creation of atmosphere is related to the light environment, color molding, space scale and texture molding (Fig. 5).

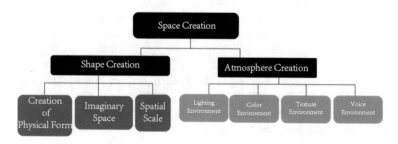

Fig. 5. The relationship diagram of space creation [8]

The creation of the atmosphere is closely related to the psychological feelings of the audience and can convey diverse emotions to audiences.

3.3.1   The Shape Creation

The China National Film Museum is mainly constructed by the creation of physical form. The central circular hall (Fig. 6) with its large circular rising space, makes people feel pretty and rich, which can promote the audience experience of the extent of the museum It also satisfies the audience's need for spatial predictability and identifiability.

The small exhibition area of the China National Film Museum is based on the normal size of human datum and not affected by space. The emotional function is biased towards the mental concentration of exhibits. The hall uses large scale space, because it belongs to a shared public space for many audiences. Therefore the space can deliver macro, power and rich emotions. The curved corridors and the setting railing give the audience the impression of a film showing.

Fig. 6. Center circular hall

3.3.2   Atmosphere Creation

The lighting (Fig. 7) adopted in the small exhibition area in the China National Film Museum is mainly environmental and contour light, while the exhibition light is focal. Environmental light cannot clearly be noticed in the space, and it easily provides a soft and comfortable feeling. The central hall is mainly environmental light, and does not use the contour lighting, so from a psychological point of view, the audience can easily get the psychological experience of being parcels, which creates a feeling of security.

The lighting of the small exhibition area (Fig. 8) is primarily direct and semi-indirect lighting. The direct lighting of the focal lights can hint at audiences to pay attention to the exhibits. The diffuse light of the ambient lighting in the hall makes the space atmosphere seem bigger, more permeable and lighter, which makes people feel comfortable when watching the movie.

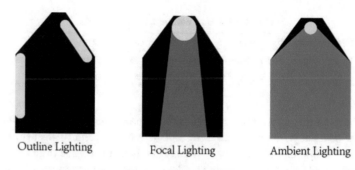

Outline Lighting   Focal Lighting   Ambient Lighting

Fig. 7. Lighting category

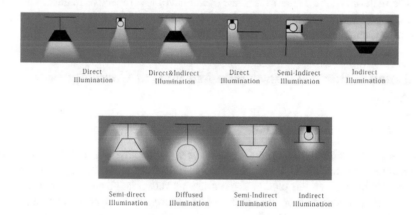

| Direct Illumination | Direct&Indirect Illumination | Direct Illumination | Semi-Indirect Illumination | Indirect Illumination |

| Semi-direct Illumination | Diffused Illumination | Semi-Indirect Illumination | Indirect Illumination |

Fig. 8. Lighting category

It is necessary for a designer to guide and awaken the audience's emotions, because it can stimulate the audience's desire, passion and other positive influences from different angles. In the atmosphere of creation and emotional transition, the status of color is especially important. Different perception of color is associated with different emotions. Red in China, means festivity, New Year, and enthusiasm. But in Europe, red represents negative emotions associated with the Middle Ages. Therefore, the colors used in museums in different regions will have different emotional affects. The culture, history, moral standards, social ethics and varying psychological feelings of the audience should be considered in the space color design process.

The China National Film Museum is unified with a black, white, and gray system. Blue, yellow and red are the auxiliary colors of the museum. For example, the color of the round hall is black as the bottom, and red and blue are the main colors projected. And red and blue have different emotional feelings (Table 1) [9].

Table 1. Table of emotional color in Asia

Color	Associative things	Psychological emotion
Red	Blood, Flame, Sun, War	Revolution, Anger, Enthusiasm, Brave
Blue	Sky, Sea, Galaxy, Dawn	Mystical, Recall, Soul, Heaven

The colors of the museum are related to revolution, war, and life. The theme of the museum is centenary history at its core and in modern century history, China has experienced the first and second world wars. Therefore, the choice of red and blue has the purpose of evoking the emotional affect of the commemorative thoughts and feelings of the audience.

The material of the museum is made of abrasive, low reflective materials and other materials which reduce the glare and interfere with the exhibits. It also creates a quiet, mysterious, soft and dim feeling for the whole space.

4 Japan Kahitsu – Kan, Kyoto Museum of Contemporary Art

4.1 The Emotional Expression

In a fast-paced industrial era, people yearn for soothing rhythms and warm emotions. To create a romantic mood, many designers extract from nature. Using visual form of the plant exhibition as the plot, designers use narrative language, the objective environment of understanding and the thoughts of concrete characters to create a romantic mood of non-rational, original ecology thus developing an idyllic exhibition space which emphasizes the design composition and the main colors evoked by Fig. 9.

Fig. 9. Kahitsu – Kan, Kyoto Museum of contemporary art

4.2   The Clear Theme of the Exhibition

Japan's Kahitsu – Kan, Kyoto Museum of Contemporary Art, is a free, quiet art museum which uses "one step in, the extension of the quiet space" as the core design discipline. This museum exhibits the collection of Kajikawa Yoshitomo, the owner of the museum, some famous photographers' works and other modern art works. It can be clearly understood from the official website that the museum wants to convey a "one by one" design principle, like the stage lighting, and the release of charm. The museum strongly wants to express a kind of emotional design by providing a positive psychological emotion to the audience.

4.3   The Construction of Space Atmosphere

4.3.1   The Shape Creation

The art museum is based on the combination of entity and virtual body. It uses the horizontal form in the closed space, and combines the line with the curve. The masculinity of the straight line is combined with the feminine sense of the curve, and then is combined with the shadow of the plant. Even if the full sunlight is sprinkled indoors, it will be transformed from surface to point. It appears that the room is not too bright or too exuberant. In the form of horizontal space, the sense of stability is used very well to convey the core spirit of the museum - to make audience feel liberated. The three-dimensional cross space form enlarges the space and increases the sense of hierarchy, while the mother and child space increases the flexibility of space. The space scale of the museum is primarily small and moderate. The main purpose is to create a sense of territoriality and privacy in every area, and works to achieve a comfortable rest space and to some extent achieve personal space and foster a sense of liberation. This also meets the demand that space provide people minimal irritation (Fig. 10).

Fig. 10. Cross space & the mother and child space

4.3.2   Atmosphere Creation

Japan's Kahitsu – Kan Museum in Kyoto presents a semi open space. Natural lighting is the main source of illumination in the pavilion, followed by concentrated lighting and environmental light when the sun is not enough. In order for visitors to achieve a sense of natural emancipation and comfort, the lightness of the museum is also in the middle and lower stages. This meets the basic lighting conditions and creates a sense of self-discipline in the forest cultivation and an ideal psychological state of harmony between man and nature. The color of the museum is mainly green light. Table 2 depicts the emotional expression of the museum. The green main color in the museum is intended to convey feelings of peace, hope, leisure, idealism, happiness and other positive emotions. The museum uses wood and other natural materials as its material. The wood is soft, giving people a warm, simple and genuine sense of nature and making them feel comfortable and stable. Natural materials are more easily accepted than artificial materials because human beings and nature have a sense of interdependence.

Table 2. Table of emotional color in Asia

Color	Associative things	Psychological emotion
Green	Steppe, Plain, Rye, Plant	Peace, Leisure, Ideal, Happiness
Yellow	Flammule, Lemmon	Wish, Develop, Upward

5   Denver Art Museum

5.1   The Emotional Expression

The Denver Art Museum creates the perception of individual creativity. In the influence of pop art, occasional art, and surrealism, we can extract a kind of "creative personality in the form of artistic conception". Negative aesthetics are standard practice, modeled against objective cognition. The pursuit of the complete liberation of thought, using

Fig. 11. Denver art museum

exaggeration and cartoon design techniques are applied in this museum (see Fig. 11). The museum displays the modern civilization of cynicism and the challenges traditional culture in the new world order though using this different effect.

5.2 The Clear Theme of the Exhibition

The theme of the Denver Art Museum is abstract painting, which transforms the painting line into a deconstructed architectural line. The design of the staircase at the Denver Museum of Art shows "skeleton abstraction." Multi-layer polylines appear in the shape of stairs. The expression of polylines and tooth lines in the abstract images has become the visual representation of the folding, subdivision, overlay and tilt of the staircase space. Abstract painting techniques and bold geometric abstraction make the entire museum into an exquisite display.

5.3 The Construction of Space Atmosphere

5.3.1 The Shape Creation

The art museum is based on the combination of entity and virtual body. The semi-open space uses the form of a polygon to express a lively, vivid spatial expression. The rising space creates wide space, while to some extent, limiting the natural light penetration to convey the abstract sense of the creation of mystery space.

5.3.2 Atmosphere Creation

The lighting of the museum is mainly environmental, and focal light is auxiliary. The practical focal light and the overall sense of ambient light results in exquisite focus on the painting. It emphasizes the effect of color through space by refraction on a single color wall. The lamplight of the museum is relatively weak, but it is strong in its emotional transmission. The color of the museum is red, but because of the different exhibition, the effect of the red is very different compared to the China National Film Museum. This demonstrates the different presentation of color emotional design in different cultures. The following chart is the difference between the psychological and emotional transmission of the same color in different geographical areas (Table 3).

Table 3. Table of emotional color in America

Color	Associative things	Psychological emotion
Red	Blood, Flame, Sun, Poppy	Anger, Barbarous, Snippy, Life

It can be seen that color emotion has diversity and territoriality. This content echoes Arthur Norman's reflective level emotional design theory. Emotional expression is often related to the audience's personal experience, field, and cultural background. The same color will often produce different psychological emotional experiences.

6   Conclusion

Based on the three case studies and literature searches, our results on the Emotional Design Process in Museums are shown in Fig. 12. We believe that can have a better understanding of how designers can improve emotional design for museum exhibition spaces throughout that entity's space by utilizing atmosphere spatial construction and exhibition design.

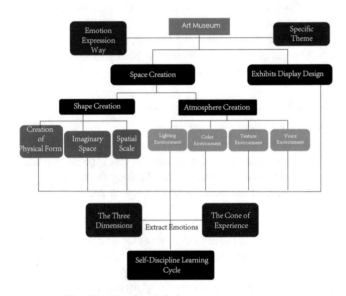

Fig. 12. Emotional design process on museum

We note there are quite a few similarities and differences in creating the exhibition space.

Each museum has its own independent color elements and strong emotional distribution of colors. These museums are more likely choose to use environmental lighting in exhibition spaces and focal lights on the display items. To be more specific, environmental light has a leading affect in uniting one atmosphere or an entire surrounding. It is fundamental for building up the sense of space and the whole emotional design.

The more use of rising space, the greater and larger the spatial sense that can be felt by viewers. This can also lower the sense of oppression due to extended periods of time staying in the museum. At the same time, people can fulfill their needs for security and identity.

However, each museum encourages different types of emotions. For instance, both the Denver Art Museum and the China National Film Museum use red as their primary hue, but the presentation of emotions is totally different. This is because of their culture differences and historical background.

Regarding the expressing of emotion by materials, Japan Kahitsu-kan Museum in Kyoto is more likely to use natural materials, which makes viewers not only feel closer and relaxed. The other two museums have generally chosen processed materials. They also have utilized scientific and technological effects throughout the entire exhibition center. This could be the result of these museums being built in modern times, with the increased availability of high-tech skills.

References

1. Hui Ma, D.: Research in Thematic Museum Entrance Designs Based on Emotion Expression. CNKI, pp. 11–24 (2016)
2. Ren Kangli, J.: The Representation with Abstract Painting and Deconstructionist Architecture for a Contemporary Visual Space. New Architecture (2012)
3. Bai Yue, D.: Emotion Design Research in the Exhibition Design. CNKI, pp. 9–40 (2012)
4. Nan Lin, D.: Research on Visual Psychology in Exhibition Space. CNKI, pp. 25–43 (2013)
5. Tomiko, Y.: Realization and abstraction operation seen in clear and logical discourse- On the Logical Cognitive Process of "Edge of Experience". Edgar DALE(2011)
6. Cao Renyu, M.: Environmental Art Design, pp. 46–62. Biejing Chemical Industry Press, Beijing(2017)
7. Yang Jing, D.: Emotional Design of Interior space. CNKI, pp. 25–43 (2008)
8. Scott Doorley, M., Scott Witthoft, M.: Make Space How to Set the Stage for Creative Collaboration, pp. 38–53. Wiley, Wiley-Blackwell, Hoboken (2012)
9. Li Jia, D.: The Color Emotion Research on the Museum Exhibition Design. CNKI, pp. 15–29 (2013)

Analysis of Changes in Personal Fashion Choices Owing to Situational Constraints

Miu Sato[1($\boxtimes$)], Takashi Sakamoto[2], and Toshikazu Kato[1]

[1] Chuo University, 1-13-27 Kasuga, Bunkyo-ku, Tokyo 112-8551, Japan
a13.43bk@g.chuo-u.ac.jp, t-kato@kc.chuo-u.ac.jp
[2] National Institute of Advanced Industrial Science and Technology, AIST,
Central-2, 1-1-1 Umezono, Tsukuba, Ibaraki 305-8568, Japan
takashi-sakamoto@aist.go.jp

Abstract. We generally choose clothes depending on the circumstance. Coordination recommendation is, therefore, made for not only the preference but also the situation. However, the selection of fashion for the situation varies from individual to individual, although in the past the coordination suitable for the situation was common. Therefore, in this research, we aim to clarify the change in fashion choice depending on the situation, whether to choose one's favorite fashion or to choose fashion according to situations.

Keywords: Fashion · Situation · Constraint · Entrainment · Selection
Change

1 Introduction

In recent years, the scale of fashion EC site markets, such as Amazon and ZOZO-TOWN, has expanded; thus, consumers can now easily purchase goods using the EC site without actually going to the stores. One can see many products at once on the EC site; however, because the number of products is enormous, it is often difficult for consumers to find the best product among them. Hence, fashion recommendations that match products to consumers' preferences are on the rise. By recommending products that match consumers' preferences, there is the advantage that they can find their favorite products. Additionally, when considering fashion coordination, one does not only consider preferences but also the place, purpose of the outfit, etc. Therefore, there are recommendations suitable for the occasion. Kawabata et al. [1] analyzed and classified the relationship between clothing lifestyle and preference clothes. As a result, it turned out that there are many people who care about time, place, occasion (TPO), and clothing norms. Therefore, when people decide fashion, they have fashion preferences and at the same time care about TPO and the situation. For example, consider the case of a university matriculation ceremony. Here, some people will adorn themselves to suit the occasion; others, however, will dress to suit their preference regardless of the occasion. Thus, even under the same circumstances, the choice of fashion may differ.

© Springer International Publishing AG, part of Springer Nature 2019
S. Fukuda (Ed.): AHFE 2018, AISC 774, pp. 250–256, 2019.
https://doi.org/10.1007/978-3-319-94944-4_28

Therefore, in this research, we aim to clarify changes in fashion choice depending on the circumstance, i.e., whether to choose based on one's preference or to synchronize with others.

2 Existing Research Related to Fashion Recommendations

The research on conventional fashion recommendation from two viewpoints has been compiled. The first is a recommendation that guesses the preference of the user. The second is a recommendation considering the TPO wearing fashion.

2.1 Recommendations Tailored to Consumer Preferences

Methods for predicting consumer preferences are generally divided into two types.

The first is collaborative filtering, which automates the "review" process whereby consumers review products based on their preferences. The recommendation target is decided through cooperation with consumers.

The second is content-based filtering, which involves choosing based on a condition, such as a consumer's favorite brand, color, and pattern. Hence, recommendations are based on the content of the search target.

Conventionally, these methods have fostered fashion recommendations tailored to consumer preferences.

2.2 Recommendation Considering TPO

Kataoka et al. recommended a coordination suitable for TPO. In this study, we used a style inherently related to TPO to learn the feature of the quantity of items used for coordination for each style. We then modeled the ratio of the number of styles for which each item is easy to use. The user inputs an image of a desired item to use for coordination along with a desired style. Thus, the one with the highest style ratio is recommended.

However, this recommendation also includes coordination suitable for TPO for those who wish to wear their favorite clothes. Therefore, based on the circumstances, we consider whether to prioritize fashion preferences or suitability for the occasion depending on the individual.

3 Differences in Fashion Choice

When generally think about the situation for the choice of clothing. Depending on the situation, one may sacrifice his/her preference in order to match the surroundings. This is a conformity behavior that emphasizes that it is not significantly different from the surrounding clothes in the situation.

Fujiwara [4] defined conformity behavior as "when asking for opinions, attitudes, and behaviors different from yourself from the surroundings, it is a mechanism that adapts to the surrounding opinions, attitudes, and behaviors while getting lost."

In addition, it says that it can be divided into "internal conformity" which accepts opinions and actions of others from the inside and can be divided into "superficial conformity" which seems to be synchronized on the surface but has a different inner surface.

Okabayashi [5] defined clothing choice as being able to express superficial conformity easily and briefly to others. Therefore, we conducted an experiment aiming to clarify what conformity behavior of clothes selection is for college students and what the background is. A questionnaire was conducted using multiple scales, with 127 clothing-related students and 144 college students. As a result, the proportion of those who chose conformity clothing and those who chose nonconformity clothing was almost half. In addition, it was found that clothing-related students are more likely to choose nonconformity clothes. Additionally, conformity behavior of clothes selection shows that there is a tendency that self-expression tends to be negative, and in nonconformity clothing selection behavior, it is thought that it reflects the strength of interest to others.

However, this research focuses only on conformity behavior of clothes selection among friends, and the situation for the choice of clothes is not considered. We believe that the conformity behavior of clothes selection differs not only from individual to individual but also in situations. In addition, because the presence or absence of conformity behavior of clothes selection varies depending on the subjects, we also target clothing-related students and other students in this experiment.

4   Experimental Method

In this research, we first selected the occasion for the outfit to be used for the experiment through a questionnaire. Thereafter, a questionnaire was conducted for each set situation regarding whether to select a favorite fashion or a fashion suitable for a situation. We also conducted a questionnaire survey on fashion consciousness of subjects. Then, changes in fashion choices were analyzed based on the obtained data.

4.1   Selection of Situations

We generally consider whether the situation is a formal place or a casual place to decide on the clothing. As a result, the degree of coordination restriction is determined, and we select an appropriate fashion. Therefore, we conducted a questionnaire to clarify the degree of constraint owing to situations. The subjects were five female college students. The situations used for the questionnaire are the following 32 situations assumed in the lives of female college students. Table 1 shows the situations used for the questionnaire.

For each situation, we conducted a questionnaire using the visual analogue scale (VAS), "whether the situation to consider coordination is formal or casual." A part of the questionnaire is shown in Fig. 1.

Table 1. Situations used for questionnaire.

No.	Situation
1	Shopping (alone)
2	Shopping (with friends)
3	Shopping (with lovers)
4	Shopping (with family)
5	Sightseeing (alone)
6	Sightseeing (with family)
7	Sightseeing (with lovers)
8	Sightseeing (with friends)
9	Theme park (with friends)
10	Theme park (with lover)
11	Drinking party (with friends)
12	Drinking party (including seniors and professors)
13	Company information session
14	Company information sessions (private clothes acceptable)
15	First date
16	Date (about the 5th time)
17	Beauty salon
18	Girls' night/Girls' day out
19	Singles party
20	Café
21	Museum
22	Afterparty of the coming of age ceremony
23	Friend's wedding ceremony
24	University entrance ceremony
25	Funeral
26	School
27	School (with presentation)
28	Live
29	Dining at home
30	Gym
31	Walk
32	Convenience store · Supermarket

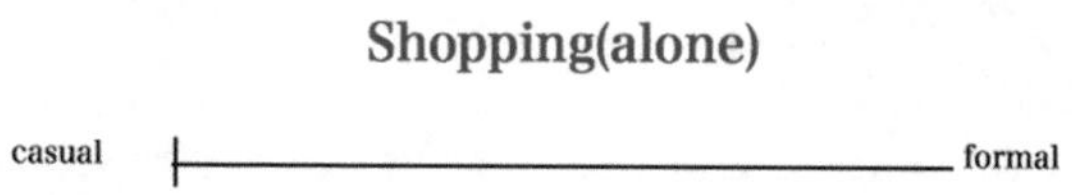

Shopping(alone)

casual |————————————————————— formal

Fig. 1. Part of the situation questionnaire.

5   Experimental Results

The degree of constraint of the situation obtained by VAS was tabulated with casual as 0 and formal as 100. Table 2 shows the average points of the five people for each situation.

Table 2. Average points of the five people for each situation.

Situation	Average score	
Girls' night/Girls' day out	0.0	**Casual**
Dining at home	4.2	
Convenience store · Supermarket	4.8	
Walk	7.2	
Live	7.6	
Gym	7.6	
Shopping (with family)	9.2	
Shopping (alone)	14.2	
Drinking party (with friends)	15.8	
Sightseeing (with friends)	17.6	
Theme park (with friends)	18.8	
Sightseeing (with family)	19.6	
Sightseeing (alone)	24.4	
Shopping (with friends)	30.0	
Beauty salon	31.2	
School	32.6	
Theme park (with lover)	33.4	
Sightseeing (with lovers)	39.2	
Date (about the 5th time)	40.4	
Shopping (with lovers)	50.0	
Café	57.4	
First date	58.0	
Museum	58.0	
Drinking party (including seniors and professors)	58.4	
Singles party	62.6	
School (with presentation)	75.2	
Company information sessions (private clothes acceptable)	85.4	
Friend's wedding ceremony	89.2	
Afterparty of the coming of age ceremony	89.4	
University entrance ceremony	95.2	
Company information session	95.8	
Funeral	96.6	**Formal**

From Table 2, an order based on the constraint degree of each situation was obtained. In a casual situation, it seems that there are many things for each person's behavior and everyday situations. Additionally, in formal situations, there are many formatted events, and it is considered that there are not many situations in daily life.

For each situation, we summarized the individual difference in the evaluation (the difference between the maximum value and the minimum value). Additionally, in order to select the situation to be used in the future experiment such that the degree of constraint is not biased, VAS 0 to 100 were converted to 11 steps. The results are shown in Table 3.

Table 3. Individual differences in evaluation values of situations and 11 levels of classification.

Levels	Situation	Individual difference	
0	Girls' night/Girls' day out	0	**Casual**
0	Dining at home	18	
1	Convenience store · Supermarket	13	
1	Walk	21	
1	Live	22	
1	Gym	18	
1	Shopping (with family)	23	
2	Shopping (alone)	26	
2	Drinking party (with friends)	41	
2	Sightseeing (with friends)	53	
2	Theme park (with friends)	35	
2	Sightseeing (with family)	37	
3	Sightseeing (alone)	60	
3	Shopping (with friends)	27	
3	Beauty salon	62	
4	School	72	
4	Theme park (with lover)	27	
4	Sightseeing (with lovers)	57	
4	Date (about the 5th time)	70	
6	Shopping (with lovers)	68	
6	Café	48	
6	First date	53	
6	Museum	44	
6	Drinking party (including seniors and professors)	40	
7	Singles party	87	
8	School (with presentation)	24	
9	Company information sessions (private clothes acceptable)	45	
10	Friend's wedding ceremony	31	
10	Afterparty of the coming of age ceremony	26	
11	University entrance ceremony	12	
11	Company information session	10	
11	Funeral	17	**Formal**

From Table 3, we can see that those with small individual differences are based on levels of 0, 1, and 11. In addition, it shows that centers have large differences among individuals. Therefore, it is considered that there is no difference in images among each subject for which the degree of constraint is clear. In a situation where individual differences are large, we believe that changes and differences in fashion choice become apparent for each individual.

For future experiments, we will use 11 situations, with a situation at each stage. The situations to be used for several stages is currently under consideration. Additionally, the results of the experiment will be announced on the day of AHFE conference.

6   Future Work

As a future prospect, we may believe it possible to improve the conventional fashion recommendation by paying attention to the differences in individual fashion choice for the situation considering fashion coordination. Moreover, by clarifying the difference in individual correspondence with constraints, it may be applicable to fields such as makeup and interior design.

Regarding Japanese fashion consciousness, Hirota et al. [6] compared Japanese and Italian female students. From this study, it follows that while Japanese chose simple and pretty fashion, Italians tended to wear something different from those who would like to express their individuality and emphasize their own personality. Therefore, it is thought that the idea of fashion selection differs depending on the cultural sphere. Therefore, in this research, we conducted experiments for Japanese female students; nonetheless, it is also necessary to make comparisons with other countries.

References

1. Kawabata, S., Kondoh, N., Kawamoto, E., Watanabe, S., Nakagawa, S.: Lifestyle, clothing lifestyle, and their relations to clothing preferences among female students. J. Text. Mach. Soc. Japan. 50(10), 285–293(1997)
2. Kamishima, T.: Algorithms of recommender systems. J. Jpn. Soc. Artif. Intell. (2016). http://www.kamishima.net/
3. Kataoka, K., Sudi, K., Kinebuchi, T.: Fashion coordinates recommendation system based on fashion styles. IPSJ SIG-CVIM Comput. Vis. Image Media. 205(25), 1–5(2017)
4. Fujihara, M.: Pilot Study on construction of conforming and social oriented behavior scale, its reliability and validity 1: from the recollection at the time of the 5th grader in an elementary school by the college students. In: Annual Report of the Faculty of Education, Bunkyo University, vol. 40, pp. 1–9 (2006)
5. Okabayashi, S.: Investigation on conformity behavior of clothes selection in university students. J. Bunka Gakuen Univ. Bunka Gakuen Junior Coll. 48, 13–22 (2017)
6. Hirota, K., Kotani, T., Ishii, T.: Fashion sense and life-style of European and Japanese female students. J. Kobe Yamate Coll. 52, 1–35(2009)

Kansei Engineering and Product Design

The Influence of the Mechanism of the Acoustic Sound Amplifier on Sound Transmission

Keitaro Sato[1]($\boxtimes$) and Wonseok Yang[2]

[1] Graduate School of Engineering and Science,
Shibaura Institute of Technology, 3-7-5, Toyosu, Koto-ku, Tokyo, Japan
cyl4218@shibaura-it.ac.jp
[2] Engineering and Design, Shibaura Institute of Technology,
3-9-14, Shibaura, Minato-ku, Tokyo, Japan
yang@shibaura-it.ac.jp

Abstract. Sound amplifiers have wide variety of shapes because of its high degree of freedom of shape. However, many existing products of sound amplifier focus on material and appearance. There are few things that emphasize the functionality of the shape conveying sound. There is the study to raise the radiation efficiency of sound by devising the shape of the acoustic horn. There are many studies of such communication aids. However, there are few studies on sound amplifier. So I clarified the influence on the sound by the shape of sound amplifier. Thus we set the acoustic measurement experiment of the sound amplifier based on the acoustic measurement of the speaker. Then we conducted experiments on 12 paper mocks and 15, 3D printer models. As the result, We were able to get tendencies of acoustic characteristics in sound amplifier. We think that this knowledge can be used for speaker design.

Keywords: Influence of shape on sound · Sound amplifier
Sound measurement · Acoustic characteristics

1 Introduction

Music was enjoyed in a limited place and time, but audio equipments have evolved to mobile along with technology development. In addition, as the distribution of music via the Internet has become widespread, audio players have also shifted to personal computers and smartphones. In particular, smartphones became the center of sound source playback by the improvement of the speaker function and the spread of the streaming distribution application advanced. In addition, the number of wireless music devices using Bluetooth and Wi-Fi has increased, and the style of reproducing music by wirelessly communicating music data from a smartphone or a personal computer has also increased. However, Bluetooth speakers have trouble with pairing and charging.

In the living space, we often listen to music directly from the speakers of the smartphone without using external speakers. However, the sound directly coming from the smartphone can't be heard depending on the surrounding environment, or the sound changes slightly and it sounds. In order to make the sound bigger and clearer, recently it has been attracting attention to sound amplifier that increase the sound coming out of

© Springer International Publishing AG, part of Springer Nature 2019
S. Fukuda (Ed.): AHFE 2018, AISC 774, pp. 259–268, 2019.
https://doi.org/10.1007/978-3-319-94944-4_29

the smartphone or making it easier to hear. This has the function of naturally improving the transmission of the sound emitted from the smartphone by giving directionality to the sound like a megaphone. There are also interior elements like ornaments. However, many existing products focus on material and appearance simply, and few things focus on the functionality of the shape conveying the sound. According to Yasuhiro Ouchi, he confirmed that the sound radiating efficiency is improved by devising the shape of the acoustic horn [1]. There are many researches on communication aids such as acoustic horn and acoustic reflector, but there are few studies on sound amplifier. Therefore, it is necessary to investigate the influence on the sound by the shape of the sound amplifier. Since sound amplifier have no electronic equipment in the internal mechanism, the contents are hollow. So sound amplifier have various forms by the degree of freedom of the shape. The purpose of this research is to clarify the influence of the shape of the sound amplifier on the sound.

2   Research Method on Acoustic Measurement

2.1   Current Status of Acoustic Measurement

According to Mr. Kenyoshiro Hino, the amount of the acoustic measurement device has dropped from tens of millions of yen to less than 500,000 yen even if it includes a measuring microphone. It is much cheaper to measure with PC's audio interface and free software. In the past, it was installed as an expensive measurement system in a standard environment such as an anechoic chamber. Now that it is inexpensive and compact, it can be said that opportunities to measure easily even in the space that actually produces sound has increased. It is easy to measure the sound at the site where sound is emitted, to use tuning and maintenance, or to measure for confirmation when the system developer listens to the actual sound [2].

2.2   Items of Acoustic Measurement

Frequency Characteristic
According to Mr. Kenyoshiro Hino, what shows how the sound pressure level changes when the frequency is changed while the input power to the speaker is kept constant is called the frequency characteristic. It is represented by a graph with the frequency [Hz] on the horizontal axis and the scale of the sound pressure level [dB] on the vertical axis [2].

Directivity Characteristic
According to Chisato Mohri, a microphone measuring the sound pressure level is placed on the extension line of the front axis of the speaker, and the straight line is set as the reference axis. Directivity characteristic that expresses how the sound pressure level varies depending on the change in angle and frequency of the speaker's front axis with respect to the reference axis while the input power to the speaker is kept constant [3].

2.3 Sound for Acoustic Measurement

Sweep Wave

According to Mr. Kenyoshiro Hino, a signal changed from a low frequency to a high frequency at a constant speed is called a sweep signal. In the measurement of frequency characteristics, it is desirable to use a sine wave signal (sweep wave) whose frequency is continuously changed. This is because there are many peaks and valleys of frequency characteristics due to the inherent resonance of the speaker's vibration system [2].

Pink Noise

According to ARI Co., Ltd., pink noise is all-band noise with equal energy per octave. Since the energy for each band called an octave band becomes uniform, the magnitude of the sound is the same in any octave band. Therefore, pink noise is often used for acoustic adjustment and measurement [4].

3 Peliminary Experiment Using Paper Crafts

3.1 Experiment Content and Purpose

There are many types of sound amplifiers, but we chose a rectangular tube for the shape to be experimented this time. And we made a hypothesis that the proportion of the pipe influences the sound. We created paper mocks of sound amplifier for the experiment. Paper craft paper was used as the material. We selected the iPhone 7 as smartphone of sound source which was the latest in the research start. We made a cube type sound amplifier based on the width 15 mm of the internal speaker of the iPhone 7. And we made 4 items each of width, height, and depth increased by 15 mm with the cube as a reference. The purpose of this experiment is to clarify the influence on the acoustics due to the shape of the tube in the sound amplifier (Fig. 1).

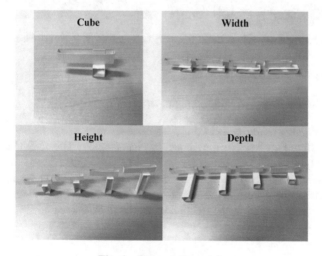

Fig. 1. Paper craft models

3.2   Plan of Experiment

Where the acoustic experiment is conducted an anechoic chamber is ideal. In recent years, the acoustic measurement system has become inexpensive and miniaturized, so that the developer experiment in the echo chamber [2].

This time we conducted an experiment in a room close to the living space where the sound amplifier is actually used. Frequency characteristic software, condenser microphone and digital sound level meter were used for the acoustic measurement system.

In order to stabilize the position to be measured, the iPhone 7 was floated using an arm type smartphone stand and the sound amplifier was installed. The sound data was reproduced in that state.

Experiment Conditions
In order to investigate the influence of the shape of the sound amplifier on the frequency characteristic, the sweep wave was reproduced with the smartphone and the microphone separated 1 m and the frequency characteristics were measured. The sweep wave has the time of 150 s and the frequency band of 20–20,000 Hz.

In order to investigate the influence of the shape of the sound amplifier on the distance attenuation, one sound pressure meter was placed 1 m away from the smartphone and 2 m apart from the smartphone to reproduce pink noise suitable for sound pressure measurement. The frequency band of pink noise is 20–20,000 Hz (Fig. 2).

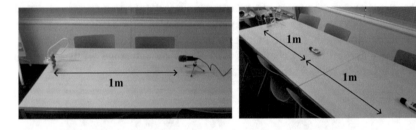

Fig. 2. Experiment conditions (Left: frequency characteristic, Right: distance attenuation)

3.3   Result

The comparison standard of width, height, depth is a cube.

Results of Frequency Characteristic
The blue graph in Fig. 3 shows the frequency characteristic. The vertical axis is db and the horizontal axis is Hz. In this time, the band of 20–20000 Hz was observed.

When the width was 30 mm, the first half of the midrange became smaller. As the height increased, the disturbed the waveforms of the low and mid-range. The peak of the sound pressure moved to the treble area as the depth increased (Fig. 3).

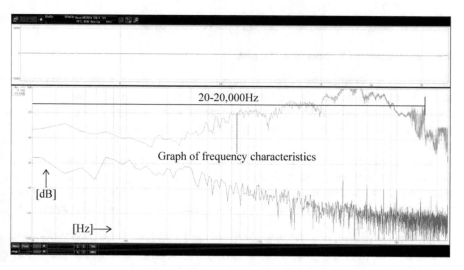

Fig. 3. Experiment result: frequency characteristic of cube

Result of Distance Attenuation

Figure 4 shows the value of the distance attenuation obtained by subtracting 2 m point sound pressure from the value of the sound pressure at 1 m and 2 m point and 1 m point sound pressure.

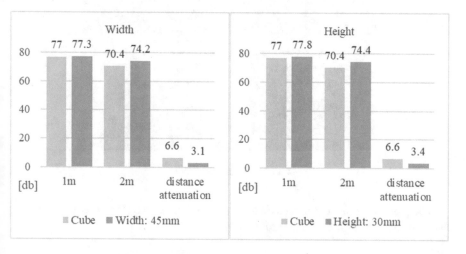

Fig. 4. Experiment result: distance attenuation

As the width increased, the sound pressure at the 2 m position increased and the distance attenuation fell. As the height increased, the sound pressure at the 2 m position increased and the damping of the distance attenuation declined. The depth did not affect much (Fig. 4).

4   Main Experiment Using 3D Printer Models

4.1   Experiment Content and Purpose

Experiments with directivity characteristic could not be performed due to the problem that the arm stand cannot change the angle exactly in the experiment using the paper craft. This problem was solved by making a model with an internal shape similar to a paper craft which stands independently without relying on an arm stand by a 3D printer.

In the model using the paper craft, the stand part and the pipe part were integrated. However, this will waste materials and time because it will print the stand part each time. Therefore, we decided to improve the efficiency of production and printing time by separating the parts with the base stand and the passage part changing the dimensions.

Five pieces of each of which the width, the height, and the depth were increased by 15 mm each on the basis of the 15 mm cubic sound amplifier was created respectively (Fig. 5).

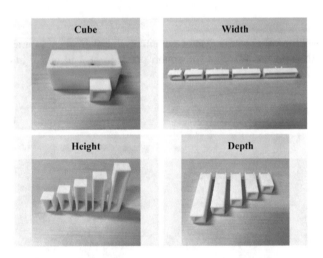

Fig. 5. 3D printed models

The purpose of this experiment is to clarify the influence on the acoustics due to the shape of the tube in the sound amplifier.

4.2   Experiment Plan

Experiment Conditions
Distance attenuation and frequency characteristic were performed under the same conditions as in the previous experiment.

All experiments were done in duplicate.

Directivity characteristics were measured using pink noise with the sound level meter and model being rotated by 30 degrees about the speaker surface at a distance of 1 m. The frequency band of pink noise is 20–20,000 Hz [2] (Fig. 6).

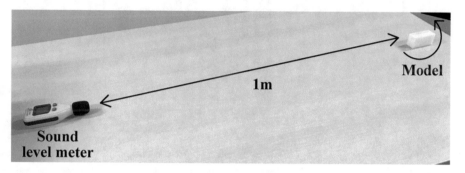

Fig. 6. Experiment conditions (directivity)

4.3   Result

Results of Frequency Characteristic
The result of the frequency characteristic raises or disturbs the waveforms of the low region and the mid region depending on the width, raises the waveform of the high region, and affects the disturbance of the waveform of the super high region. Increase the waveforms of the low, mid and high regions depending on the height, and influence such as to disturb or raise the waveform of the super high region. The waveform of the low range rises and falls due to the depth, raising the waveform of the middle range, lowering the waveform of the latter half of the middle range, and raising the waveform of the high range and the super high range.

Result of Distance Attenuation
Figure 7 shows the value of distance attenuation obtained by subtracting 2 m point sound pressure from 1 m and 2 m point sound pressure and 1 m point sound pressure.
As a result of the distance attenuation, the sound pressure increases as the width increases, and the distance attenuation tends to increase. As the height increases, the sound pressure decreases and the distance attenuation tends to decrease. As the depth increases, the sound pressure decreases and the distance attenuation tends to decrease (Fig. 7).

Result of Directivity Characteristic
Figure 8 shows the sound pressure when the angle is directed to 0°, 30°, 60°, 90°. As the result of the directivity characteristic, as the width increases, the sound pressure at the front becomes larger and the sound pressure at the side becomes smaller, so that the directivity characteristic tend to become larger. As the height increases, the sound pressure at the front decreases and the sound pressure at the side increases, so that the directivity characteristic tend to become smaller. As the depth increases, the sound pressure at the front decreases and the sound pressure at the side decreases, so that the directivity characteristic tend to increase (Fig. 8).

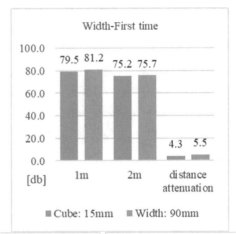

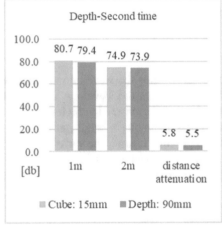

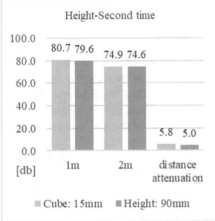

Fig. 7. Experiment result: distance attenuation

4.4 Analysis

The results of model of paper and 3D printer showed different trends. This is considered to be related to the difference in material between paper and plastic, and the difference in model accuracy. Therefore, the results of 3D printers with high model precision are considered to be effective this time.

Analysis of Distance Attenuation
As the width increases, the sound pressure increased and the distance attenuation tended to increase. It is thought that this is because the internal resistance is reduced by increasing the width.

As the height increases, the sound pressure decreased and the distance attenuation tended to decrease. It is thought that this is because the sound is dispersed upward as the height increases.

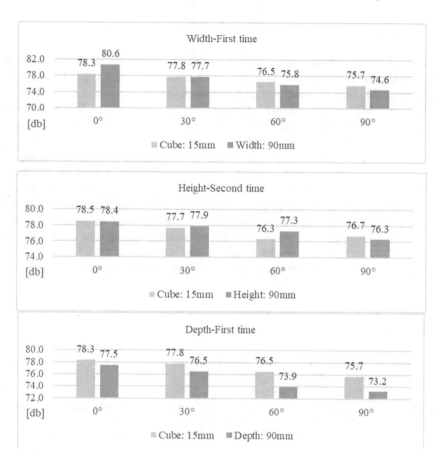

Fig. 8. Experiment results: directional characteristic

As the depth increases, the sound pressure decreased and the distance attenuation tended to decrease. It is thought that this is because the distance to reduce the momentum of the sound is increased internally as the depth increases.

Analysis of Directivity Characteristic
As the width increases, the greater the front sound pressure and the smaller the side sound pressure. So the directivity characteristic tended to increase. It is thought that this is because the larger the width, the closer the noise meter blocks the sound.

As the height increases, the sound pressure in the front decreases and the sound pressure on the side increases. So the directivity characteristic tend to decrease. This is because sounds are more likely to be dispersed to the left and right.

As the depth increases, the front sound pressure decreased and the side sound pressure decreased. So the directivity characteristic tended to increase. This is because the distance that reduces the momentum of the sound increases as the depth increases, and it is considered to be because the sound is released at a position farther from the sound level meter.

5 Conclusion

During the experiment the sound amplifier itself was vibrating up and down. Since the angle at which the smartphone is placed is perpendicular to the sound amplifier, the vibration of the built-in speaker is transmitted vertically. Also the direction in which the shape of the tube extends is also perpendicular to the built-in speaker. The internal shape unique to this sound amplifier and the elements of each dimension combine to give different influences to the sound.

In this research, we chose the subject to be investigated as the square pipe, and made the hypothesis that the proportion of the pipe influences the sound. For verification, we designed a cube type sound amplifier based on the width of the built-in speaker of the smartphone. Based on this model, the models with increased width, height, and depth at equal intervals were created and acoustic measurement experiments were performed. And we clarified the acoustic characteristic of each dimension respectively. If we use the results of this research for sound amplifier and speaker design, we can support predicting sound by geometry and choosing a shape to create the desired sound.

References

1. Ouchi, Y., Yamazaki, Y.: Acoustic characteristics of acoustic horn with annular opening. J. Jpn. Acoust. Soc. **73**(1), 5–11 (2017)
2. Kenyoshiro, H.: Audio and measurement. JAS J. **53**(3) (2013)
3. Mori, C., Kondo, A., Yagi, M.: Development of student experiment teaching materials on acoustic engineering. Bull. Kagawa Tech. Coll. **3**, 131–137 (2012)
4. ARI Co., Ltd.: White Noise, Pink Noise. http://www.ari-web.com/service/kw/sound/noise.htm
5. Okiboshi, T., Oikawa, Y., Ohuchi, Y., Yamasaki, Y.: Sound collection hearing aid system using curved reflector plate. In: Conference papers of the Japan Acoustical Society Research Presentation (CD-ROM), vol. 2013, issue: Fall, Page: ROMBUNNO.1-2-2 (2013)
6. Mitsuishi, M., Furihi, K.: A Basic study on horn speaker for automotive multi-way system. Technical report of IEICE, EA, Appl. Acoust. **111**(270), 55–60 (2011)
7. Yoshida, T., Yokoi, M.: Introduction of acoustic experiments for students in engineering. Osaka Sangyo University papers. Natural Sciences ed. J. Osaka Sangyo Univ. **128**, 1–19 (2017)
8. JIS standard: speaker for sound system. http://kikakurui.com/c5/C5532-2014-01.html
9. Sound fan: Mirai speaker. https://soundfun.co.jp/
10. Cloud Funding Makuake: No power speaker "Listen". https://www.makuake.com/project/listen/
11. Basic knowledge of noise. http://www.city.gifu.lg.jp/secure/6589/soukiso.pdf
12. Shino, S.: Psychology of the sound world. Nakanishiya Publishing, Tokyo, June 2014
13. iPhone Mania: Volume Comparison of iPhone built-in speakers of the past ~ From first generation to iPhone 6. https://iphone-mania.jp/news-60547/
14. iPod LOVE: Built-in speaker of "iPhone 8" is slightly higher volume than "iPhone 7". https://ipod.item-get.com/2017/09/iphone_8iphone_7_1.php

Comparison of Relaxation Effect from Watching Images Between 2D and Head-Mounted Displays

Yoshiki Koinuma$^{(\boxtimes)}$ and Michiko Ohkura

Shibaura Institute of Technology,
3-7-5, Toyosu, Koto-ku, Tokyo 135-8548, Japan
ma17041@shibaura-it.ac.jp,
ohkura@sic.shibaura-it.ac.jp

Abstract. Stress is a major problem in modern society. Therefore, much research has focused on its reduction. Previous research has reported that people are relaxed by watching 2D video of natural environments. In recent years the virtual reality (VR) market has begun to grow, and the amount of such types of relaxation continues to increase. However, research has failed to clarify the difference of relaxation effect obtained by watching relaxing video between a 2D display and a HMD. In this research, we presented relaxing video and compared the relaxation effects of two types of displays. We used a questionnaire (POMS2), bio-signals (ECG), and salivary α-amylase tests to evaluate the relaxation effects.

Keywords: Virtual reality · Relaxation · 2D display · Head mounted display
Kansei engineering · ECG · Salivary α-amylase test

1 Introduction

Since stress is a major problem in modern society, much research has addressed how to relieve it, including sports and forest bathing [1]. Previous research reported that 2D nature videos also relax people [2]. The virtual reality (VR) market is also expanding. The availability of inexpensive, head-mounted displays (HMDs) has increased, and their future potential looks promising [3]. Thus, various VR images and applications are being proposed and provided [4], and the amount designed for relaxation is growing. However, the difference of relaxation effects between 2D display and HMD has not been clarified.

In this research, we presented relaxing video and compared their relaxation effects by two different types of displays. We used a questionnaire (POMS2), bio-signals (ECG), and a salivary α-amylase test to evaluate the relaxation effects.

S. Fukuda (Ed.): AHFE 2018, AISC 774, pp. 269–276, 2019.
https://doi.org/10.1007/978-3-319-94944-4_30

2   Evaluation Experiment

2.1   System and Content

We created a computer graphics video of a forest to relax our participants. The video included bird songs and the sounds of flowing rivers, presented to participants by headphones. A screenshot of our video is shown in Fig. 1.

Fig. 1. Screenshot of relaxing video

We compared two display methods. One presented the video on a liquid crystal display (LCD) and the other presented it on a HMD. The outline of our experimental systems is shown in Fig. 2. We used EV2450 (23.8 in.) for the LCD and Oculus rift CV 1 for the HMD. Participants can look around by moving their head while they are watching video by HMD. While watching the video on LCD, they can manipulate the viewpoint with direction keys instead of moving their head.

We performed a questionnaire and a salivary α-amylase test three times: after entering the room, after the stress task, and after watching the video.

2.2   Experimental Method

The following is our experiment's procedure:

(i) The experiment was explained to the participants who also give their informed consent.

(ii) The experimenter placed the measuring instruments on the participants.

(iii) The experimenter recorded their ECGs and set the resting state section (1 m).

(iv) The participants performed a stress task: a mental calculation task (5 m).

(v) Participants watched the 8-min video. Then we repeated (iv) and (v) twice.

(vi) The experimenter stopped recording the ECGs
(vii) The experimenter removed the measuring instruments.

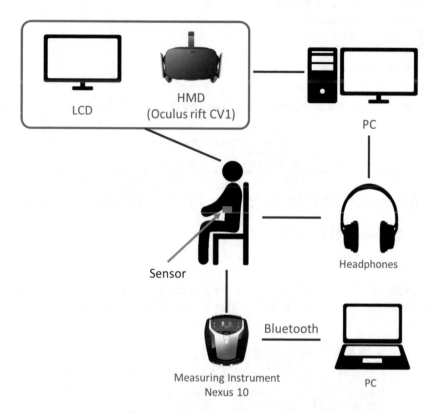

Fig. 2. Outline of experimental system

We measured the ECGs of the participants during the experiment.

After (iii), (iv), and (v), we did a questionnaire and a saliva α-amylase test. The order of the display method was counter-balanced.

2.3 Evaluation Method

We use the Profile of Mood States, 2nd ed., (POMS 2) [5] for our questionnaire. POMS 2 assesses the mood states of individuals and quickly assesses transient, fluctuating feelings, and enduring affect states [6]. This tool has been widely used in previous researches on stress relief and relaxation [7]. POMS 2 has both full-length and short versions. We used the short version (POMS 2, ed., adult short version) because it provides answers relatively quickly. POMS 2 yields seven scale scores and total mood disturbance (TMD).

The following are the seven scales:

- AH: anger-hostility
- CB: confusion-bewilderment
- DD: depression-dejection
- FI: fatigue-inertia
- TA: tension-anxiety
- VA: vigor-activity
- F: friendliness

The physiological indices for our analysis are shown in Table 1.

Table 1. Physiological indices

Biological signals	Physiological indices	Relationship to feelings of relaxation
ECG	LF/HF	Decreases when feeling relaxed
Saliva α-amylase test	Amylase value	Decreases when feeling relaxed

3   Experimental Results

3.1   Outline

We performed an experiment with six male students in their twenties. The experiment took about one hour per participant. Figure 3 shows an experimental scene.

Fig. 3. Experimental scene

3.2   Questionnaire Results

We performed five questionnaires:

- after resting state section,
- after stress task (2D or HMD),
- after watching the video (2D or HMD).

Figure 4 shows the averages of each scale and the TDM.
We analyzed the questionnaire results and found no statistical differences.

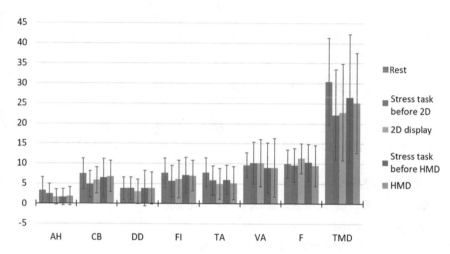

Fig. 4. Average of each scale and TDM

3.3   Saliva α-amylase Test Results

We performed a saliva α-amylase test five times (after the resting state section, after the stress task (2D or HMD), and after watching the video (2D or HMD)). Figure 5 shows the saliva α-amylase values for each participant and their averages. We performed a paired t-test on the following pairs:

- After stress task (2D) and after watching the video on a 2D display;
- After stress task (HMD) and after watching the video on HMD;
- After watching the video on a 2D display and after watching it on HMD.

We found the following significant differences:

- After watching the video on HMD < after stress task (HMD) ($p < 0.05$)
- After watching the video on HMD < after watching it on 2D display ($p < 0.01$).

This result shows that the saliva α-amylase value was significantly lower after watching the video on HMD than on a 2D display and after a stress task (HMD).

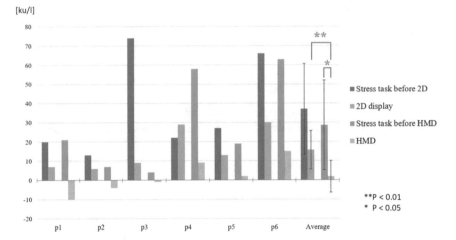

Fig. 5. Saliva α-amylase values for each participant and their averages

Participants seemed to feel more relaxed when watching the relaxing video on HMD than on a 2D display.

3.4   ECG Results

For the ECG analysis, we used the following sections:

- resting state section
- watching the video section (2D display or HMD).

We considered the results during the resting state as the basis and calculated the differences between the resting state section and watching the video section for analysis. Figure 6 shows the LF/HF for each participant and their averages. The paired t-test results show significant differences between the two display methods:

after watching video on HMD < after watching it on 2D display ($p < 0.01$).

LF/HF was significantly lower while watching the video on a HMD than on a 2D display. Therefore, participants seemed more relaxed when watching the relaxing video on HMD than on a 2D display; this result agrees with the saliva α-amylase test result.

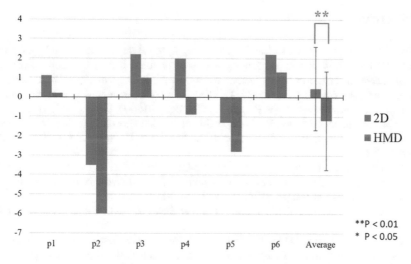

Fig. 6. LF/HF for all participants and their averages

4 Discussion

In our experimental results, participants seemed to feel more relaxed when watching the relaxation video on HMD than on a 2D display. Therefore, we obtained a higher relaxation effect by watching video on a HMD than on a 2D display.

5 Conclusion

We experimentally clarified the differences of relaxation effects while watching relaxing video between a 2D display and a head-mounted device (HMD) and obtained the following conclusions:

- The saliva α-amylase value was significantly lower after watching the video on HMD than watching it on a 2D display.
- LF/HF was significantly lower while watching the relaxing video on HMD than on a 2D display.

From these results, when watching relaxing videos, we found that watching on HMD provides a higher relaxation effect than watching on a 2D display.

Acknowledgements. We thank the students at Shibaura Institute of Technology who participated in our experiment.

References

1. Park, B.J., Tsunetsugu, Y., Kasetani, T., Kagawa, T., Miyazaki, Y.: The physiological effects of Shinrin-yoku (taking in the forest atmosphere or forest bathing): evidence from field experiments in 24 forests across Japan. Environ. Health Prev. Med. **15**(1), 18–26 (2010)
2. Kawakubo, A., Yoshioka, A., Oguchi, T.: Effects of motion pictures and sounds of the natural environment on stress reduction. Rikkyo Psychol. Res. **2015**(57), 11–19 (2015). (In Japanese)
3. Worldwide Shipments of AR-VR Headsets Maintain Solid Growth Trajectory in the Second Quarter, According to IDC, 5 September 2017. https://www.idc.com/getdoc.jsp?containerId=prUS43021317
4. After mixed year, mobile AR to drive \$108 billion VR/AR market by 2021, Digi-Capital, January 2017. https://www.digi-capital.com/news/2017/01/after-mixed-year-mobile-ar-to-drive-108-billion-vrar-market-by-2021/#.WpZHd-jFKM8
5. Heuchert, J.P., McNair, D.M.: Profile of Mood States 2, Multi-Health Systems (MHS), Toronto (2012)
6. Heuchert, J.P., McNair, D.M.: POMS-2 Manual, Multi-Health Systems (MHS) (2012)
7. Tsujiura, Y., Toyoda, K.: A basic examination of psychological and physical responses to a video of forests –the possibilities of therapy using a video of forests–. Jpn. J. Nurs. Art Sci. **12**(2), 23–32 (2013). (in Japanese)

Influence of Fixation Point Movement on Visually Induced Motion Sickness Suppression Effect

Naoki Miura[1]([⊠]), Ujike Hiroyasu[2], and Michiko Ohkura[1]

[1] Shibaura Institute of Technology,
3-7-5, Toyosu, Koto-ku, Tokyo 135-8548, Japan
al14087@shibaura-it.ac.jp,
ohkura@sic.shibaura-it.ac.jp
[2] National Institute of Advanced Industrial Science and Technology (AIST),
AIST Tsukuba Central 6, 1-1-1 Higashi, Tsukuba 305-8566, Japan
h.ujike@aist.go.jp

Abstract. Previous studies reported that rendering the fixation point is effective for suppressing visually induced motion sickness (VIMS). This study investigates why the fixation point has a VIMS suppression effect. Gazing-specific-place and gazing-specific-object are two properties of static-fixation points. To clarify which property is responsible for the VIMS suppression effect of fixation points, we employ a moving-fixation point and employ 3D CG content experienced by participants through a head-mounted display (HMD). This study compares the following: no-fixation-point, static-fixation-point, and reciprocating-fixation-point conditions. Each condition is evaluated by a simulator sickness questionnaire (SSQ) and subjective evaluations. Both the SSQ and subjective evaluation results show the VIMS suppression effect and compare it under the reciprocating-fixation-point condition and with the no-fixation-point condition. The gazing-specific-object property might cause a VIMS suppression effect.

Keywords: Visual induced motion sickness · Fixation point
Head-mounted display · Kansei engineering · Cognitive studies

1 Introduction

Motion sickness is a common by-product of exposure to optical depictions of inertial motion. This phenomenon, called visually induced motion sickness (VIMS), has been reported in a variety of virtual environments. Because of the remarkably rapid progress of virtual reality technologies, their applications are becoming more familiar in our daily life. Therefore, VIMS should be scrutinized by cognitive studies. Even though a previous study reported that rendering the fixation point suppresses VIMS [1], the reason remains unclarified.

Gazing-specific-place and gazing-specific-object are two properties of static-fixation points. To clarify which is responsible for the VIMS suppression effect of fixation points, we employ moving-fixation points. Since moving-fixation points have

© Springer International Publishing AG, part of Springer Nature 2019
S. Fukuda (Ed.): AHFE 2018, AISC 774, pp. 277–288, 2019.
https://doi.org/10.1007/978-3-319-94944-4_31

the latter property but not the former property, they can be used to separate the effects of the two properties. If the VIMS suppression effect of fixation points is due to the gazing-specific place property, the suppression effect should disappear when the fixation point is moving.

2   Method

We employed 3D CG content that was experienced by participants through a head-mounted display (HMD). We prepared two types of fixation-point movements on the VIMS suppression effect: (a) sine wave reciprocating and (b) square wave reciprocating. For each movement type, we compared the following conditions: no-fixation-point, static-fixation-point, and reciprocating-fixation-point. Each condition was evaluated by a simulator sickness questionnaire (SSQ [2]) and subjective evaluations on sickness. SSQs were conducted before and after the experience. We conducted subjective evaluations during the content experience.

3   Preliminary Experiment

We presented the 3D CG content to the participants, instructed them to gaze at the fixation point if it was present, and to follow it with their eye if it moved. We prepared two fixation-point movements and conducted Preliminary Experiments 1 and 2 for each movement type. In experiment 1, the fixation point's movement type was sine wave reciprocating (Fig. 1). In experiment 2, it was rectangular wave reciprocating (Fig. 2). The frequency of these reciprocating movements was set to 0.4 Hz, and the amplitude was set to the four conditions shown in Table 1.

Fig. 1. Sine wave reciprocating

3.1   Evaluation Method

Our evaluation method was comprised of the following two items:

(1) SSQ: SSQ was conducted before and after the content experience. Values were obtained by subtracting the value before the content experience from the value after it.
(2) Subjective evaluation of sickness: During the content experience, subjective evaluations on five grades from 0 to 4 were conducted every 30 s, 15 s after starting the content.

Fig. 2. Rectangular wave reciprocating

Table 1. Experimental fixation-point conditions

Conditions	Fixation-point existence	Amplitude
1	No	–
2	Yes	0°
3	Yes	3°
4	Yes	12°

3.2 Equipment

We used the following equipment in our experiment:

Head-mounted display: The 3D CG content was presented by FOVE 0 (FOVE, Inc.) which is an eye-tracking type of HMD (Fig. 3). By using FOVE, we can observe the participant's eyes (Fig. 4).

Response box: For answering subjective evaluations on sickness, a response box (Fig. 5) was used. In the experiment, participants responded by pushing its button while experiencing the content with a HMD.

Figure 6 shows an experiment scene.

Fig. 3. FOVE 0

3.3 3D CG Content

The content used in the experiment was a six-minute 3D CG image created by Unity, a stationary image for the first minute and a moving image in the landscape for the subsequent five minutes. Figure 7 shows an example of the content.

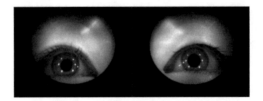

Fig. 4. Image of participant eyes

Fig. 5. Response box

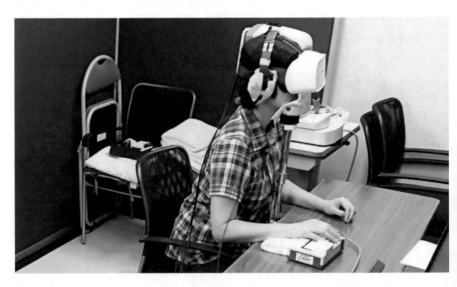

Fig. 6. Experimental scene

Fig. 7. Example of content screen (with fixation point)

3.4 Experiment Participants

Preliminary Experiment 1 had 30 participants (16 males, 14 females, ages 20–47). Preliminary Experiment 2 had 32 participants (16 males, 16 females, ages 20–46). However, the participants who experienced physical problems during the experiment were excluded from the analysis. As a result, in both experiments, we analyzed 18 participants.

3.5 Experiment Procedure

   I. Participant answers the SSQ before the content experience.
  II. Experimenter puts the HMD on the participant.
 III. Participant experiences the 3D CG content.
 IV. Experimenter removes the HMD.
  V. Participant answers the SSQ.
 VI. Participant takes a 30-min break after answering the SSQ.
VII. (I) to (VI) are repeated for the remaining conditions.

The preliminary experimental conditions of the fixation point in Table 1 were counterbalanced for condition order.

3.6 SSQ Results

Figures 8 and 9 show the average values of the SSQ Total Scores of the participants. To compare the results between those who became very sick and those who did not, the participants were classified as high and low score groups based on SSQ Total Scores under the no-fixation-point condition. In both experiments, there were five participants in the high score group and 13 in the low score group. The following is the multiple comparison results:

- No-fixation-point < Sine wave reciprocating (12°) (p < 0.05)

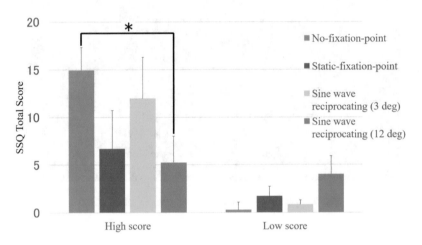

Fig. 8. Preliminary Experiment 1: average SSQ Total Score

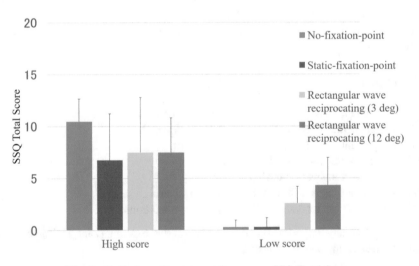

Fig. 9. Preliminary Experiment 2: average SSQ Total Score

3.7   Subjective Evaluation on Sickness Results

Figures 10 and 11 show the average values of the subjective evaluations on sickness of the high score group. The following are the multiple comparison results:

- Sine wave reciprocating (12°) > three other conditions ($p < 0.05$)
- No-fixation-point < static fixation point ($p < 0.05$)
- No-fixation-point < rectangular wave reciprocating (12°) ($p < 0.05$).

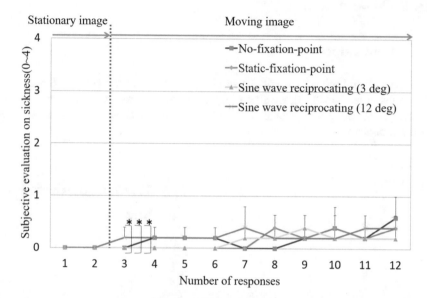

Fig. 10. Preliminary Experiment 1: average values of subjective evaluation of sickness of high score group

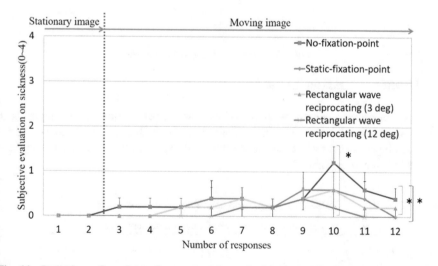

Fig. 11. Preliminary Experiment 2: average values of subjective evaluation of sickness of high score group

3.8   Problems of Preliminary Experiment

Most participants did not feel VIMS after they experienced the content. Even under the no-fixation-point condition, the proportion of participants with low SSQ Total Scores (less than 5) was 72%.

4   Content Improvement

Based on our preliminary experiment results, we improved the content. Since image rotation increases VIMS [3], we alternately added pitch or roll rotation (amplitude: 8°, frequency: 0.2 Hz) to the content per 30 s. Figure 12 shows the timeline of the improved content.

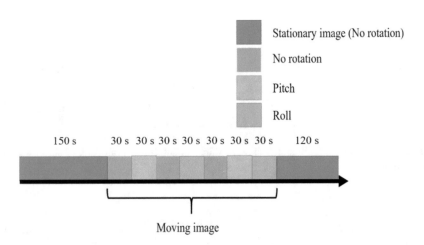

Fig. 12. Timeline of improved content

5   Evaluation Experiment

Next, we experimented with the improved content. The evaluation method, equipment, and procedure were identical as in the preliminary experiments. The frequency of the reciprocating movements was set to 0.4 Hz, and the amplitude was set to the three conditions shown in Table 2.

Table 2. Experimental conditions of fixation point

Conditions	Fixation-point existence	Amplitude	Movement type
1	No	–	–
2	Yes	0°	–
3	Yes	12°	Experiment 1: sine wave
			Experiment 2: rectangular wave

5.1   Experiment Participants

Seventeen people (7 males, 10 females, ages 21–50) participated in Experiments 1 and 2. After removing the participants who became very sick, eight participants remained in Experiment 1 and 13 in Experiment 2.

5.2   SSQ Results

Figures 13 and 14 show the average values of the SSQ Total Scores of the participants who were classified in high and low score groups based on SSQ Total Scores under the no-fixation-point condition. In Experiment 1, there were five participants in the high score group and three in the low score group. In Experiment 2, there were five participants in the high score group and eight in the low score group. The following are the multiple comparison results:

- No-fixation-point < static fixation point (p < 0.05)
- No-fixation-point < rectangular wave reciprocating (12°) (p < 0.01)

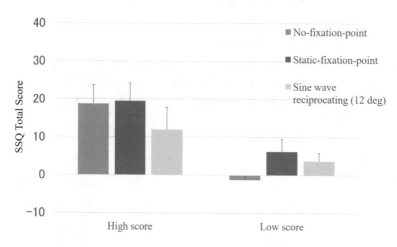

Fig. 13. Experiment 1: average SSQ Total Score

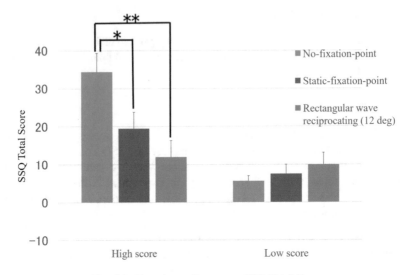

Fig. 14. Experiment 2: average SSQ Total Score

5.3   Subjective Evaluation on Sickness Results

Figures 15 and 16 show the average values of the subjective evaluations on the sickness of the high score group. The following are the multiple comparison results:

- No-fixation-point < rectangular wave reciprocating (12°) (p < 0.10)

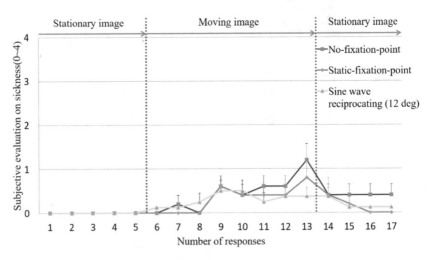

Fig. 15. Experiment 1: average value of subjective evaluations on sickness of high score group

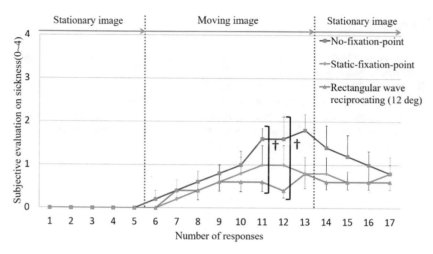

Fig. 16. Experiment 2: average value of subjective evaluation on sickness of high score group

6 Discussion

In our experiments, because the participants showed resistance to VIMS under the reciprocating-fixation-point condition, this result suggests that a moving-fixation point can suppress VIMS. In addition, since it only has the gazing-specific-object property, this property suppresses the VIMS effect.

7 Conclusion

We clarified which property results in the VIMS suppression effect of fixation points by experimenting with moving-fixation points and experimentally obtained the following conclusions:

(1) The gazing-specific-object property contributes to the VIMS suppression effect.
(2) VIMS can be suppressed without always gazing-specific-place.

Future work will collect more data and investigate the VIMS suppression effect of the gazing-specific-place property.

Acknowledgements. We thank all the participants in our experiments.

References

1. Webb, N.A., Griffin, M.J.: Optokinetic stimuli: motion sickness, visual acuity, and eye movements. Aviat. Space Environ. Med. **73**(4), 351–358 (2002)
2. Kennedy, R.S., et al.: Simulator sickness questionnaire: an enhanced method for quantifying simulator sickness. Int. J. Aviat. Psychol. **3**(3), 203–220 (1993)
3. Ujike, H., Yokoi, T., Saida, S.: Effects of virtual body motion on visually-induced motion sickness. In: 26th Annual International Conference of the IEEE on Engineering in Medicine and Biology Society, pp. 2399–2402. IEEE Press (2004)

A Kansei Engineering Based Method for Automobile Family Design

Huai Cao$^{(\boxtimes)}$ and Yang Chen

School of Mechanical Science and Engineering, Huazhong University of Science and Technology, Wuhan, People's Republic of China
caohuai@hust.edu.cn, 1959511079@qq.com

Abstract. Product family design is a widely applied design strategy for maintaining a brand image. Especially in the field of automotive styling has developed a mature family design. However, there are also many problems with family design. The lack of change in family design not only limits the designer's play, but also makes consumers feel boring. This paper presents a method based on Kansei engineering for automobile family design. The objective of the method is to retains the brand identity without using rigid family elements by directly study the semantics of the design features.

Keywords: Kansei engineering · Automobile styling design
Semantic differential method · Feature

1 Introduction

1.1 Background

Product family design is a design strategy widely applied in industrial design. An important quality of a strong brand is the presence of a clear, well-defined brand identity [1]. Many successful brands are using family design, such as BMW's kidney-shaped grille, MacBook Pro's frosted aluminum case. Product family design is that the same brand but different products often use the similar or even the same design. The benefits of this approach are the continuation of successful designs that help to create a consistent brand image and saving on development and manufacturing costs. But on the other hand, family design has also been accompanied by the lack of innovation.

Family design is the most mature automotive design field. For example, in the car design, family design usually expresses in a series of similar design elements between different models of the same brand, such as grilles and lights. Different types of some brands are nowhere similar. Many consumers are bored and confused, and their purchase decisions are affected. Low-end models and high-end models using the same design. Although it may help short-term sales of low-end models, but this approach will certainly lead to long-term brand value devaluation.

© Springer International Publishing AG, part of Springer Nature 2019
S. Fukuda (Ed.): AHFE 2018, AISC 774, pp. 289–300, 2019.
https://doi.org/10.1007/978-3-319-94944-4_32

1.2   Car Design Process

From the designer's point of view, car styling is a emotional and complex job that can not be accurately described by words or numbers. General production car design process includes design research, early concept design, design deepening, design freeze, details and product design. Throughout the design process, the design's presentation includes two-dimensional sketches and renderings, as well as digital three-dimensional digital models and clay models. Among them, the most decisive role in the final design stage is the early concept design, followed by a series of processes are in close contact with the early sketch, as well as the effect of the three-dimensional model to deepen and optimize the initial design. In the early concept design phase, the design team often set a fuzzy design goal (such as a paragraph of text, a series of pictures, even audio and video, etc.) and basic engineering requirements to the designer, and then the designer went according to his own understanding to find inspiration, complete the early design. The rapidity of the sketch, the initial, the imprecision of the inaccuracy, and the sketch of the ambiguity can stimulate creativity [2].

Because the early concept design had a huge impact on the entire design process, this paper chose to make influence to the design process at this stage. More specifically, this approach is to inspire designers by a series of filtered features of the designer during the early concept design. Instead of telling the designer directly as a common design approach, which brand design elements are not changeable and which ones can be changed.

1.3   Expression of Car Design

In the field of automotive styling, no matter what the design goals are, expressions are similar. The main expression is the proportion, stance, volume, surface, graphic design and so on. Because the design method proposed in this paper is aimed at the early concept design (mainly sketching and rendering), and to make the automobile family design more easily to study, this paper focuses on 2-dimensional perspective. From a 2d perspective, the car styling design is divided into silhouette, layout, light and shadow (volume and features), materials and details from primary to secondary. Remove the effect of material, 2d car styling can be simplified to silhouette, layout, feature. In the mass-produced and same-class car design process, there is little space for designers to free play due to engineering and safety constraints in the silhouette levels and layout levels except for feature levels. In addition to proportion and stance, family design DNA mainly expresses in a series of family features. For example, Audi's hexagonal grille, BMW's crease line. The design method proposed in this paper is to start with the feature level of study on how to innovate automobile family design.

2   Method

The Kansei engineering is defined as "translating the customer's Kansei into the product design domain" [3]. From a Kansei engineering point of view, the goal of automobile family design is to create a brand-specific feeling and impression for consumers, not necessarily all cars using a series of highly similar and concrete design elements. Further speaking, different design elements can be used, but only to give consumers a unified feeling and impression.

Therefore, the key to solve the rigid problem of family design elements is to break the boundaries of family elements and add more possibilities in the early concept design. In order to achieve this goal, the method adopted in this paper is to extract each individual feature collected from the picture of car design field and non-car design fields to establish a feature library, and to evaluate and classify these features through the semantic difference method in Kansei engineering, and then inspired designers with filtered features according to the design goals in the early concept design stage, not just the classic family elements of the historical model design.

In the process of innovative design, how does this article recognize the family DNA of a car brand? The semantic expression of products can enhance the sense of value brought by the function of products, and affect users' cognitive emotion through semantic description, expression, marking and recognition [4]. The method presented in this paper is based on the semantic differential measurement of many individual features of the brand's cars. In this paper, we put these semantic measurement results of many features of a single brand into a rose chart called the Brand Feature Graph (BFG), as a brand's family design metrics.

In addition, this paper also classifies the features according to the similarities of the results of Kansei measurements of different features. This article refers to these categories as similar features. Similar to BFG, the Kansei measurement of all features in each similar feature class is made a rose chart called similar feature graph (SFG).

In summary, this design method can use the intentional image deconstructed from different brands or even different fields to build a feature library, and the designer can be not limited to the inherent family elements, which is beneficial to discovering more possibilities. At the same time, the semantic analysis can understand the consumer's feeling, re-recognize the design features and family DNA, so that the design of the car to maintain a certain brand image between different models can also avoid excessive duplication.

2.1   Feature Library

The feature library contains many features as objects, each feature has a unique ID, citation relation, Kansei engineering evaluation, and other attribute values (Fig. 1 shows). These features are also categorized in two ways, that is, the brand and the Kansei characteristics. Designers can select multiple features by brand, pick features by preference of SFG, select one feature and then find features that the SFG are close to or opposite. And this is still a dynamic library can be updated as the number of surveys increases, new features are added and updated.

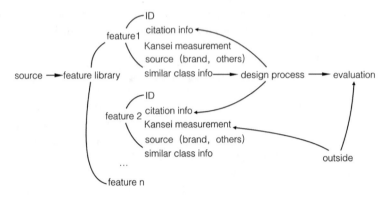

Fig. 1. Data flowchart

2.2   Design Process

Step 1: Feature extraction

Collect inspiration images from different sources (cars, products, art, nature, etc.) and edit the images to get sharp and clear features.

Step 2: Feature evaluation

Determine several adjective pairs as attributes of each feature. Use semantic differential method to take Kansei measurements for obtaining a series of attribute values corresponding to each feature.

Step 3: Feature classification

Classify all the features by brand. Analyze the attributes of the existing brand design including the features and obtain the BFG. Study the relationship between the BFG and the design philosophy of the brand.

According to the method of cluster analysis, the features are classified into multiple classes by similarity to obtain the SFG.

Step 4: Early concept design

Determine the proportion of models. Combined with the brand's existing design philosophy and the BFG in step 3. Set a series of adjectives as a design goal. According to the design goal to select the appropriate similar class. Do sketches and renderings inspired by the filtered features

Step 5: Design evaluation

Take Kansei measurement of complete design by semantic differential method to determine if the design meets the design goals (Fig. 2).

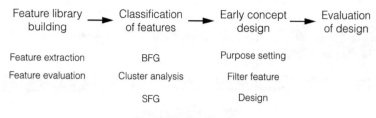

Fig. 2. Method process

3 Experiment

The experiment in this paper takes the design of car body side view as an example. We use semantic differential method in Kansei engineering to make Kansei measurements of individual features and side view designs. This paper has conducted two questionnaires of the semantic differential methods. The respondent mainly are 18–25 years-old college students. The majority has design background.

The first survey of semantic differential method conducted Kansei measurements of individual features. The purpose is to analyze brand image and analyze the characteristics of similar features.

The second survey of semantic differential method conducted a Kansei measurement of the complete side-view design using the features in the survey 1. The purpose is to prove the significant impact of individual features on the complete design, further illustrating the feasibility and effectiveness of the proposed design method.

3.1 Experiment Process

Step 1: Experimental selected from five brands, two models of each brand a total of 10 models of the side picture as a design material. We edit the images to get the clear feature with clear light and shadow and the outline. There were 36 extracted features, each feature has a unique ID in the upper left corner (Fig. 3 shows).

Step 2: Determine the five pairs of adjective of Kansei measurements: dynamic - balanced, powerful- elegant, sharp - soft, thick - light, pure - complex. Due to the large number of features, we place 36 features in two questionnaires that each contains 18 features. We also conducted a questionnaire of both questionnaires by using a five-level Likert scale.

Step 3: Screen valid questionnaire results. We obtained 36 features of Kansei measurement results.

First of all, according to the brand these features will be divided into five classes. Kansei measurements of each brand are analyzed to obtain the Brand Feature Graph (BFG).

Then, using the system classification method, according to the similarity, the result of cluster analysis of the 36 features is obtained and multiple similar feature classes are obtained. The features in each class are analyzed to obtain the similar feature graph (SFG).

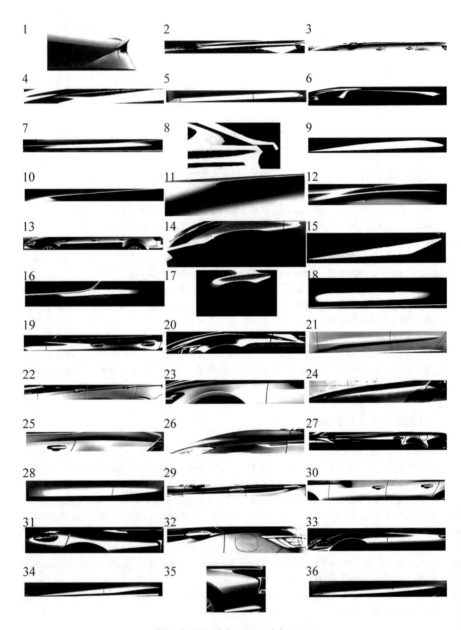

Fig. 3. The 36 extracted features

Step 4: We select the 2018 Audi A7 model from the 10 cars in Step 1 as a original design; select a number of coordinated features from all 36 features to make 3 new side-view designs (Fig. 4 shows); the 3 new designs are made by changing the main features and retaining outline and graphic design of 2018 Audi A7.

No. 0 design is the original design, using the features of no. 21, no. 23, no. 25, no. 30;
No. 1 design used no. 11, no. 15;
No. 2 design uses no. 13, no. 28;
No. 3 design uses no. 22, no. 24, no. 36.
Step 5: This paper makes a Kansei measurement of the Audi A7 original design and
3 new designs by using the five adjective pairs in step 2.
Step 6: Comparisons and analysis the Kansei measurement of individual features
and the complete side-view design.

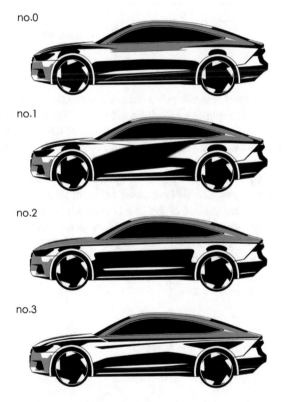

Fig. 4. The original design and 3 new designs

4   Results and Analysis

The experiment has conducted two online questionnaires of Kansei measurement. The
first questionnaire studied a total of 36 features, and each of the two questionnaires
received 30 valid responses. The second questionnaire received 40 valid responses.

4.1   Kansei Measurement Results of the Features

Analyzed from the mean, it shows that dynamic and purity is the semantic preferences of the overall car styling features, while the other three adjective pairs tend to be neutral. From the analysis of variance, dynamic - balanced, sharp - soft, thick - light showing a greater difference, but strength - the difference is very small variance (Table 1 Shows).

Table 1. Kansei measurement results of the features

	Dynamic-balanced	Powerful-elegant	Sharp-soft	Thick-light	Pure-complex
Mean	2.714	2.946	2.911	3.011	2.788
Variance	0.222	0.086	0.219	0.138	0.194

4.2   Analysis of the Brand Feature Graph

Survey 1 studied the local features rather than the complete design. Is there a correlation between the design philosophy of the brand and the local features? From the Brand Feature Graph, we can clearly see that the results of Kansei measurements of the features of the five brands still show significant differences among the five different brands. Not only the existence of differences, compared to the overall design philosophy of the brand, we can see that there is a significant correlation between the whole and the local. For example, the BFG of Audi is the most easily recognizable, most balanced, elegant, soft and concise of the five brands, which is consistent with Audi's quest for balance and simplicity (Fig. 5 shows). Another example of BMW's BFG shows powerful, sharp, light, complex, and the pursuit of sporty, sculpture sense of the BMW design philosophy is consistent. (Fig. 5 shows) Further, the BFG can be used as a measure of the design effect of the family design. That is, the local features reflect the overall design effect to some extent.

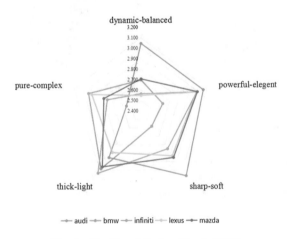

Fig. 5. The Brand Feature Graph (BFG)

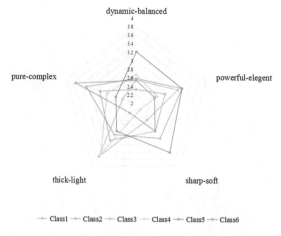

Fig. 6. The Similar Feature Graph (SFG)

4.3   Analysis of the Similar Feature Graph

Figure 7 is the tree clustering map. According to the results of Kansei measurements, we systematically cluster the 36 features by using the method of cluster analysis with. If all the samples are divided into six classes, as the red line shown in Fig. 7.

The SFG are shown in Fig. 6, each class of which consists of features that have similar Kansei measurement results.

Six similar classes contain 10, 5, 3, 5, 11, 2 features respectively. Kansei measurements of different similar classes also showed significant differences. The BFG shows the brand's Kansei attributes and the SFG shows the sentimental attributes of the similar-feature class. Observing the shape of a feature in each of the similar classes reveals the fact that similarly-shaped features have similar Kansei measurements, features with differences in shape may also have similar Kansei measurement results.

4.4   Analysis of the Complete Side-View Design

The Table 2 contains results of the original design (no. 0) and the three new design (no. 1, no. 2, no. 3) and excerpts the Kansei measurements of the features applied in the all 4 design.

In this paper, the Pearson correlation coefficient is used to measure the similarity between the individual features and the complete design. The Pearson correlation coefficient is in the range of $[-1, 1]$. The larger the absolute value is, the stronger the correlation is.

In this paper, the Pearson correlation coefficients of the all side-view design of numbers 0, 2, and 3 with their respective main features are calculated, respectively 0.7224, 0.8602, 0.7896. The results have shown a significant correlation.

However, we see that no. 1 does not show the same semantic similarity with the main features. After discussing with many professional designers, this paper argues that the reasons for this are mainly the differences of the marginal processing methods and

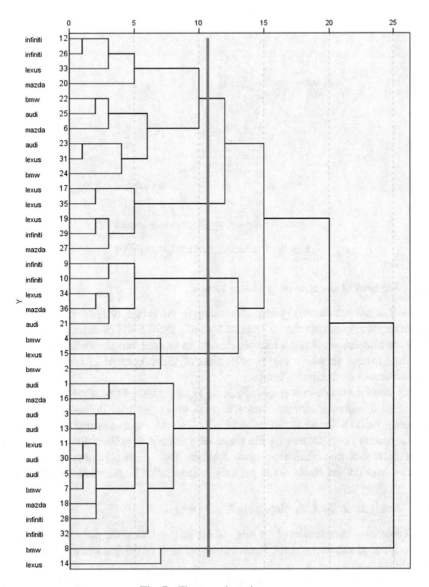

Fig. 7. The tree clustering map

the interference of the surrounding features, which make the characteristics of individual features and complete designs vary greatly.

Table 2. Kansei measurement results of the designs and main features

Design	Main feature	Dynamic-balanced	Powerful-elegant	Sharp-soft	Thick-light	Pure-complex
No. 0	30	3.03	3.23	3.50	3.23	2.48
No. 1	11	2.45	2.70	2.53	3.00	3.20
No. 2	13	3.38	3.28	3.50	2.98	2.45
No. 3	22	2.55	2.80	2.75	3.03	3.23
Feature ID		Dynamic-balanced	Powerful-elegant	Sharp-soft	Thick-light	Pure-complex
11		3.20	3.27	3.30	2.73	2.43
13		3.70	3.20	3.73	2.73	2.73
22		2.50	2.83	2.87	3.07	2.93
30		3.50	3.17	3.47	2.83	2.47

5 Conclusion

In this paper, the study of automobile family design based on Kansei engineering is carried out at the feature level. In order to get rid of the unchanging family design elements, this paper proposed a family design method using features with similar semantic as the inspiration of the early concept design. It also provides a way to get design features from any image. Through experiments, the whole method was verified.

The results of the Survey 1 show that not only the similar shape features but also the different shaped features can reach similar semantics, that is, they bring similar feelings and impressions.

Through the analysis of the BFG, there are significant differences between the different attributes of the automobile brands and are highly relevant to the brand's design philosophy.

The results of the survey 2 directly demonstrate that there is a significant correlation between the semantics of local features and the semantics of the complete design, that is, to bring similar feelings and impressions.

The results of the survey 2 indirectly explanation when we have established a large enough feature library, especially the classes of similar feature, we can design cars retained family DNA with the change of features.

The method proposed in this paper is still deficient.

First, The morphological changes and combination of features in the design process both affect the semantics of the overall design and depend on the designers' subjective play. For example, the main feature of the complete side-view design in survey 2 is identical to the same feature in survey 1 as the difference in Kansei measurements (sharp and gradual) and the surrounding secondary features result in significant differences in Kansei measurements.

Second, Likert scaling is a bipolar scaling method, measuring either positive or negative response to a statement. Bipolar scale may be too simple to complex feelings about design features. The scaling ignores the middle option of "both A and B". Such as "powerful and elegant", "sharp and soft" and etc.

References

1. McCormack, J.P., Cagan, J.: Speaking the Buick language: capturing, understanding, and exploring brand identity with shape grammars. Des. Stud. **25**, 1–29 (2004)
2. Purcell, A.T., Gero, J.S.: Drawings and the design process. Des. Stud. **19**(4), 389–430 (1998)
3. Nagamachi, M.: Kansei Engineering as an ergonomic consumer-oriented technology for product development. Appl. Ergon. **33**, 289–294 (2002)
4. Wikström, L.: Methods for Evaluation of Products' Semantics. [PhD Thesis], pp. 25–36. Chalmers University of Technology, Göteborg (1996)

Examining the Relationship Between Lean Adoption and Housing Finance in Ghana

I. Salifu Osumanu, C. O. Aigbavboa[(✉)], and D. W. Thwala

Department of Construction Management and Quantity Surveying,
University of Johannesburg, Johannesburg, South Africa
ibosumanu@yahoo.com, {caigbavboa,didibhukut}@uj.ac.za

Abstract. The paper investigates the relationship between lean adoption and housing finance in Ghana. Ghana currently has 1.7 million housing deficit and this is attributable to high population growth rate and urbanization. The deficit remains one of the most socio-economic challenges facing the country today. Given the huge unmet housing needs of Ghana, there is the need for an appropriate financing model to be developed to secure funding for housing construction. The objective of this study is to provide a description of the factors that affect house financing within the context of Ghana and provides lean strategies for reducing construction cost. The study is designed with a quantitative methodology which 120 respondents will be selected to answer questions relating to how lean principles could be used to reduce construction cost. The findings will serve as policies to the real estate industry and government. Recommendations will be offered to real estate companies, the banks and government to enable them develop a comprehensive financing model.

Keywords: Lean adoption · Housing finance model · Affordability
Appropriate technology

1 Introduction

Ghana currently has a huge housing deficit of 1.7 million units which needs a well coordinated system between Government and the private real estate developers to address. For Ghana to reverse this trend and make huge impact, it is believed that the right combination of lean strategies and prudent cost management will reduce construction cost for real estate companies to develop affordable housing units that are within the reach of the average Ghanaian.

The insufficiency of housing has produced several squatters and slums in the major towns and cities of Ghana. There is 5.4 million slum dwellers and it is anticipated to reach 7.1 million by 2020 (Ghana Statistical Service 2010).

These figures suggest that the annual supply of housing units in Ghana falls short of demand between 65%–70% of the national requirement. The 2010 population census recorded a population of approximately 25 million with a total of 4.9 million housing units. This confirms the fact there was a deficit of 1.5 million housing units in 2010 which has rising to 1.7 million in 2014.

© Springer International Publishing AG, part of Springer Nature 2019
S. Fukuda (Ed.): AHFE 2018, AISC 774, pp. 301–311, 2019.
https://doi.org/10.1007/978-3-319-94944-4_33

The successful implementation of Lean manufacturing principles, the Toyota model to the real estate industry will enable estate companies carry out projects with new technology leading to reduction in cost of production and low cost of housing (Johansen and Walter 2007).

Globally, Lean concept is used in most advanced countries such as United States of American, United Kingdom and the Netherlands. This has achieved remarkable success in the first year of implementation. The USA and UK experienced 17% and 37% improvement in their operation leading to low cost of construction products (Koskela 2004). However, lean adoption in some developing countries such as Indonesia, Kenya, Egypt, South Africa and Ghana is yet to yield some gains (Weru 2015).

This paper will contribute to theoretical and empirical literature on the adoption of lean, it will serve as an implementation framework to guide lean adoption in the real estate industry, it will assess the operational cost and house financing for the real estate industry, it will also be used as benchmark for assessing the operational and financing model for the real estate industry and to provide recommendation for bridging the housing deficit of Ghana.

1.1 Problem Statement

The problem this paper is seeking to address is from two broad areas; firstly, there are high levels of waste and poor performance in the real estate industry. This has occasioned the high prices of housing in Ghana. Poor performance in the real estate industry has also contributed to undue delays and cost overruns. According to (Ekanayake and Ofori 2000), construction materials constitute between 55–60% of a total cost of a house. Studies conducted in the Netherlands revealed that the level of material waste is 10% of total materials brought to site for use. Secondly, studies done by (Johnson and Brennan 2006) and (Common et al. 2000) in the Netherlands and UK revealed that real estate industry is a bit reactive in the adoption of lean manufacturing principles. In the case of most developing countries such as Ghana, the application of the traditional system to the real estate industry is dire and might stifle project completion. This system produced a situation where projects became very expensive when completed (Dulaimi and Anamas 2001). It is in this view that this paper finds it comparative to carry out this research to examine the nexus between lean adoption and housing financing in Ghana.

1.2 Research Objective

The general objective of this paper is to examine the nexus between lean adoption and housing financing in Ghana:

The paper has the following specific objectives;

1. To establish the effect of lean adoption on housing financing in Ghana.
2. To establish the relationship between lean adoption and effective procurement practices in Ghana.
3. To assess the relationship between lean adoption and cost control measures in Ghana.

1.3   Research Questions

The paper attempted to answer the following research questions.

1. What is the effect of lean adoption on housing financing in Ghana?
2. What is the relationship between lean adoption and effective procurement practices in Ghana?
3. What is the relationship between lean adoption and control measures in Ghana?

1.4   Research Hypothesis

H1: There is a positive relationship between the adoption lean and housing financing in Ghana.

H2: There is a positive relationship between lean adoption and effective procurement practices in Ghana.

H3: There is a positive relationship between lean adoption and cost control measures in Ghana.

2   Literature Review

The literature review focused on the state of real estate in Ghana, the state of housing in Ghana, theories of transformation, Finance Theories, value generation theories and the attempt to explain the effect of lean adoption on housing financing in Ghana.

2.1   Theoretical Framework

The theoretical framework for this study was built around the lean model to explain the theoretical framework for the study. Assist in explaining the financing relationship of real estate projects. Theories that serve as useful guidelines for addressing challenges facing the housing sector in Ghana (Koskela 2000). The real estate industry in Ghana, the Theory of production as applied in construction industry, value generation, financing theories from the perspective of demand and supply relationship.

2.2   Transformation Theory

The theory of transformation is expressed in terms of inputs that are converted into outputs in the production process. These inputs comprise capital, machinery, labour, capital and building materials. According to Koskela and Howell (2002), the underlying theory of contemporary project management in construction is founded on the conventional transformation theory which defines the project scope and delivery system. Most Managers in the real estate industry make use of similar deterministic models such as CPM, Gant Chart and WBS. These models are very important for calculating project costs, delivery time and profit maximization.

2.3   Value Generation Theory

The Value generation theory posits that project management follows a methodical delivery process. According to (Koskela 2000) value is created when there is an exchange of service or good whilst production is recognized as the creation of goods and services to achieve optimal returns to the customer. Under this theory, customer value creation is significant and that more resources need to be mobilized to transform imbalanced customer expectations to quantifiable levels during the house construction process. Its goal is to reduce waste and costs so that the construction project will speedily be directed to meet customer expectations (Lukowski 2010).

2.4   Flow View Theory

The flow view theory focuses on waste management and examines production system perspectives with many components such as material, labour, equipment and information. The flow begins with the purchase of raw materials and ends with a finished product. The theory was developed by Koskela (1992) to essentially eliminate waste as well as reduce non-value adding activities within the construction process. The Flow View theory is an extension and application of lean manufacturing principles to aid value addition and waste removal (Bertelsen 2004).

2.5   Conceptual Framework

The conceptual framework for the study was based on the relationship amongst procurement technology improvement, planning and decision-making, waste elimination which will eventually contribute to an overall effective cost control of real estate companies. The conceptual framework therefore depicts a direct relationship amongst the lean adoption variables.

Lean Adoption Variables

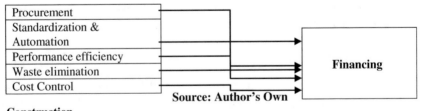

Source: Author's Own

2.6   Empirical Literature

The empirical literature focused on the application of lean techniques and its effect on financing. The implementation of lean techniques has significantly reduced waste and improved performance in construction projects (Weru 2015). Since independence,

Ghana has traditionally relied on manual architectural drawings with limited standardization and automations. According to (Brownhill and Rao 2002), one of the aims of lean adoption is to reduce waste, improve cost and enhance housing financing. The lack of standardization and automation makes it difficult for real estate companies to achieve operational efficiency. The presence of procurement irregularities as well as wrong product specifications contribute to high construction costs (Picchi and Granja 2004).

2.6.1   The State of Real Estate in Ghana

Since independence, Ghana traditionally has a limited real estate market with few companies, residence and investors. Trends in demography, income and availability of retail credit also point to increased demand for residential housing over the next decade. With the population increasing sharply since 2000 and urban residents growing from 43.8% of the overall populace in 2000 to 50% in 2010, Ghana will need an estimated two (2) million new housing units by 2020. Real estate services in Ghana accounted for only 1.78% of Ghana's GDP in 2010. Available statistics from the Ghana Statistical Service indicate that Ghana has a huge housing deficit of 1.7 m.

2.6.2   Factors Affecting Housing Finance in Ghana

Incomplete Development. One of the challenges with self-financing of new homes by owners is the long duration it takes to complete the development. Within serviced plots, owner-financing undermines the uniformity and beauty of the estates due to large stretches of uncompleted houses. It is also observed that some developers fail to fully complete the projects and do not keep their promises to home buyers. Common observations include open drains, poor infrastructure, deplorable roads, etc.

Poor Management. Poor management of housing communities to guarantee good maintenance and preserve the investment values of housing units continue to be a challenge for homebuyers. Deterioration of road network, sanitation system and poor waste disposal are some of the common characteristics of most housing developments in Accra and the surrounding areas.

Quality Issues. One of the main issues with estate development is the low quality standards. Some developers use poor quality fittings and fixtures which deteriorate or break down soon after such properties have been acquired. Home-buyers therefore do not gain optimal value after few years of purchasing houses.

Access to Finance. Access to finance for home buyers continue to a major challenge in Ghana. In deciding to seek finance for finance from banks to buy a house, potential buyers are influenced mostly by interest rates. The cost of borrowing has deterred many individuals and corporate entities from going for loans from the commercial banks.

3   Methodology

The study was designed as a quantitative research in which random sampling technique was adopted to select 120 respondents from the real estate industry in Ghana to answer questions relating to the adoption of lean and financing. The study used both primary and secondary data sources. The researcher largely used secondary data for the study. These were obtained from the published articles, journals and other relevant published research on lean adoption. Primary data were collected through a self-administered questionnaire. The work surveyed a sample of the targeted population and the minimum desirable sample size was determined using the following model (Sapoka 2006). Multiple regression was used to analyze the relationship between lean adoption and profitability of real estate companies in Ghana at 5% level of significance. The data gathered was analyzed statistically through correlation and multiple regressions with SPSS to generate the results. Multiple Regression was used so that the researcher will be able to analyze the relationship between lean adoption and financing of real estate companies in Ghana.

3.1   Model Specifications

The model used for analyzing the relationship between lean adoption and real estate financing. They comprised dependent, independent and some control variables (Yin 2009).

$$LAPit = \beta_0 + \beta_1 Pit + \beta_2 SAit + \beta_3 PEit + \beta_3 WEit + CCit \, \acute{\epsilon}it \qquad (1)$$

Where

Dependent Variable
Financing

Independent Variables

1. Procurement
2. Standardization & Automation
3. Performance Efficiency
4. Waste Elimination
5. Cost control

e   = Stochastic error or disturbance term.
t   = Time dimension of the variables
$\beta 0$ = Constant or Intercept.
B1–B4 = Coefficients to be estimated or the coefficients of slope parameters. The expected signs of the coefficients (tested at 5% level of significance) are such that B1–β5 > 0.

4   Data Analysis

The researcher analyzed the data based on the 120 questionnaires distributed to respondents from the real estate industry in Ghana out of which 100 were received for analysis. Multiple regression analysis, descriptive statistics, and correlation were used to present the results of the study. The data analysis began with a multiple regression which sought to examine the relationship between housing finance and lean adoption, followed by descriptive statistics and correlation the discussion of findings follows.

Table 1 showed the relationship between house finance and lean adoption. From the data analysis, it was found that the lean adoption was significantly different from or it had a significant positive correlation with housing finance with a score of R = 73%, R2 = 71%, P = .000. The R is the coefficient of determination and it indicates the extent of correlation between lean adoption and house finance. From Table 3, a score of 71% indicates that the two variables are positively associated and is taken to be the best measure of the results because it has factored into the computation the margins of errors. The R2 explains a refined level of association between the two variables. The P-value is the probability of default or the alternate hypothesis and should not exceed 5% or 0.05 for the results of the study to be valid. From the analysis above, p-value of 0.00 implies that the hypothesis was tested below 5% level of significance. This significant level is within the accepted correlation level of 0–5%. This implies that when lean is adopted, it has the tendency to improve performance efficiency, waste elimination, and cost control.

Table 1. Multiple Regression Analysis of lean adoption and housing finance in Ghana

Source	SS	Df	df MS		Number of obs	100
					F(5, 100)	1.51
Model	1671.33	5	196.33		Prob > F	0.2401
Residual	22136.4	98	1794.91		R-squared	0.73
					Adj. R-squared	0.71
Total	4829.74	103	1991		Root MSE	14.032
House finance	**Coef.**	**Std. Err.**	**T**	**P > t**	**[95% Conf.**	**Interval]**
Standardization & automation	−0.3124	0.24098	−1.2962	0.02	−0.6352	0.78912
Performance efficiency	0.0733	0.4398	2.56	0.05	−0.122567	2.12776
Waste elimination	0.66433	0.43999	3.67	0.06	−0.74561	0.66532
Cost control	0.01234	0.42333	2.90	0.03	−1.562078	0.22421
Procurement	0.12889	0.38802	0.33217	0.01	−0.656122	1.10084
Cons	10.0855	0.74116	13.61	0	−1.11278	1.3289

Source: Output from SPSS

Table 2. Correlation matrix of house financing market and apartment/house cost

Lead adoption (X)	Correlation coefficient
House finance (Y)	
Standardization & Automation	.34*
Performance efficiency	.78*
Waste elimination	.73*
Cost control	.79*
Procurement	.45*

Source: Output from SPSS Version 15

Table 3. Descriptive Statistics of lean adoption and house financing in Ghana

Variables	Obs	Mean	Std. Dev.	Min	Max
House finance	100	122.5	14.8167	45	100
Standardization & Automation	100	116	12.9719	56	88
Performance efficiency	100	78	8.99451	44	56
Waste elimination	100	93	10.2325	56	65
Cost control	100	99.5	12.337	45	77
Procurement	100	72.5	12.2769	39	53

4.1 Correlation Analysis

Correlation analysis was used to ascertain the association between lead adoption and housing finance in Ghana. From the analysis, it was evident that there was a significant positive correlation between lead adoption and house financing. This implied that the lack of market for houses has positively influenced its pricing. Capital costs, labour, land, infrastructure and building materials collectively having impact on apartment or housing costs. The correlation analysis of house finance market and apartment cost have been presented Table 2. The analysis indicated that lead adoption variables such as performance efficiency, cost control and waste elimination have effects on housing finance as they recorded correlations of 78%, 79% and 73% respectively.

The descriptive statistics showed that lean adoption could have a positive effect on housing finance. A possible explanation is that lean factors could help improve the funding when cost is internally controlled, procurement streamlined, when standards are well followed by the real estate companies. In effect, when waste is eliminated, whilst performance efficiency is boosted, it will serve as a morale booster for banks to finance projects of real estate companies.

4.2 Discussion of Findings

This study indicates that a major issue facing the real estate sector in Ghana is how to raise finance from the banks, development partners and allied institutions to construct affordable houses. Potential real estate developers view this as a major headache since there are virtually no huge funds for the construction of houses. The other financing model is through project financing where the underlying asset is used as collateral for loan. Real estate companies in Ghana have not benefited from project based funding.

From the discussion, there was a moderate positive significant difference between lean adoption and housing finance through better management of waste, cost and operational efficiency, all other things being equal.

This means that the hypothesis developed is partly supported by the results. Thus, *there is a significant positive difference between the house finance market and effective and efficient operation of the real estate sector.*

The findings of the study support the hypothesis that lean adoption leads to performance efficiency, cost control and waste elimination. For real estate companies to get funding from the banks, they must be good managers of costs and internal operations as well as demonstrate that they have what it takes to repay these loans. In addition, the house finance market is underdeveloped and there it takes a long time for real estate developers to get buyers for their houses. The findings from the study showed that the right combination of government policy, private sector involvement in the real estate sector will go a long way to solve the challenges facing developers. The study similarly found out most real estate developers in a bid to get finance, fall on friend and microfinance institutions that grant them short term loans at high interest rate.

4.3 Theoretical Implications

This study contributes to literature in three different ways. First, the transformation theory explains that all factors of production must be employed in their optimal form to help real estate companies to perform better. This will help real estate companies to achieve operational efficiency and to reduce cost. Second, the value generation theory indicates that real estate companies must offer value to their customers. This call for financial resource mobilization to augment house construction. To eliminate waste, real estate companies must adopt lean principles to meet customer expectations, automate their processes for better results. Third, the flow view theory posits that production systems of real estate companies must be streamlined so that production can go uninterrupted.

4.4 Managerial Implications

The study has several implications for managers. First, for real estate managers to be able to bridge the 1.7 million housing deficit in Ghana, access to long-term financing is necessary. Secondly, managers must adopt lean principles to streamline their operations to be competitive, relevant and profitable. Third, there is the need for a coordinated response to the house finance problem.

5 Recommendations

The study has given a clear picture about the benefits of adopting lean and its effect on house financing in Ghana. It is therefore recommended that issues of training in mortgage financing be taken seriously as one of the surest means to help potential real estate companies in Ghana to an avenue to raise funds. The following recommendations are being offered in this direction:

1. It is recommended that the real estate developers set up a committee to adopt lean and train their members.
2. The government must set up a seed fund for the development of houses in Ghana.
3. The government must offer tax incentives to banks that offer mortgage financing to real estate developers.
4. Venture capital could be considered as an alternative financing model for real estate companies.

5.1 Conclusion

From the findings in chapter four and the summary in this chapter, it can be concluded that the real estate developers are lean adoption have effect on housing finance in a variety of ways. The adoption of lean could help the real estate company to streamline its operations, control costs and improve upon the overall efficiency of the real estate companies. Banks and microfinance institutions that lend to real estate companies examine the financial reports of real estate companies to investigate how efficient they are. Also, the study found out that lean adoption could contribute to the overall financial health of real estate companies as it enables the companies to follow certain processes. It was established from the study further that there was a significant positive relationship between lean adoption and a vibrant housing market and strong real estate sector. The study further revealed that there should be a strategic partnership between the real estate companies and the financial institutions to grant loans to developers at reasonable rate so that they can go about their construction without any let or hindrance.

5.2 Recommendation for Future Research

The study focused on key lean factors that affect housing finance within the real estate industry in Ghana. It has been shown from the findings of the study that when real estate managers are able to manage their internal dynamics such as cost, eliminate waste and improve procurement processes, banks, development partners and other institutions find real estate companies or projects attractive for funding. In the absence of these two financing models, real estate companies tend to go to friends and family for small financing. Whilst the researcher has dealt with some of the issues and challenges within the house financing market, there are still some pockets of grey areas that must be looked at. There is therefore the need for research in the area of venture capital.

Time and resource constraints did not permit the researcher to look into some of these dynamics and to prosecute the topic to its logical conclusion. Nevertheless, future research should look at the relationship between lead adoption and venture capital financing.

References

Ahmed, S., Forbes, L.: Modern Construction: Lean Project Delivery and Integrated Practices. CRC Press, New York (2011)

Demeter, K., Matyusz, Z.: The impact of lean practices on inventory turnover (2011)

Ghana Statistical Service, Population Census Report (2010)

Gyadu-Asiedu, W.: Assessing construction project performance in Ghana. Modelling practitioners and clients" perspectives. A Thesis Submitted to the Technology, University of Eindhoven, Faculty of Architecture, Planning and Building, Eindohoven, Netherlands (2009)

Womack, J., Jones, D.: Lean Thinking Banish Waste and Create Wealth in Your Corporation, 2nd edn. Free Press (2010)

Johnston, R.B., Brennan, M.: Planning of organizing: the impacts of theories of activity for management of operations. Omega **24**(4), 367–384 (1996)

Koskela, L.: Making do - the eighth category of waste. In: 12th Annual Conference of the International Group for Lean Construction, 3–5 August 2004, Helsingor, Denmark (2004)

Luther, R.: Construction Technology Centre Atlantic (2005). http://ctca.unb.ca/CTCA1/sustainableconstruction.html. Accessed 1 Oct 2009

Larson, T., Greenwood, R.: Perfect complements: synergies between lean production and eco-sustainability initiatives. Environ. Qual. Manag., **13**(4), 27–36 (2004)

Lopes, J.: Construction in the economy and its role in socio-economic development (2012)

Ofori, G. (ed.): New perspectives on construction in developing countries, vol. 6, pp. 40– 71. Spon, Abingdon (2011)

Marzouk, M., Bakry, I., El-Said, M.: Application of lean principles to design (2011)

Shah, R., Ward, P.T.: Defining and developing measures of lean production. J. Oper. Manag. **25**, 785–805 (2007)

Picchi, F., Granja, A.: Construction sites: using lean principles to seek broader implementations. In: Proceedings of International Group of Lean Construction, 12th Annual Conference, Copenhagen, Denmark, 3–5 August (2004)

Weru, M.: Lean Manufacturing Practices and Performance of Large Scale Manufacturing Firms in Nairobi, Kenya (MBA Project), University of Nairobi (2015)

Boakye, C.K.: Addressing Ghana's housing and job creation challenge (2010). http://www.ghanaweb.com/GhanaHomePage/features/

Shah, R., Ward, P.T.: Defining and developing measures of lean production. J. Oper. Manag. **25**(4), 785–805 (2007b)

Development of a Kansei Engineering Artificial Intelligence Sightseeing Application

Shigekazu Ishihara[1(✉)], Mitsuo Nagamachi[2], and Toshio Tsuchiya[3]

[1] Department of Assistive Rehabilitation, Hiroshima International University,
555-36, Kurose-Gakuendai, Higashi-Hiroshima 739-0043, Japan
i-shige@he.hiorokoku-u.ac.jp
[2] International Kansei Design Institute,
5-37-17, Aga-Kita, Kure 737-8506, Japan
mitsuo.nagamachi@gmail.com
[3] Faculty of Economics, Shimonoseki City University,
2-1-1 Daigakucho, Shimonoseki 751-8510, Japan
tsuchiya@shimonoseki-cu.ac.jp

Abstract. Kansei engineering methods and text mining methods were applied to the web application for assisting sightseeing travel at Kure city. Kansei engineering methods used here were a questionnaire and multivariate analyses for research inter-relations between aims of travels and interests. Text mining was done on logged tweets on Twitter. The number of tweets mentioned on Kure was around 3400 to 3700 tweets per day. The text mining reveals latest events and people's interests in the daily basis. With Twitter text mining to conventional Kansei engineering methods, both general Kansei on sightseeing and rapidly changing interests are kept reflecting to the inference rules of the web-application.

Keywords: Kansei engineering · Sightseeing · Text mining · Inference

1 Introduction

This research was conducted for analysis Kansei, purpose and preference of Kure-city visitors. Kure city is located southeast of Hiroshima city, and it faces inland sea. The city has more than 100 years history of the Navy port and shipbuilding. Recently, industrial heritage and related an animation movie attract more visitors. Attracting and satisfying visitor demands is indispensable for the city.

It is more demanded to suggest visiting places and attractions in their limited visiting time. In this research, we have applied Kansei engineering research methods to build sightseeing assisting web application.

Kansei engineering methods were well been developed and applied to a large number of products and services [1–5]. Traditional Kansei engineering method has four stages. 1. Collecting appropriate Kansei words, including general and domain-specific expression words. 2. Kansei evaluation with a large number of Kansei word pairs. 3. Multivariate analyses. In Kansei analysis, principal component analysis or its derivative methods are used to extract Kansei structure. Multiple regression analysis and its

© Springer International Publishing AG, part of Springer Nature 2019
S. Fukuda (Ed.): AHFE 2018, AISC 774, pp. 312–322, 2019.
https://doi.org/10.1007/978-3-319-94944-4_34

derivative methods are used for revealing Kansei and design elements relations. Various statistical tests are also used. 4. Derived relations between Kansei and design are rewritten as a set of inference rules that are used for making products with the Kansei system that helps product designers.

Although these traditional methods are well established, it takes a certain effort to keeping renewal of evaluation and has an analysis of the data. In addition, sightseeing places and attractions are changing. Some attractions are tightly related to seasons and there are many particular events those held on a short time. Recent days, SNSs are very powerful to spreading event information. Thus, in this research, we utilized both traditional Kansei engineering methods and text mining methods on numerous tweets mentioned Kure on Twitter.

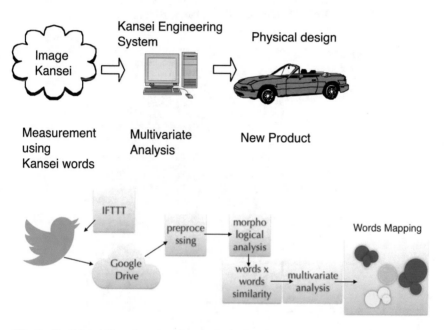

Fig. 1. Traditional Kansei engineering method (above) and Twitter text mining (lower)

2 Kansei Survey and Analyzed Result

We asked visitors to participate in questionnaire study at Warship Yamato-museum (the largest warship at WWII). 500-yen ticket for souvenir purchase was issued for participants. There were 165 valid answers. 14 participants were over 66 years old.

The questionnaire used for this research contains five sections; Profile (male or female, age group, prefecture), Purpose of this travel, Interests & hobbies, Aims & expectations on travels, a Main vehicle of travels.

2.1   Kansei Survey Analysis and Results: Aims and Expectations of Travels

Questions of trip purposes and hobby are analyzed with principal component analysis. The figure shows the principal component loadings for trip purposes. Lines show higher correlations (top 10% of correlations).

Under the age of 65. The result of participants under age of 65; Activities are densely associated with each other. Upmost part corresponds to physical activities (Cycling, Fishing, Climb). Upper right, there are experiences (Workshop, Sea bathing, Relaxation, Shopping). Up middle part has Kure specific activities (visit Anime pilgrimage – anime shooting places, Navy site). Lower right side relates to the historic sites (World heritage, Historic site, Temple, Industrial heritage). Lower middle is local culture (Gourmet, Museum, Visit TV play shooting places, Bloom watch) (Fig. 2).

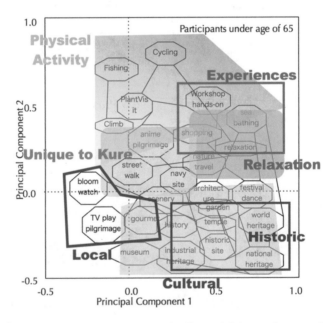

Fig. 2. Analysis result: principal component loadings of "aims and expectation of travels" Participant under the age of 65. Activities and aims are tightly released each other

Over the age of 65. The result of participants over 65 shows separations between 3 groups of activities. Upper left corresponds with experience of self-involvement (Climb, Workshop, Street walk, Gourmet, Festival). Upper right part is visual experiences (Bloom watching, Garden, Nature, Plant visit, Museum). Center down part is historical places and relax (Historical site, Temple & shrine, Navy site, Industrial heritage, Architecture, Relaxation, Shopping, Scenery). These separations suggest differences of physical strength and activeness of elder visitors (Self-involvement vs. Passiveness). The center down part is located in between of them. There are two connections between groups and suggest elder visitors preferences; associations between Festival – Navy Site and History visit – Famous Architecture. Festival visitors could be encouraged to visit Navy sites. History visit participants may interest more on architectures (Fig. 3).

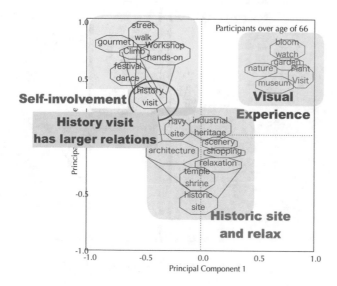

Fig. 3. Analysis result: "aims and expectation of travels" Participant over the age of 65. It revealed apparent clusters and had limited variations

2.2   Kansei Survey Analysis and Results: Interests and Hobbies

Questionnaire section of "Interests and hobbies" are also analyzed with PCA and shown dominant correlations. Younger participants result shows the three closely gathered clusters. At the uppermost, TV game and Manga & Animation are very closely tied. Right side, there is creative activities cluster consists of Painting, Haiku & calligraphy, Pottery. Down left, there is relax & enjoyment cluster of Relaxation, Gourmet, Travel, Gardening, Nature Visit, Spa, and Shopping.

Although the result of participants over 65 does not show apparent clusters, some key activities emerged. One of them is Walk & Jog, which correlated to Flower visit, Museum, Music, and Gardening. "Walk & Jog" implies to a certain level of physical strength to do these visiting activities. At right middle, Museum, Photo, Music, Movie, Pottery, Gardening and History are densely tied with higher correlations. These activities are a similar level of physical strength required for daily living. The sport has eight links to other activities; it is one of the hub activities correlates with other activities. Similar to Walk and Jog, it correlates to higher physical activities of climbing, also correlates to ADL level of activities of Shopping, Cooking, Movie, Photo, and Pottery. Haiku & Calligraphy has a unique correlation with Sport.

The participants under 65 show groups of interests and hobbies. The participants over 65 show decisive activities of Walk & Jog and Sports. The level of physical strength shows vague distinctions on activities (Fig. 4).

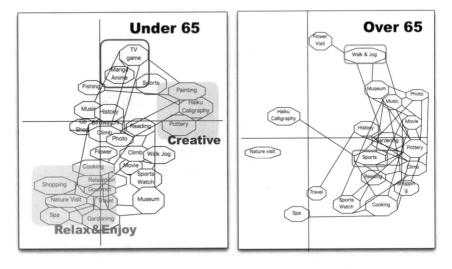

Fig. 4. Analysis result: "Interests and Hobbies"

3   Twitter Text Mining on Kure

3.1   Twitter and Its Uniqueness for Research

Twitter has several distinctive features from other SNS. One is its short length writing as the "tweet". Length of the tweet has been limited to 140 characters since its beginning. In November 2017, Twitter officially extends its limitation to 280 characters in European languages. Japanese, Chinese, and Korean are still limited to 140 characters [6].

Because of its short length per a post, users tend to make post "tweet" in real-time and at the site, compared with other SNSs.

The second is its short, concise writing style. Initially, Twitter does not have the feature of adding photos and movies. These services were provided by third party companies, and later, Twitter acquired them and integrated their services. Still, Twitter community keeps its tradition to communicate mainly in short text, rather than photo and movies.

For our research, the third and most significant feature of Twitter is its spreading function, "Retweet." Retweet, often shortened as RT, is a function of spreading other person's tweet to one's followers. With this function, the tweet evoking attention spreads to a broad population in very short time. Thus, the retweeted tweet by many users appears many times in Twitter log. We can use the number of RT of the same tweet as a sort of measure of curiosity and attentiveness. Overall text mining process was shown in Fig. 1.

3.2   Twitter Text Mining Methods

Tweet Collection. From 19th December 2016 to 27th July 2017, we have collected tweets those mentioned "Kure (呉)". Every one minute, searching command for "Kure" was executed from IFTTT (ifttt.com). IFTTT is a web service site to provide linkages between various web services and automatize their execution. The tweets hit to the search command were stored in Google drive spreadsheet.

Preprocessing on Gathered Tweet Data. For example, on 19th December 2016, there are 3553 tweets those contain "Kure". Roughly, from 2600 to 4700 tweets are collected per a day, during the period. Among these tweets, there are many tweets not related to Kure city. For example, the character "呉" share with ancient Chinese nation "wú." Since Chinese historical text "Sangoku-shi *(The Records of the Three Kingdoms)*" is very popular for Japanese people in longtime and there are many derivatives in novels, in a manga and even in computer games. We build automatic filtering script for these Sangokushi related tweets.

Text Mining Methods. At first, tweets are decomposed into Morpheme, and identified its word class. This process, *Morphological Analysis* was done with MeCab, Japanese morphological analyzer developed by Dr. Taku Kudou when he belonged to Nara Institute of Science and Technology (http://taku910.github.io/mecab/).

Then, we utilized word to word co-occurrence approach to analyzing tweets. The basic idea is as follows; There is a word, which appeared in a tweet. In the same tweet, there is another word. These two words have co-occurrence. If co-occurrence is frequent, these two words have strongly related. To quantify this idea, we used Jaccard coefficient. Jaccard coefficient s_{ij}, between $word_i$ and $word_j$ is defined as

$$s_{ij} = \frac{p}{p+q+r} \tag{1}$$

Where, p is the number of positive for both words (occurred $word_i$ and $word_j$ in the same tweet), q is the number of positive for $word_i$ and negative for $word_j$ (occurred $word_i$ in a tweet). r is the number of negative for $word_i$ and positive for $word_j$ (occurred $word_j$ in a tweet).

With Jaccard coefficient, similarity measures between words pairs were made into a similarity matrix. On this matrix, further multivariate analysis techniques are applied.

In this paper, we present two analyzing results. One result is on tweets mostly made on 19th February. That was from log file #121 + #122; those have 4,000 tweets (2000 tweets per file), the log from 23:02 of 18th February 2017 to 02:21 20th February 2017. These 4000 tweets were made during 27 h and 19 min. The combined file has 135,962 words. 56,440 words were used for analysis. Omitted words were non-words such like Twitter IDs, link URLs, symbols, signs, and Emoji.

Another result is on 27th and 28th July, those are tweet log file #400 and #401. These 4000 tweets were made during 21 h and 12 min, from 09:14 27th July 2017 to 06:26 28th July 2017. Filtered and combined file has 190,064 words of 3760 tweets. 84,403 words were used for analysis.

Cluster analysis was used to determining appropriate word classification. That merging methods were Ward and Group Average method.

On processing 19th February data, minimum appearance number was set to 60. It means words appeared at least 60 times were accounted for processing. Appearance frequency was expressed as a diameter of a bubble in MDS plot. Colors shown in MDS map are corresponding 6 clusters as shown in Fig. 5. On 27th July data, appearance number was set as also 60. Cluster analysis shows 7 clusters were appropriate, thus corresponding 7 clusters were shown in MDS plot.

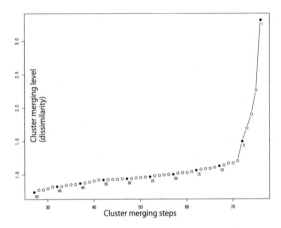

Fig. 5. Dissimilarity plot along with cluster merging steps (19th February data). From 50 clusters solution to 6 clusters, dissimilarity increases linearly. There is a jump between 6 to 5 clusters. This case, 6 cluster solution is recognized the appropriate result.

Figure 6 shows the result of 19th February. The plot shows the result of Multidimensional Scaling (MDS) on similarity matrix. Computing of similarity matrix, hierarchical cluster analysis and MDS were done with KH Coder [7, 8]. MDS optimization method was Kruskal. Optimization was assigned to 2-dimensional space.

3.3   Twitter Text Mining Results

19th February Tweets Analysis. At the center of the plot, there are two "Kure". This was caused by the morphological analyzer, that has divided it into two different part-of-speech (POS) tag. These two Kure points same Kure. Kure was located at very center of the plot, that means Kure is related to all words.

Including Kure, there are many words in orange. Near Kure, there are very tightly related "Hiroshima" and "Navy." Hiroshima is the large city next of Kure. Kure city was developed as shipbuilding site and port of Navy in modern age Japan. In orange color cluster, verbs like "go", "see", "like", "watch". Nouns are "book", "event" and "sea". Since these verbs and nouns are fundamental, thus they seem to relate general activity and sightseeing. "Curry" seems characteristic to Kure city, since Curry was spread to Japanese popular dishes, from Navy. Many restaurants in Kure serves

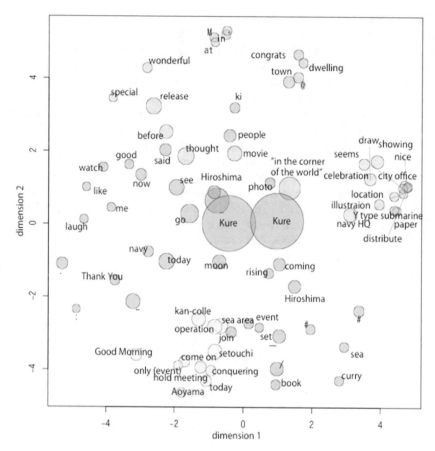

Fig. 6. MDS plot of words of tweets made on 19th February 2017.

different curry dishes to relate with Navy vessels. Each vessel has own curry recipe and they have cooperated with restaurants to serve their dish.

All words belong to yellow cluster in the bottom are related to the event of "Kantai-Collections". Originally it was an on-line sea-battle game of old navy vessels and warships, but ships are personified to girls. Anthropomorphism is not new in human history, in Japan and in China, such methods attract great number of players. Now, anime, manga and costume-play enthusiasts frequently have gatherings. Since Kure is the old navy city, this kind of events keep attracting young visitors.

Words of green-blue relate to the movie "In the corner of the world". The animation movie has great success in 2017, and in 2018, it was featured in many movie theaters in not only in Japan but also in US, France, Italy, Germany, UK and South-east Asian countries. The movie tells the story of Hiroshima girl who married with clerk man who is working at the navy office in Kure. Repeated massive airstrikes on Kure and A-bomb dropped on Hiroshima force destroyed ordinary family's life. Green words are applause to the movie. Words of purple are tweets noticing cerebration since 100-day past from the movie's release.

27th and Morning 28th July Tweets Analysis. This plot also has two "Kure." One belongs Green-Blue cluster. Green-blue cluster words related to tweets on "In the corner of the world" and Kure airstrike on the same day, 28th July of 1945. Cruiser "Aoba" and "Haruna" suffered bombing raid, then they swamped. The original tweet was made with colored photography of damaged cruiser "Haruna". It was one of series of neural network based colored historical photographs by Prof. Hidenori Watanabe of Tokyo University [9]. The tweets were largely retweeted. Swamped "Haruna" appeared in a scene of the "In the corner of the world", based on the same photograph. Left of Green-blue cluster, there is a small Light-Green cluster. These words show the similar tweets to above mentioned one. The tweet also mentioned today of 72 years ago, almost no counterattack due to lack of fuel, by the cause of the naval blockade.

The Red cluster at the bottom shows the tweets from Sunao Katabuchi, director of the "In the corner of the world". The movie draws the detail of Kure air-raid. The air-raid warning was issued, then "Suzu", the heroine of the movie, and her family was evacuated to a self-made shelter. Amber cluster of left middle also mentioned air-raid.

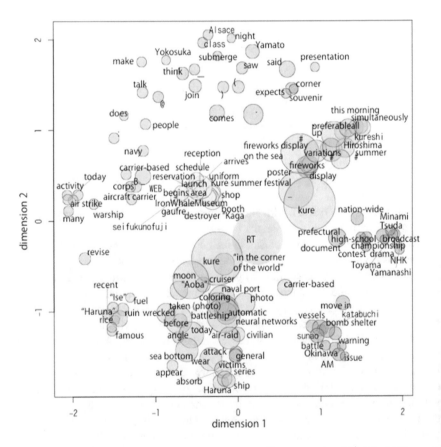

Fig. 7. MDS plot of words of tweets made on 27th and morning 28th July 2017

Air attack task force 38th of US Navy and B-24 bombers have departed to air-raid on Kure, this day 72 years ago.

Another Red cluster in right-middle was mentioned Kure-Mitsuta high school was honored as the first place at original radio documentary contest held by NHK.

Yellow cluster near the center is an announcement of Gaufre which package features destroyer Kaga (at WWII), that on sale shop at Iron Whale (Submarine) Museum. It also goes on sale at Kure summer festival.

The Purple cluster of right-hand corresponds fireworks display on the sea and its posters, which will be held next day. Those tweets were made by Kure city tourism official. Words of Blue cluster are verbs those commonly used in other tweets (Fig. 7).

4 Inference Rules for Sightseeing Web Application and Remarks

There are several implications to incorporating Kure sightseeing web application.

Under age of 65, visitors travel aims and expectations have interrelated and diversified. 6 different groups of them were observed (physical activities, experiences, relaxation, unique to Kure, cultural, historical (part of cultural)).

Over age 65 people have clear distinctions between expectations (Self-involvement/Visual experience/Historical visit and Relax). The result suggests differences in physical strength and activeness of elder visitors (Self-involvement vs. Passiveness).

"Interests and Hobbies" also have differences between age groups. Under 65, people show 3 strong correlative interest groups of relax and enjoy/creative/TV game and anime. Over 65 does not show such strong correlative structure, but physical strength suggests as a principal factor. Analyzed results were built into Kansei Engineering Artificial Intelligence helper application for Kure city sightseeing.

We have implemented Kansei survey results as recommendation inference rules in the sightseeing travel web application.

Twitter tweet mining shows the large difference of tweet topics between a day in February and in July. This mining result is reflected in daily recommendation of destinations and events. The mining was half-automated and manually reflected to the web application now, but we are trying fully automated renewal.

The URL of Kure sightseeing web application is https://kureikonavi.appspot.com and operated in beta testing. Japanese, English, Chinese, Simplify Chinese, and Korean languages are supported (Fig. 8).

Fig. 8. Kure sightseeing Web application, Kure-iko Navi. https://kureikonavi.appspot.com

References

1. Nagamachi, M.: An image technology expert system and its application to design consultation. Int. J. Hum. Comput. Inter. **3**(3), 267–279 (1991)
2. Nagamachi, M., Senuma, I., Iwashige, R.: A study of emotion technology. Jpn. J. Ergon. **10** (2), 121–130 (1974). (in Japanese)
3. Nagamachi, M., Lokman, A.M.: Innovations of Kansei Engineering. CRC Press, Boca Raton (2011)
4. Nagamachi, M. (ed.): Kansei/Affective Engineering. CRC Press, Boca Raton, Florida (2011)
5. Nagamachi, M., Lokman, A.M.: Kansei Innovation: Practical Design Applications for Product and Service Development. CRC Press, Boca Raton (2015)
6. Rosen, A.: Tweeting Made Easier, Twitter official blog (2017). https://blog.twitter.com/official/en_us/topics/product/2017/tweetingmadeeasier.html
7. Higuchi, K.: A two-step approach to quantitative content analysis: KH coder tutorial using anne of green gables (part I). Ritsumeikan Soc. Sci. Rev. **52**(3), 77–91 (2016)
8. Higuchi, K.: A two-step approach to quantitative content analysis: KH coder tutorial using anne of green gables (part II). Ritsumeikan Soc. Sci. Rev. **53**(1), 137–147
9. Watanabe, H.: Reboot memories – memory inheritance based on communication emerged by flowing records. Ritsumeikan Heiwa Kenkyu (Ritsumeikan Peace Study) **19**(1), 1–12 (2018)

The Impact of Design Semantic on the User Emotional Image in the Interface Design

Ren Long[(⊠)] and Jiali Zhang

School of Mechanical Science and Engineering,
Huazhong University of Science and Technology, Wuhan, China
longren@hust.edu.cn

Abstract. Interface design, as a medium of communication between human and computer, transmits information through design semantics and makes the user generate corresponding emotional changes, which affects the user experience. This paper explores the shape semantic and color semantics of the interface design by referring to the product semantics, and by using the theory of Kansei engineering, analyzes the influence of design semantics on the users' emotional images. Experimental results show that the design semantics affect the user's emotional image, and the influence of color semantics on the user's emotional image in the interface design is greater than that of the shape semantics. Therefore, in the interface design process, according to different product positioning, select the appropriate graphic elements and color collocation, can accurately convey the information expressed by the designer, and match the emotions generated by the user.

Keywords: Interface design · Design semantic · Kansei engineering

1 Background

In recent years, with the rapid development of computer science, more and more various types of software, interface design has gotten more and more attention by software developers and users. Today's interface design not only need to meet the functional needs of users, but also to have a unique emotional foundation to win the user's identity, which requires designers to consider a variety of factors in designing interface and use a variety of theories to guide the design. In the perceptual age with increasing emotional needs, consumers put forward higher requirements for digital products to meet people's psychological needs. It's a question worth considering that how designers can reflect the feelings of users in the design and design products that meet the emotional expectations of people.

© Springer International Publishing AG, part of Springer Nature 2019
S. Fukuda (Ed.): AHFE 2018, AISC 774, pp. 323–333, 2019.
https://doi.org/10.1007/978-3-319-94944-4_35

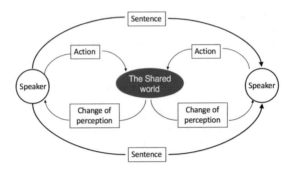

Fig. 1. The interaction of people-people

2 Significance

In the process of human interaction, people communicate with each other through the sentence and then understand each other by the sematic after that they change emotion (Fig. 1).

There are many similarities between dialogue process of "human-machine" and "human-human": the user interacts with the computer interface through input orders, after obtaining the user's order, designers use the computer program to process the user's input and the user get visual feedback of the program result through the organization of the interface graphic elements, which complete an interaction [1]. In this process, the organization of graphic elements play a role of language in the dialogue of "human-human" (Fig. 2). Therefore, the interface design is also interactive design. The communication between interface and users is made through the semantics which is expressed by the interface. Interface design provides visual stimulation through semantics, which enables users to visually associate with the contents of the interface, so as to change the user's emotional images. Users have different emotional experience on the interface, therefore affecting the user's stay time on the internet.

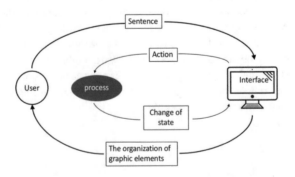

Fig. 2. The interaction of people-interface

3 Shape Semantic and Color Semantic of Interface Design

As an interdisciplinary research subject, Interface design includes cognitive psychology, linguistics and so on. As a branch of linguistics, product semantics (especially the shape semantic and color semantic) plays an important guiding role in the design. The method is also applicable to the interface design [2]. This paper is based on the use of product semantics, try to introduce the principles of Kansei engineering into the steps of user interface design, and provide a strong basis and guidance for the whole process which turn perceptual user data into rational design results. Discussing the problem about the impact of shape semantic and color semantic on user's emotional images in the interface design. experimental data is used to show the important role of shape semantic and color semantic in the interface design, and their influence on user's emotional images are compared in this paper.

In the process of human communication, the meaning expressed through different sentences enables the listener to produce different perceptions and emotions. In the interface design, the process of using the design elements to form an interface is equivalent to the process of using vocabularies to construct sentences in the linguistics. The designer constructs the image of the interface through the use of words and graphics, and the layout of the design elements, to make users produce perception and emotion on the interface. Gestalt psychologist Rudolph Anheim (1954/1985) believes that even the simplest lines can express emotions or feeling [3]. Through modeling semantics, interface design can reflect the grade, quality, fun, etc. of the website, allow users fell the belongingness and identity, thus establishing the characteristic of the interface.

Color is the most abstract language and a completely subjective cognitive experience. People's perception of color will lead people to produce some emotional psychological activity. The color semantics in the interface design comes from the visual impact and physiological stimuli of color to human beings. And from this people generate a wealth of empirical associations and physiological associations, resulting in a specific psychological experience. Therefore, in the interface design, color semantic is particularly important. The interface design which has good user experience, better characteristic and function through using the color semantic to resonate powerfully with user, to hold the user's psychology, recognition and understanding.

4 Experiment Design

Questionnaire method is used to do the experiment in this paper. And then statistical method is used to analysis the collected data to compare the different influence on human emotional images of each sample.

In the interface design, the combination of words, titles, pictures, etc., will form a variety of lines and shapes on the page. The combination of these lines and shapes constitutes the overall artistic effect on the interface. In the early collection process of interface layout, a lot of interface layout are found, and generally they can be divided into three types: straight line (rectangle), arc (circle) and the combined use of both. We chose a thematic activity of the largest e-commerce platform in China as the interface prototype. In order to better compare the differences between the shape elements, we re-create the interface with straight line (rectangle) and arc (circle). Combining these two interfaces with 7 coloring schemes: red, yellow, green, blue, purple, black, white. In order to ensure that the information of text and picture will not interfere the result of user emotional image, we blurred all the text and picture, only appear monochrome block with different transparency. At last, identified 14 samples (Fig. 3).

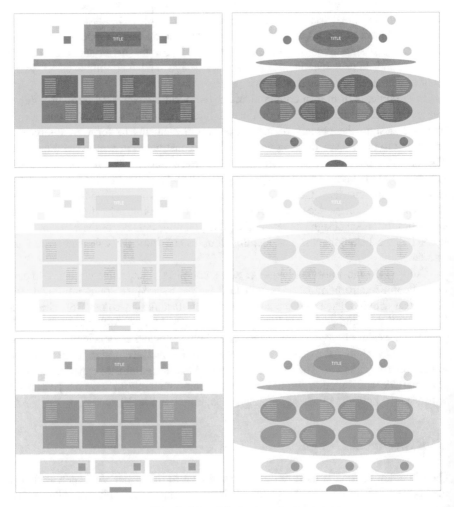

Fig. 3. Materials for experiment

Fig. 3. (*continued*)

According to the six types of basic emotional prototypes of emotional psychology, we make a Emotional Image Questionnaire of interface with Likert scale using love, surprise, fear, happiness, sadness and anger as option of emotional images. 100 college students (20–24 years old) participated in the study, all of them from universities and non-design industries.

Experimental Process. Participants use the same computer display, watch interface in the order of the samples, fill in the questionnaire after 5 s for each interface, and then choose one of the emotional images that best match the psychological response according to the subjective feelings after watching the interface.

5 Statistics and Analysis of Results

5.1 Percentage Statistics Analysis of Emotional Images of User on Different Interfaces

According to the percentage value of different emotional image from choice, we found the emotional images of user on red interface is affected by the shape change, but the red interface will give people more happy and angry emotional images. The yellow interface is less affected by shape change, and always give people more joy and surprise emotional images. The green interface will be affected by the change of shape. The green interface of arc (circle) structure will generate more happy emotions, and the green interface of the straight line (rectangle) structure will generate more surprised emotions. Blue color interface is basically not subject to the impact of changes in shape, and always give user more emotions for the happy, surprise. The emotional image of purple is not clear, the emotional percentage value is close, and is less affected by the changes of shape. Black interface will also be affected by the shape change, especially in the happy mood. The white interface is less affected by changes in shape. In general, red and green are more affected by changes in shape, while yellow and purple are the least affected by the shape.

5.2 Line Chart of User Emotions on Different Interfaces

According to the line chart of emotional images of each sample, we found when the shapes are the same, the emotional images will react differently with the change of color, and there are obvious differences in every color, only purple shows less difference (Figs. 4 and 5).

However, when the color is same, it can be clearly seen that the trend toward emotional images is almost the same of different shapes, the difference of each emotional is small.

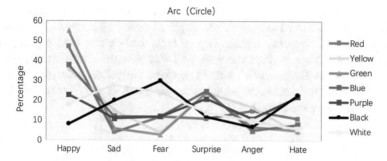

Fig. 4. The user emotions of different colors on arc (circle) interface

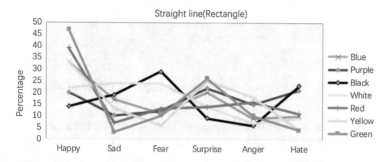

Fig. 5. The user emotions of different colors on straight line (rectangle) interface

The red color shows the highest emotional happiness, especially in the arc structure. Probably because red is commonly used to celebrate holidays in China, the red interface creates a happy emotional image. So the red color is especially suitable for festive and happy theme interface (Fig. 6).

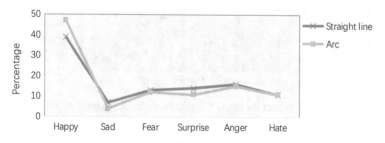

Fig. 6. The user emotions of different shapes on red interface

In the yellow color interface, it can be clearly seen that both the straight-line shape and the arc shape show happier and more surprised emotional images. The reason may be that yellow in China often represents good luck, the emperor, etc., to give people a happy emotional image. At the same time, yellow is often used for warnings, due to its maximum stimulus for people's vision. Therefore, the yellow color can be used in consumer's business interface to stimulate consumption or the interface which need shows lively, happy emotion (Fig. 7).

Fig. 7. The user emotions of different shapes on yellow interface

The green color interface is somewhat similar to the yellow color interface, and the emotional image brought to users is also focused on both happiness and surprise. However, the emotional image of angry in the green interface is obviously less than the yellow interface. The reason may be green is often used to represent the forest, peace, people have a harmonious and natural image. Therefore, in the interface design which need reflects the natural theme, both arc shape and green color are good choices (Fig. 8).

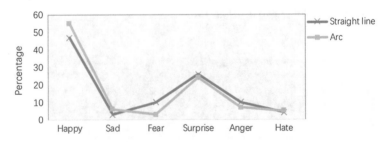

Fig. 8. The user emotions of different shapes on green interface

The blue color interface is shown as a happy and surprisingly intense interface, especially with an arc structure. Comparing with other colors, the emotional images of sadness, fear and distasteful on blue interface are slightly higher, probably due to people often associate blue with the sky, the sea and so on, these kinds of images are always not certainty for people, so will also bring negative feelings (Fig. 9).

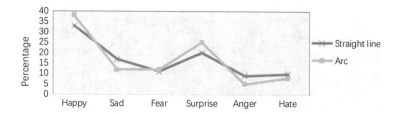

Fig. 9. The user emotions of different shapes on blue interface

Emotional images of the purple color interface is not obvious, but it can be seen that the emotion image of happiness, surprise and hate are higher in intensity (Fig. 10).

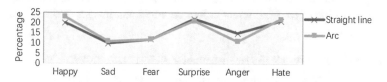

Fig. 10. The user emotions of different shapes on purple interface

In different shape structures, the fear of black interface, emotional images of fear and hate are most obvious, especially in the straight line interface. This shows that in the interface design requires the careful use of black square, large module black interface is easy to make users discomfort (Fig. 11).

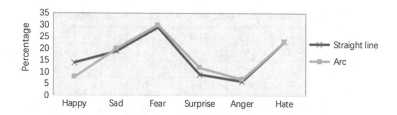

Fig. 11. The user emotions of different shapes on black interface

White color interface reflects the emotional image of sad is highest, especially the arc structure of the white interface. Emotional images of surprise, angry, and hate with lower choose suggests that white is easy to fit into the interface design of different types and styles, but the use of too many white colors can also make people sad and fear (Fig. 12).

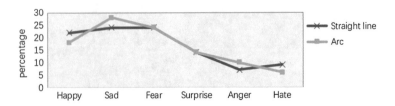

Fig. 12. The user emotions of different shapes on white interface

5.3 Effects Compare of Shape Semantic and Color Semantic on User Emotional Images

We compared the variance of the effect of different shapes on emotional Images with the variance of the effect of different colors on emotional images (Tables 1 and 2).

Table 1. Variance of the effect of different shapes on emotional images

	Red	Yellow	Green	Blue	Purple	Black	White
	8	3	8	5	3	6	4
	3	2	3	5	1	1	4
	1	2	7	1	0	1	0
	3	1	2	5	1	3	0
	1	1	3	4	4	1	3
	0	0	1	2	1	0	3
Variance	6.88889	0.91667	6.66667	2.55556	1.88889	4	2.88889

Table 2. Variance of the effect of different colors on emotional images (Straight line)

	6	−8	6	19	25	−14	0	13	19	14	27	33	19	6
	−6	4	−10	−3	−12	10	−4	3	−6	−14	−7	−16	−2	−9
	7	3	2	1	−16	−4	−5	−6	−23	−1	−2	−19	−18	−17
	−11	−12	−6	−8	5	−1	5	3	16	6	4	17	11	13
	−2	6	7	1	10	8	9	3	12	1	−5	4	3	9
	7	7	1	−10	−12	0	−6	−17	−19	−6	−17	−19	−13	−2
Variance	49	53	37	89	215	62	30	86	281	77	185	395	164	110

In this paper, the greater the variance, the greater the impact of variables. It can be clearly shows in the table, the change of color semantic has a greater impact on the user's emotional image than the shape semantic on the interface.

6 Conclusion

Research shows that design semantics affect the participants' perception and emotional imagery of the interface design. In the same shape interface with different colors, participants' perceived color semantics is much higher than the shape semantics. It's shown that color semantic plays a leading role in the user's perception of the interface design. In contrast, the impact of color on visual perception is far greater than the morphology, confirming the opinion that the impact of the Color semantics in the interface design on the user's emotion is faster and stronger than that of modeling semantics. Therefore, when designing the interface, designers need to consider the impact of different design semantics on users. According to product positioning and user emotions, choose the best psychological elements of the user and color elements to design, in order to design the interface which can match the user emotion. This study is expected to play a part of guiding role in future interface design. Whether the interface design semantics causes emotional image of the interface are different because of cultural differences between countries, it still needs to do further research.

References

1. Preece, J., Rogers, Y., Sharp, H.: Interaction Design: Beyond Human-Computer Interaction. Wiley, Hoboken (2002)
2. Wenting, H.: Research on Humanized Interface Design Based on Semiotics. Hefei University of Technology (2010)
3. Arnheim, R.: Art and Visual Perception: A Psychology of the Creative Eye. University of California Press, Berkeley and Los Angeles (1954)
4. Cai, Z.: The Research of User Interface and Interaction Design of Dimension Based on Kansei Engineering and Color Psychology. Nanjing University of Aeromautics and Astronautics (2009)
5. Yamamoto, K.: Kansei Engineering - The Art of Automotive Development at Mazda. The University of Michigan, Ann Arbor (1986)
6. Nakamura, T., Sato, T., Teraji, K.: Arrangement of color image words into the non-luminous object color space. J. Color Sci. Assoc. Jpn. 50(17), 1740–1747 (1994)
7. Wang, M., Kuo, Y.-S., Lai, A.C.: Primary school students' emotional images of combinations of shapes and colors. J. Hum. Soc. Sci. 3(1), 81–94 (2007)
8. Chen, X.: Application of Kansei engineering in color design of modern products. Theory Res. 16, 182–183 (2009)
9. Sun, D., Sun, H.: Color Psychology (2017). ISBN 9787542656605

Understanding Young Chinese Consumers' Perception of Passenger Car Form in Rear Quarter View by Integrating Quantitative and Qualitative Analyses

Chunrong Liu$^{(\boxtimes)}$, Yang Xie, Yi Jin, and Xiaoguo Ding

Shanghai Jiao Tong University, Shanghai 200240, China
cheeronliu@sjtu.edu.cn

Abstract. The characteristics of young Chinese consumers' perception of passenger car form in rear quarter view are explored by integrating quantitative and qualitative analyses to help designers understand young consumers' aesthetic appreciation. Based on investigation into consumers' judgment on similarity between car forms, the classification and the perceptual map are obtained by cluster analysis and multidimensional scaling, respectively. The distribution and form change of 78 samples in the perceptual map are analyzed qualitatively. The findings show that (1) the characteristics of consumers' perception of passenger car form in rear quarter view can be mapped by dimension reduction onto the horizontal and the vertical dimensions in the two-dimensional perceptual mapping space; and (2) consumers' perception of and evaluation on form differences can be interpreted by three types of form design features, which include the local feature in rear view, the local features in side view, and the overall transitional features between the rear and side of passenger car form.

Keywords: Passenger car form in rear quarter view · Consumer research
Perceptual mapping · Quantitative and qualitative analyses

1 Introduction

Product form is a carrier of communication between designers and users as consumers. Designers embody their deep emotions and psychological thinking to life and arts in product form design and users perceive product form as the objective of cognition [1]. With product form features, users make emotional response to product form when touching and perceiving them [2]. The passenger car product, carrying many connotations, is a carrier not only to convey the spirit of innovative design but the expression of the consumers' yearning for a better life, not only a simple product.

Focusing on the innovation and generation of aesthetic form in product design and automobile styling and design fields, various studies have been conducted by researchers, involving in a huge quantity of topics including form feature [3–7], passenger car's interior [8], dashboard design [9], brand's genes [10, 11] of passenger car form (in front view [12, 13], front quarter view [14, 15], side view [16]), relationship between passenger car form and traditional culture [17], reflective attribute of

© Springer International Publishing AG, part of Springer Nature 2019
S. Fukuda (Ed.): AHFE 2018, AISC 774, pp. 334–343, 2019.
https://doi.org/10.1007/978-3-319-94944-4_36

automobile styling surface [18], style features [19, 20], image cognition [21, 22], cognitive relationship between the dynamic feeling and form features [20], cognition differences between automobile designs [23] and between designers and users [24], consumers' purchase decisions and preferences [14, 25–28], consumers' latent desires for product form [29], feeling quality of products [30], form attractiveness evaluation [31, 32], semantic properties [33], emotion toward car styling [34, 35], application of anthropomorphism in car styling innovation [36], automobile product experience [37] and so on. These studies attempt to understand the characteristics of consumers' perception of product and automobile form and to explore the perceptual differences between consumers and designers.

It has been proven that when Chinese consumers observe and perceive passenger car form, the rear quarter view is an important point of view that reflects the design intent and product family features, although passenger car form in front quarter view is the greatest concern [15, 25]. Meanwhile, some studies have discovered that form of the taillight cluster in rear view is one of the key points of car form design [3], and the form of rear autobody has a great influence on the forms of other areas [24]. These findings show the importance of passenger car form in rear quarter view in designing whole autobody form. In this study, the characteristics of young Chinese consumers' perception of passenger car form in rear quarter view are explored to help designers understand young consumers' aesthetic appreciation of passenger car form in rear quarter view.

2 Quantitative Analysis Procedure

2.1 Stimuli-Preparation Process

Passenger car is in main force role in current Chinese car market [38]. This study focuses on the triple-compartment passenger cars with engine displacements from 1.5L to 2.4L, involving in twenty-five main car brands of production cars in Chinese market including FAW, Dongfeng, SAIC, GAC, Changan, Chery, BYD, FAW-Toyota, FAW-VW, FAW-Mazda, FAW Audi, SAIC-VW, GAC-Honda, Dongfeng-Honda, Dongfeng-Nissan, Dongfeng-Peugeot, Dongfeng-Citroen, Dongfeng-Renault, Beijing-Hyundai, BBAC (Beijing Benz), Changan-Mazda, Changan Ford, BMW-Brilliance, SAIC-GM Buick and SAIC-GM Chevrolet.

Totally, 129 pictures of passenger car form in rear quarter view of motioned-above brands are collected from websites such as online forums. All pictures are selected in a way ensuring they were photographed visually at as same viewing angle as possible to reduce experimental error in the coming-up investigation phase, and are processed with the same wheel hub by blurring radially so as not to be in a specific wheel hub form. Meanwhile, in order to avoid in the coming-up investigation phase the possibility that car body color distracts the subject from the passenger car form presented in the pictures, all these pictures are converted into black-and-white color mode, and passenger car form in each picture is presented centrally on the white background while the brand logo and license plate in each picture is removed (as shown in Fig. 1). Finally, all 129 samples in pictures of passenger car form in rear quarter view are marked randomly with serial numbers of V1, V2, ⋯, V128, V129.

Fig. 1. An interface of the grouping task tool

2.2 Pilot Experiment of Grouping Task

In order to reduce the number of form samples, a pilot experiment is conducted at the early stage of the study. Ten subjects from design professional background are invited to evaluate the similarity between one hundred and twenty-nine samples of passenger car form in one hundred and twenty-nine pictures, respectively. Every subject is kindly asked to estimate on the similarity between any one pair of passenger car form in rear quarter view among all samples in pictures, and to classify samples of passenger car form according to his/her own judgment on form's similarity in a relatively quiet environment during experiment with an interactive grouping task tool developed by the first author's research team for similarity judgment (as shown in Fig. 1).

One piece of similarity matrix data is generated by the grouping task tool when a subject finished his/her trial and evaluation. As result of pilot experiment, ten effective pieces of similarity matrices data are acquired.

An averaged similarity matrix is obtained by averaging above ten original similarity matrices. This averaged similarity matrix is then analyzed by hierarchical cluster analysis method and a dendrogram as result is plotted.

By observing and analyzing the dendrogram, it is reasonable to classify one hundred and twenty-nine samples of passenger car form into 9 categories according to the related cluster analysis principle [39] that it is ideal for all samples to be classified as 'evenly' as possible into categories. Then another cluster analysis in K-Means method is carried out by setting group value nine. According to the distance of a form sample to the center of the corresponding category that it belongs to, those with relatively closer distance values in its corresponding category are selected. Totally, 78 form samples displayed in pictures are remained for next formal experiment and marked randomly with serial numbers of V1, V2, ⋯, V77, V78, respectively.

2.3   Formal Experiment of Grouping Task

In formal experiment phase, 48 young subjects are invited to participate in similarity judgment experiment. These subjects, consisting of 25 male consumers and 23 female consumers with age of 25 to 35, are kindly asked to estimate the similarity between the 78 samples of passenger car form in rear quarter view in a relatively quiet environment during experiment with the interactive grouping task tool mentioned above. Finally, 48 pieces of effective data are acquired, with which an averaged similarity matrix is calculated by averaging all forty-eight similarity matrices from forty-eight subjects.

This averaged similarity matrix is then analyzed by hierarchical cluster analysis method and a dendrogram as result is plotted. According to related principles of cluster analysis method [39], it is found reasonable to classify seventy-eight samples of passenger car form into 7 categories. Furthermore, a cluster analysis with K-Means method is completed with group value seven, resulting in seventy-eight form samples being classified into 7 clusters which are named G1, G2, G3, G4, G5, G6 and G7 as an identifier to each category for the next research, respectively.

2.4   The Perceptual Mapping

By transforming the motioned-above averaged similarity matrix in formal experiment of grouping task, a dissimilarity matrix is obtained. With multidimensional scaling method using the dissimilarity matrix, subjects' perceptual map of passenger car form in rear quarter view is plotted (as shown in Fig. 2. In this figure, a polyline with segments in the same color is illustrated for every category to connect circles that mark the positions of form samples in the same category in the perceptual map, and the horizontal and vertical axes are outlined to facilitate the description below).

Analyzing the distribution of form samples in each category in the perceptual map, it can be observed that the perceptual map presents visually the characteristics of subjects' perception of passenger car form in rear quarter view as follows: (1) the form samples V75 to V41 in the first category (G1) are mainly distributed in the periphery except the bottom-right corner area; (2) the form samples V21 to V67 in the second category (G2) are mainly distributed in the central area along the horizontal dimension; (3) the form samples V47 to V70 in the third category (G3) are mainly distributed in the central area along and closer to the horizontal dimension than samples in G2; (4) the form samples V24 to V50 in the fourth category (G4) are mainly distributed in the periphery except the area close to the right end of horizontal axis; (5) the form samples V46 to V33 in the fifth category (G5) are mainly distributed in the periphery except the upper-right corner area; (6) the form samples V4 to V74 in the sixth category (G6) is mainly distributed in the periphery closer to the central area than samples in G5; and (7) the form samples V9 to V27 in the seventh category (G7) is mainly distributed along and close to the vertical dimension.

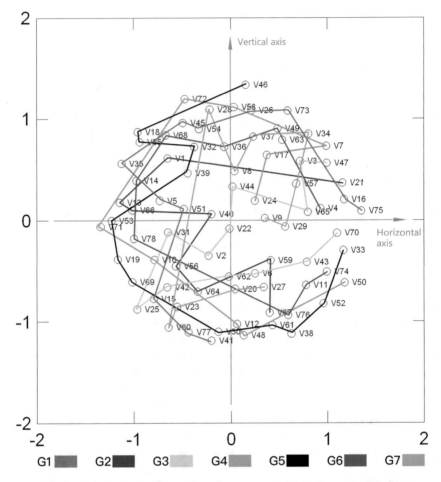

Fig. 2. The distribution of samples of seven categories in the perceptual map

3 Qualitative Analysis Procedure

The characteristics of consumers' perception of passenger car form in rear quarter view are embodied in the distribution and its regularity in form changes in the two-dimensional perceptual map plotted by multidimensional scaling by means of dimension reduction. To set up qualitative analysis and explore the regularity in form changes in the perceptual map, seventy-eight pictures corresponding to form samples are printed and placed on a physical plate in the corresponding locations as presented in the perceptual map, and are observed and analyzed with help of related car styling and design knowledge. It is found that along the horizontal and vertical dimensions in the perceptual map, the distribution of seventy-eight forms implies observable regularity in form changes of passenger car form in rear quarter view as follows.

3.1   Form Change Along the Horizontal Dimension

From the left to the right end along the horizontal axis in the perceptual map, the transitional feature of the rear bumper form to the rear side wall is embodied as follows: for the majority of the samples of passenger car form scattered and located onto the left end of horizontal axis in the perceptual map, the outlines of rear bumper form extend to the rear side wall as illustrated in the left figure in Fig. 3, while for the majority of those scattered and located onto the right end of horizontal axis in the perceptual map, the outlines of rear bumper form extend to the tail-light area as illustrated in the right figure in Fig. 3.

Fig. 3. The transitional features of rear bumper form

3.2   Form Changes Along the Vertical Dimension

By comparing the majority of the samples of passenger car form scattered and located onto the upper end of vertical axis with the majority of those scattered and located onto the lower end of vertical axis in the perceptual map, the regularities of the morphological transitional features of passenger car form are discovered as follows.

Firstly, the visual inclining angle of the C-pillar as viewed from the rear quarter view shifts from a larger angle as illustrated in the left figure in Fig. 4 to a smaller angle as illustrated in the right figure in Fig. 4.

Fig. 4. The features of C-pillar's inclining angle

Secondly, the beltlines appearance shifts from a larger bulging degree as illustrated in the left figure in Fig. 5 to a smaller bulging degree as illustrated in the right figure in Fig. 5. In addition, the beltline's rake angle shifts from a greater degree as illustrated in the left figure in Fig. 6 to a smaller degree as illustrated in the right figure in Fig. 6.

Thirdly, the transitional features between the auto beltline and the upper outline of the taillight shift from an incoherent state as illustrated in the upper figure in Fig. 7 to a coherent state as illustrated in the lower figure in Fig. 7.

Fig. 5. The features of beltline's bulging degree

Fig. 6. The features of beltline's rake angle

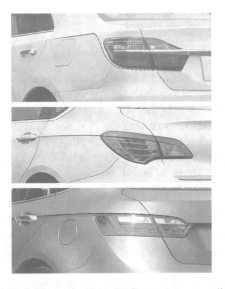

Fig. 7. The transitional features between beltline and upper outline of the taillight

Fourthly, the local form of the license plate area shifts from the open morphological state as illustrated in the left figure in Fig. 8 to the closed morphological property as illustrated in the right figure in Fig. 8.

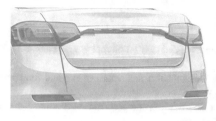 

Fig. 8. The form features of license plate area

4 Conclusion

In China, passenger car form has been the fourth most important factor among fourteen main factors that affect consumer's purchase decision [40]. It is very necessary for automakers to understand the perceptual identification of passenger car form of young Chinese consumers and form's dissimilarity and variance exposed in their morphological cognition in the fiercely competitive Chinese market of passenger car, where passenger cars' consumers tend to be younger obviously [41] and young consumers begin to be in main force role. In this study, the young Chinese consumers' perceptual characteristics of triple-compartment passenger car form are investigated by quantitative cluster analysis and multidimensional scaling methods, and by qualitatively examining the potential clues to morphological shifts in the perceptual map.

The findings show that the characteristics of consumers' perception of passenger car form in rear quarter view can be mapped onto the horizontal and the vertical dimensions in the perceptual map by dimension reduction, presenting the differential form features and gradual changes along the horizontal and vertical dimensions in the perceptual map. In conclusion, young consumers' perception of and evaluation on the differences of passenger car form in rear quarter view can be interpreted by both overall and local form features in three types, involving in (1) one local feature in rear view, i.e., the form feature of license plate area; (2) three local features in side view, including the features of C-pillar's inclining angle, beltline's bulging degree and beltline's rake angle; and (3) two overall features in rear quarter view of transition between the rear and side surfaces, including the transitional feature of rear bumper form to its adjacent autobody section, and the transitional feature between beltline and upper outline of the taillight.

These findings imply it can be helpful for a carmaker to distinguish its own passenger car form in rear quarter view from competitors by intentional design differentiation for targeted young consumer population with both overall and local form features in rear quarter view while keeping sound coordination in passenger car for in rear quarter view.

References

1. Liang, Q.: Correlation between car styling elements and user emotional image. Pack. Eng. **37**(20), 14–19 (2016). (in Chinese)
2. Zhao, D.: Research framework of emotion classification and value of product modeling. Pack. Eng. **37**(20), 1–8 (2016). (in Chinese)
3. Zhao, J., Tan, H., Tan, Z.: Car Styling: Theory, Research and Application. Beijing Institute of Technology Press, Beijing (2010). (in Chinese)
4. Hu, W., Zhao, J., Zhao, D.: Study on styling image of vehicle based on form feature lines. China. Mech. Eng. **20**(4), 496–500 (2009). (in Chinese)
5. Zhao, D., Zhao, J.: Automobile form feature and feature line. Pack. Eng. **28**(3), 115–117 (2007). (in Chinese)
6. Zhao, D.: The acquisition and representation of the knowledge on automobile form features. Master's thesis, Hunan University, Changsha (2007). (in Chinese)
7. Li, W.: Research on prototype features and form design of automobile headlights. Master's thesis, Hunan University, Changsha (2015). (in Chinese)
8. Jindo, T., Hirasago, K.: Application studies to car interior of Kansei Engineering. Int. J. Ind. Erg. **19**, 105–114 (1997)
9. Zheng, X.: a study of passenger car dashboard assembly styling and its design. Master's thesis, Shanghai Jiao Tong University, Shanghai (2016). (in Chinese)
10. Zhang, W.: A Car Styling-Based Study: The Design Methodology Based on Brand DNA. Beijing Institute of Technology Press, Beijing (2012). (in Chinese)
11. Hu, W., Chen, L., Liu, S., et al.: Research on the extraction and visualization of vehicle brand form gene. Mach. Des. Res. **27**(2), 65–68, 79 (2011). (in Chinese)
12. Wang, D.: Research on front-part styling design of SUV based on Kansei Engineering. Master's thesis, Jilin University, Changchun (2015). (in Chinese)
13. Zhao, L.: Optimization Design of Front Face Based on Kansei Engineering. Master's thesis, Northeastern University, Shenyang (2012). (in Chinese)
14. Liu, C., Zhu, X.: Young consumers' perception of form style of passenger cars. Pack. Eng. **37**(24), 6–10 (2016). (in Chinese)
15. Liu, C.: Design Strategy Development for Product Innovation. Shanghai Jiao Tong University Press, Shanghai (2015). (in Chinese)
16. Huang, D.: Research on side-view forms of sedans based on the form dynamic theory of visual perception. Master's thesis, Shanghai Jiao Tong University, Shanghai (2015). (in Chinese)
17. Huang, Z.: Modern car design and traditional culture. Art Des. (Theory) **1**, 184–186 (2010). (in Chinese)
18. Liang, Q., Zhao, J.: Classification and reflective attribute of automobile styling surface. Pack. Eng. **36**(4), 43–45, 50 (2015). (in Chinese)
19. Min, G.: Research on image of dynamic car styling. Master's thesis, Hunan University, Changsha (2007). (in Chinese)
20. Zhou, W., Zhao, J.: Dynamic Car styling based on feature comparison model. Pack. Eng. **37**(20), 50–53 (2016). (in Chinese)
21. Zhao, D., Jing, C.: The credibility evaluation methods of vehicle modeling based on the designer-user image cognition model. Pack. Eng. **36**(12), 78–82, 122 (2015). (in Chinese)
22. Hu, T., Zhao, J., Zhao, D.: Imagery cognition differences between designers and users automobile modeling. Pack. Eng. **36**(24), 33–36 (2015). (in Chinese)
23. Zeng, X., Zhao, D., Zhao, J.: Evaluation method for car styling rendering based on cognitive differences. Pack. Eng. **38**(6), 177–181 (2017). (in Chinese)

24. Hu, T., Zhao, J., Zhao, D.: Pattern study of design imagery processing and aesthetic cognition. Zhuangshi **2**, 104–105 (2015). (in Chinese)
25. Liu, G., Huang, D., Liu, C.: Preference of car form feature based on DEMATEL method. China Pack. Ind. **24**, 34–36 (2014). (in Chinese)
26. Jiang, C.: Styling and development of heavy truck based on the study of consumers' preferences. Master's thesis, Shanghai Jiao Tong University, Shanghai (2013). (in Chinese)
27. Tan, Z., Zhao, J.: User perceived preference of automobile styling. Pack. Eng. **37**(20), 9–13 (2016). (in Chinese)
28. Huang, Z.: Research on evolution of automotive design style. Master's thesis, Shanghai Jiao Tong University, Shanghai (2010). (in Chinese)
29. Chang, H., Lai, H., Chang, Y.: Expression modes used by consumers in conveying desire for product form: a case study of a car. Int. J. Ind. Erg. **36**(1), 3–10 (2006)
30. Lai, H., Chang, Y., Chang, H.: A robust design approach for enhancing the feeling quality of a product: a car profile case study. Int. J. Ind. Ergonomics **35**(5), 445–460 (2005)
31. Chang, H., Lai, H., Chang, Y.: A measurement scale for evaluating the attractiveness of a passenger car form aimed at young consumers. Int. J. Ind. Ergonomics **37**(1), 21–30 (2007)
32. Chang, H., Chen, H.: Optimizing product form attractiveness using taguchi method and TOPSIS algorithm: a case study involving a passenger car. Con. Eng. Res. App. **22**(2), 135–147 (2014)
33. Wang, B., Li, B., Hu, P., et al.: Semantic-oriented shape exploration for car styling. In: 2014 5th International Conference on Digital Home, pp. 368–373. IEEE Press, Guangzhou (2014)
34. Wan, X., Che, J., Han, L.: Car styling perceptual modeling based on fuzzy rules. App. Mech. Mater. **201–202**, 794–797 (2012)
35. Li, R., Dong, S.: Styling aiding methods based on emotional words and car styling prototype fitting. Pack. Eng. **37**(20), 25–29 (2016). (in Chinese)
36. Yusof, W.Z.M., Hasri, M., Ujang, B.: Innovation in form generation: anthropomorphism for contextual collaboration in car styling. In: Proceedings of the 2nd International Colloquium of Art and Design Education Research (i-CADER 2015), pp. 189–204. Springer, Singapore (2016)
37. Jing, C., Zhao, J.: The automobile modeling evaluation from the perspective of product experience. Pack. Eng. **35**(22), 17–21 (2014). (in Chinese)
38. Report on Local Brands in China's Car Market. http://www.motortrend.com.cn/umbraco/surface/Article/ArticleDetail/5279. (in Chinese)
39. Zhang, W.: Advanced course of statistical analysis in SPSS. pp. 235–251, Higher Education Press, Beijing (2004). (in Chinese)
40. Yang, F.: Study on influence of passenger car's form on consumers' purchase decision. Master's thesis, Shanghai Jiao Tong University, Shanghai (2016). (in Chinese)
41. The Influence of Slowing Growth of Chinese Macroeconomics in 2016 on Automobile Sales and Characteristics, Demand and Capacity of Consumer Populations. http://www.chyxx.com/industry/201612/476907.html. (in Chinese)

Research on Female University Students' Winter Clothing Colors in Guangdong, China

Huajuan Lin[1(✉)], Xiaoping Hu[1], and Wenguan Huang[2]

[1] Faculty of Fashion and Accessory Design, Guangzhou Higher Education Mega Centre, South China University of Technology, Panyu District, Guangzhou 510006, People's Republic of China
519085157@qq.com, huxp@scut.edu.cn
[2] Guangdong Overseas Chinese Vocational School, Tianhe District, Guangzhou 510520, People's Republic of China
1129900107@qq.com

Abstract. This paper takes the 210 female college students at the age of 18–23 in Guangzhou as the research subject, observes the colors of their winter clothing and makes ranking statistics, so as to analyze the reasons for selection of different colors. The tendency of dress color selection for female college students is investigated. Finally, it summarizes the characteristics of dress color for female college students in Guangdong, and analyzes the factors that may affect color selection, such as native place, age, specialty, personality, skin color, figure, temperament, popularity, personal preference and clothing consumption. It is important to study the physical, psychological and environmental factors that affect dress selection for female college students, which is of great value for the product design and marketing of related clothing brands. It can also provide guidance on the misunderstandings and related problems in the dress color for the female college students.

Keywords: Female university students · Clothing colors · Research Guangdong

1 Introduction

Colors are being increasingly used in every aspect of people's lives and gaining more and more attention. In terms of clothing, colors reflect individuals' temperament and style, serving as a means to express feelings and establish personal style.

Guangdong is a province with huge economic contribution to China. Guangzhou, the provincial capital, is a world-class metropolis and a major city in China. Located in South China, Guangzhou features a subtropical monsoon climate. Without snowfall, winter in is very different from that in North China. Hence, there is a significant north-south difference in clothing. As Guangzhou is leading China's fashion trend, university students in Guangzhou, mainly born in 80s and 90s, showcase the changing fashion trend.

In China, as the number of university students increases continuously, this population have become a major drive to societal changes. In particular, female university

© Springer International Publishing AG, part of Springer Nature 2019
S. Fukuda (Ed.): AHFE 2018, AISC 774, pp. 344–354, 2019.
https://doi.org/10.1007/978-3-319-94944-4_37

students value personal style, novelty and fashion. They are sensitive to colors and have high acceptance to new trends. This is why when new trends of colors emerge, they can immediately utilize and demonstrate them. The most influential factor on the brightness of their clothing color is climate; the pureness of their clothing colors is impacted by the subtropical monsoon climate and the metropolitan environment of Guangzhou. The speedy spread of fashion information helps female university students to stay abreast of fashion trends. While personal style was demonstrated in this research, there is also convergence across individuals in terms of take-up of trendy colors, lacking sufficient professional guidance. Female university students are going through physiologically and psychologically changing periods. This, and their respective life habits, regions of origin and cultural backgrounds, mean that choice of clothing colors is influenced by various factors. However, due to similarities in physiological factors, age, occupation and educational background, there is also a converging trend among them on clothing colors selection, especially after they move to live in Guangzhou. This research on university clothing colors adopted observation and questionnaire methods, as well as market research, data analytics and graphic demonstration. With collection and sorting of a large amount of material and information, this research concluded that female university students in Guangdong as a population have distinct clothing color selection as compared to other groups.

In terms of observation, random sampling was collected by random photographing. Target population was university students aged 18–30 in 10 universities in Guangdong. This research aims to find out the characteristics and trend of clothing colors of female university student in winter. This will provide empirical data for organizations doing fashion research, design and prediction in China. Color samples collected in this research were entered into NCS for analysis.

In terms of universities selection: a comparative analysis was made between art universities and comprehensive universities on clothing perception and judgement. While a comprehensive range of majors were included, analysis was done mainly on students majoring science, liberal arts and art.

2 Analysis of the Results

2.1 Analysis of Color Frequency Ranking

No. 1: black

In this color survey, black has become the most frequently used color. In the total number of 455 colors, black has ranking the first in the hue rankings with the absolute advantage of 180 colors and a ratio of 39.6%. Black and white are almost invariable mainstream colors, especially black which is often referred to as the most secure color and is widely applied in campus or workplace. And black is also recognized as a temperament color, which is a good helper to improve the intellectuality. Black is so popular with school students, which is related to the following factors. Firstly, black belongs to the classic color for winter clothing, which has the effect of light absorption and warmth retention in thermodynamics. Therefore, many people have been used to choosing black as the color for winter clothing so for a long time. Secondly, since the global economic crisis in 2008, black has become an important carrier for the society that has been plagued by the economic crisis to express inner feelings due to its complex psychological meaning of adamancy, durable, lost and so on. Therefore, in recent years, black has been at the forefront of the international color trend. For example, many top international brands, and even brands that have always been famous for their gorgeous colors, such as Christian Lacroix, Dior, Kenzo and so on are all using black. So black a great influence. For those who are in school, the choice of black can obviously reflect their ability to follow the trend. Thirdly, as a neutral color, black is also the easiest color to coordinate with other colors. For those who are not good at collocation, black is the most secure color. Black also has a good effect on other colors as well as resistance to dirty.

Based on the stable, good-looking, classic and easy- matching features of black, it can be imagined that black will not come out of the fashion color stage in the future. More importantly, after entering the winter, black is still called the main color in the campus as always.

No. 2: Earth Color Series

The earth color series is ranking second in the survey with the number of 110 applications and the ratio of is 24%. The color of the earth can also be called natural color, mainly referring to those colors with gray and medium and low purity such as brown, beige, olive, stone green, etc. In a word, the color of the earth is a very wide-ranged color. There are four reasons why earth color can take such a large proportion in the school clothing in the winter of 2017. Firstly, the earth color belongs to the derivative of the design concept of environmental protection in recent years. Secondly, the earth color with gray is closer to the environmental tone of the autumn and the winter, which is also the reason for the large proportion of the earth color in the autumn and winter season. Thirdly, the earth color is more suitable for Chinese skin color. It is not as flamboyant as the bright color, and it shows the feeling of maturity, elegance and intellectuality. So, it can show the implicit, introverted and unpublicized personality of Chinese people. At the same time, it is also in line with the specific identity of college students. Fourthly, the earth color is a color that will not be outdated or wrong in any occasion. It is easy to match with other colors and is not easy to create conflicts, irritation, and impact.

The earth color for women's clothes in school in winter is mainly of low brightness. However, the moderate and high earth color will be more suitable for the spring and summer season as the temperature increases with the seasonal changes, and the cream, beige, skin color will be more popular in recent years.

No. 3: Blue

The survey results of campus color shows that blue, as a color for garment or accessories, is ranking third with a total of 91 colors and a proportion of 20%. In this season, most of the blue has low saturation, so it looks dark. The reasons why blue is widely recognized among students are as follows: firstly, blue jeans take a large proportion, thus making blue rank first. It can be easily seen from the survey that the durable blue jeans almost become the essential clothing for each university student, thus greatly enhancing the proportion of application of blue in the campus. Secondly, in addition to plain denim blue, the electric blue, Klein blue, sapphire blue, blue peacock with richness or glossiness also become the most popular color in the current international clothing fashion. In recent years, in the releasing conference held by famous brands such as Giorgio Armand, Balenciaga, Louis Vuitton, Versace also use these novel and avant-garde blue. For those students who are interested in fashion, choosing colors such as electric blue and other fashionable colors can not only satisfy their psychological demands of leading the trend, but also make them attractive in the crowd, so as to achieve the purpose of publicizing personality. Thirdly, according to *Research Report on Color Preference of Post 80s and 90s*, about 30% of the young people use blue as their favorite color, and blue ranks second. It can be seen that young people are fond of blue. Fourthly, blue is often the coldest color in the vision. It gives people a quiet and calm feeling and is well coordinated with the atmosphere of the winter. There is also another factor that cannot be ignored. Blue can foil the Oriental yellow grey skin color.

Today, Denim Blue is the most indispensable color in campus clothes, and the new blue in spring and summer will add "endless vitality" to this season.

No. 4: White

Through investigation, it is found that the white as the mainstream is not widely used in this season compared with black which is non- coloured. There are only 46 for white. The white in winter makes people feel elegant, noble, pure, light. In the dreary season, white is like a snowflake, which brings bright and lightness to the winter clothes. White is considered the least aggressive color. For inexperienced students, white is a pure, longing and sacred color, which is a good self-adjustment and protection. However, the white is easy to get dirty, difficult to keep clean, and becomes too obvious with high brightness in the dark clothes. Taking into account the advantages and disadvantages, only half respondents will choose the white of the big block, and the other half will choose white to embellish, so that there is no conflict and it is also a moderate expression of the yearning for this color. Therefore, in this survey of campus clothing in this season, we found that white showed a moderate trend and was neither overused nor disappearing on the stage of color.

No. 5: Red

Of the 455 colors extracted from this survey, red appears 44 times in total and is ranking fifth with a proportion of 9.67%. In general, the main color of the campus in autumn and winter in 2017 is the light red. Firstly, red is like a fire in winter, thus becoming an important choice for people to keep warm. In other clothing color surveys of fixed groups, red has always been the most widely used color in winter. Secondly, in Chinese traditional culture, red represents joy, happiness and enthusiasm. Since ancient times, Chinese people have a stronger sense of recognition about red than other ethnic groups. According to an Investigation Report on Color Orientation of Chinese Urban Residents, 72.6% of respondents say they like red, which indicates that Chinese people have deep feelings for red. There are many festivals in winter, which provides an opportunity for the clothes color of the students. Thirdly, due to the economic crisis in recent years, many brands launch red clothing with a highly visual appeal in the clothing conferences, so as to stimulate consumers' desire to purchase.

In fact, loving red is not a patent for Chinese people. Grosse, a German anthropologist has pointed out, "red is the color of all nationalities." Red is divided into warm hue and cold hue. The cold dark red, ruby, and sexy peach red are popular in western countries. In addition, light pink is also popular with young women. For college students, the red series is in line with their youthful and aggressive spirit. Therefore, the popularity of the red series will not decrease among students.

No. 6: Yellow

In this survey, yellow appears 13 times and is ranking seventh with a proportion of 2.85%. Firstly, from the perspective of hue, blue, purple, orange, green, red, black and white colors have a debut and enjoyed popularity for a time since twenty-first Century. According to the periodic law of the development of the color itself, the popularity of the Yellow will be able to account for the days. In recent years, red are often seen in the release conferences of many big brands, such as Versace Giorgio Armani, Dior, Chanel, etc. It can be seen that the Yellow tune is bound to be popular in the next few years. Secondly, yellow is the color of the sun, which represents warmth and brilliance. Therefore, the appearance of the Yellow tune in winter is also a practice in the application of clothing color. In fact, in addition to Daisy yellow, India yellow, egg yolk, light yellow and so on, it also contains cream-colour, malt yellow and other earth colors.

With the arrival of spring and summer, in addition to the popularity of yellowish yellow, such as lemon yellow, light yellow, etc., which have high saturation, the pink and yellowish yellow are also popular with college students. After orange, the yellow is the most promising color for new round of fashion color in the international fashion field.

No. 7: Green

As a symbol of nature, green is always loved by people. It is a symbol of peace, quietness and vitality. In today's society, green also symbolizes the environmental protection and hope in the design. This season is exactly the opposite of last year when green is very popular. Does this indicate that the army green and olive green which are popular in recent years will gradually withdraw from the popular stage? There are two main reasons. Firstly, the army green and olive green with the characteristics of solemnity, stability, sinking, and knowledge, which were popular last year, have small users. Young people are more willing to show vigor, youth and vitality. So, there are not many respondents who choose green. Secondly, in recent years, the army green, olive green and other colors has been popular for a long time, resulting in a certain degree of aesthetic fatigue. So it can not satisfy the curiosity of young people.

No. 8: Orange

In the survey, orange has appeared only 4 times, and is ranking eighth with a proportion of 1%. Firstly, in psychology, bright and brilliant orange is the most warm color of all colors. In theory, orange is most suitable for winter as well as red. Secondly, orange is very glamorous and hopping in the crowd, and it is easy to attract attention. Therefore, for those who pursue personality, it is no doubt a little attraction for college students. Thirdly, in recent years, the orange tone has been highly praised in the field of international fashion. The fashion industry hopes to arouse people's yearning for a better life with this strong optimism spirit, so as to guide consumers to get out of the shadow of the economic crisis.

In all color series, orange is always the most dazzling. Orange usually gives people a lively feeling, which can remove the depression. Whether you choose to use it as a major hue or a bright spot in dull clothes, you can inject unlimited energy into your clothes in a moment, so as to make you can stand out in the crowd. Bright and sweet orange is a positive energy storm that sweeps through the streets. The dark pink orange brings new powder wax, and the base color of the corals is more unique. Copper orange gives people a positive outlook, which is both temperament and wild. It has a sense of cool of metal and warmth. The most concerned copper color is upgraded to copper orange with warmness in the new season.

In this research, we have not found the interviewees who choose the purple tone for the time being. But in the whole-body collocation, usually a little purple can improve the sense of color and the sense of hierarchy as a whole. Perhaps due to this reason, purple is more used in clothing color matching. The purple with other color tendencies is used in coloring for large pieces of clothing. In addition, the purple has rich layers and extremely personality. It is also very harmonious with the colors with the high quality common pigment, such as blue or red. But purple is not common among young campus interviewees, which is obviously related to the unmanageable characteristics of purple.

2.2 Psychological Analysis of Clothing Aesthetic

1. Factors influencing the color selection of clothing for female college students in Guangzhou

The choice of clothing color can express the micro psychological characteristics of the wearer, which is mainly influenced by the native place, age, profession, personality, skin color, figure, temperament, popularity and personal preferences of the wearer. The chart has listed some of the factors that the female college students are most concerned about in the choice of clothing. It is not difficult to see from proportionate that nowadays when female college students choose clothes, they pay more attention to personal suitability, and most of them want to improve their image through clothing and color matching. In this survey, 79% of the respondents believe that appropriate clothing colors can change a person's temperament and adjust a person's body defect. It is precisely because of this reason that women college students have higher and higher demand for clothing color.

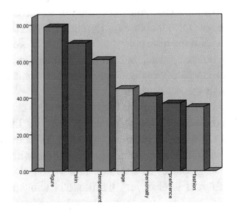

The ranking of factors influencing female college students' choice of clothing color

Figure	Skin	Temperament	Age	Personality	Preference	Fashion
79	70	61	45	41	37	35

In winter there are 64.6% of the respondents who chose color with low brightness and high saturation, such as blue, black, and brown; 29.6% of the respondents choose black and white in the non-color system, and 23% of the respondents choose middle color. From the results of the survey, it can be found that the color selection of female students in Guangzhou will be influenced by the factors such as the environment, the season, etc. But they also have fixed clothing color, which is mainly determined by the campus environment in which they are located.

2. We can understand the degree of female students to follow fashion from the aspects of native place, grade, specialty and monthly consumption. Whether the current female college students follow the fashion color depends largely on whether the fashion color is suitable, whether it is matched with their own personal conditions and other factors. But there are also some junior girls who treat popular color with their personal preferences as a standard. But with the increase of color knowledge and age, this standard can also be gradually changed, which also illustrates the importance of color knowledge education for female college students.

3   Study on the Clothing Color of Female College Students in Guangdong

The standard of the choice of clothing color for female college students in Guangzhou: At present, the female college students often choose the basic color when choosing clothing color. For spring and summer clothing, they mostly choose color with high brightness and low saturation, or bright color series with high saturation, such as rice yellow, light blue, and so on. These colors are brighter, which can bring a good sense of stability and a light pleasure. In autumn and winter, they mainly choose color with low brightness and high saturation or non-colored hue, and choose more traditional and conservative colors such as black, white, opaque color etc. It reflects the clothing characteristics of female college students, which are not only natural but also pragmatic.

The relationship between clothing color and fashion color of female college students. At present, female students know the fashion and color trend through a variety of channels and are good at grasping the fresh elements. They have rich imagination. As long as the colors can reflect the personality, the temperament, the promotion of the image, they can win their love. The novelty and variability of the colors meet the desire of female college students to attract people's attention by displaying their personality. Therefore, female college students have become an important group of fashion consumption in university campuses.

The clothing color characteristics of female college students in Guangzhou: Youth, knowledge, random and other elements have become an important standard for female college students to choose the color of clothing, which can be reflected in the following aspects:

1. The personal physical condition plays a major role in the color selection of clothing for female college students. Female college students are very sensitive to the change of seasonal clothing color. For spring and summer clothing, they mostly choose color with high brightness and low saturation, or bright color series with high saturation, such as rice yellow, light blue, and so on. In autumn and winter, they mainly choose color with low brightness and high saturation or non-colored hue, which are traditional and conservative.

2. Female college students will be influenced by regional environment and season in Guangzhou when choosing clothing color. However, they still have relatively fixed clothing color, which is mainly determined by their university campus environment. Female college students have active thinking, and they accept new things quickly. They generally believe that the popular color represents the sense of fashion in a period, but it is not suitable for them. They seek to match popular color and non-popular color in a reasonable way, oppose blindly following popular color, and advocate individualized choice of clothing color. To sum up, the unique temperament and charm of female college students endow the group with unique aesthetic characteristics of clothing color. As a whole, the clothing of female college students has the features of pleasing, fashionable and beautiful in color, and more rhythmic the collocation. Therefore, when clothes designers should pay attention to the color preference of this group in design, so as to highlight the personal charm and the overall style of female college students.

References

1. Lu, S., Li, N.: A preliminary study on the color collocation of clothing for Chinese young women. Art Des. (Theory) (2010)
2. Jing, L.: The regulation of popular color on human emotion. Wuhan Textile University, Master's thesis (2012)
3. Wei, C.: A survey on the color of women's clothes in Universities in winter. Chin. Clothing (2012)

Creating an Affective Design Typology for Basketball Shoes Using Kansei Engineering Methods

Alexandra Green[⊠] and Veena Chattaraman

Department of Consumer and Design Sciences, Auburn University,
Auburn, AL 36849, USA
{aag0042,vchattaraman}@auburn.edu

Abstract. Research has shown that for athletic shoes, visual attributes such as color and style can be more important than ergonomic or technical attributes in consumer purchase decisions. Previous studies have also shown that psychological feelings and emotions are in fact tied to products based on individual design characteristics that create a 'gestalt' feel for the product. Kansei engineering is one method commonly used in product development to gain a better understanding of emotions and their linkages with specific design characteristics, which can then be used to design products that communicate the desired 'feel'. The current study posits that the design characteristics of shoes and the emotions that they elicit can be statistically grouped together, creating Kansei/affective design types that have applications for product development, marketing, and mass customization. An exploratory study using male millennial athletes revealed four affective design types for basketball shoes, which are associated with differing design characteristics.

Keywords: Basketball shoes · Athletic footwear · Kansei engineering
Affective design · Design typology

1 Introduction

"How can we make you better?" is the question Nike asks LeBron James, one of the greatest basketball players of our generation, when designing his shoe every year [1]. Professional athletes choose their gear based on how well it will improve their performance, but they place importance on personal expression through style and aesthetic attributes as well. Design characteristics such as colorway, the combination of colors in the design [2], mesh patterns, ankle coverage (high top, mid top, or low top), material combinations, and strap feature (presence or absence of an additional closure strap) come together to create the complete product form of basketball shoes, which can have an impact on how the athlete perceives the shoe. For professional basketball players, who are often given free sneakers in exchange for publicity or as part of an endorsement deal, the shoes that elicit the most excitement, interest, and potential for performance improvement are the ones that are likely to be chosen.

Previous research has shown that psychological feelings and emotions can in fact be tied to consumer products such as digital cameras and athletic shoes based on their

© Springer International Publishing AG, part of Springer Nature 2019
S. Fukuda (Ed.): AHFE 2018, AISC 774, pp. 355–361, 2019.
https://doi.org/10.1007/978-3-319-94944-4_38

individual design characteristics [3, 4]. Kansei/affective engineering, developed by Nagamachi [5], is one method commonly used in product development to gain a better understanding of those emotions and use them to design products. Emotions and the design characteristics of a product that elicit them can be statistically grouped together, creating Kansei/affective design types, which can further be analyzed to understand how consumers psychologically and behaviorally react to differing affective design types, such as modern, powerful, or nostalgic (retro).

Studies have shown that for athletic shoes, visual attributes such as color and style are more important than ergonomic or technical attributes when athletes are making purchasing decisions [6], but no research can be found that breaks down athletic shoes into their individual design characteristics to understand how color and style impact emotions and perceptions. Previous research has focused solely on the attribute of color related to style perceptions of athletic shoes [4, 7], but other important attributes of basketball shoes, such as ankle coverage and strap feature, have been ignored. Shieh and Yeh [4] found that consumers attach semantic meanings, such as the perception of modernity, simplicity, or formality, to the combinations of colors in athletic shoes; however, these authors did not explore the contribution of other shoe features in their study. Hence, limited previous studies have explored consumer response to the design characteristic of color in athletic shoes, and almost none can be found that focus on 'athlete' perceptions. Do varying design characteristics of athletic shoes elicit specific feelings for athletes? This study sought to answer this question to provide insight for basketball shoe developers to better design products that will speak to athletes, professional and recreational, alike.

1.1   Kansei Engineering

Kansei/affective engineering, defined as a method of product development that attempts to gain an understanding of consumers' feelings and translate them into design characteristics, originated in Japan and has been used in a variety of industries spanning from automobiles to digital cameras and more [3, 5]. In Japanese, the word Kansei translates to "a consumer's psychological feeling...regarding a new product" [5, p. 4]. Nagamachi, the creator of Kansei/affective engineering, published a seminal work on the methodology of Kansei in 1995 in order to help product developers shift to a more consumer-focused mindset. For example, Nagamachi [5] explains that a customer may view a product they want to purchase as "luxurious" or "strong," but product developers may not understand the exact design characteristics that led to these judgements. Kansei/affective engineering methods aid product developers in pinpointing design characteristics and deriving their emotional or semantic meaning from consumers, allowing them to design products that are likely to elicit the desired emotional response.

The study by Shieh and Yeh [4] operationalized the product form of sports shoes based on their exterior colors using Kansei words. An environmental scan of the sport shoe market combined with consultations with experts led Shieh and Yeh [4] to 20 Kansei word pairs that can be widely applied to all sports shoes, covering the scope of what a consumer may see when they view a sports shoe product. Examples of these word pairs include simple-complex, modern-retro, obtrusive-modest, and compliant-rigid [4]. In their study, Shieh and Yeh [4] had respondents view stimulus images of sports shoes in different colorways and rate each shoe based on the 20-word pairs presented as semantic differential scales, as suggested by Nagamachi [5]. The researchers found that there was a strong correlation between the adjective pairs of modern-retro and like-dislike, while the most liked colorway was red/white followed by black/white and white/white [4]. In other words, shoes that were perceived to be modern and/or had the primary colors of red, black or white were highly liked by the participants. It was also discovered that specific colors and combinations could be linked to Kansei words; shoes with combinations of three (one) colors scored the highest ratings on the complex (simple) adjective, while the color green as a primary color resulted in high ratings for retro and white was perceived as the most formal. The above findings show that changes in a single design characteristic such as color can change the semantic effect of the product form through eliciting different Kansei words and feelings. In the exploratory study reported in the subsequent sections, we use a sample of male millennial athletes to examine the relationship between the design characteristics of basketball shoes, such as colorway, strap feature, and ankle coverage and Kansei/affective design types.

2   Methods

This study employed a web-based survey using 100 images of basketball shoes as stimuli. The product form of basketball shoes based on the design characteristics of colorway, strap feature, and ankle coverage was investigated through the lens of Kansei words to create Kansei/affective design types. The sample for this study consisted of 170 Millennial men (mean age 21 years) who play on intramural basketball teams at five large American universities from across the country: Duke University, The University of Tennessee, The University of Oregon, The University of Massachusetts, and The University of Maryland.

2.1   Stimuli

Stimuli was selected for this study using methods commonly used in design research [8] in which a group of two-dimensional images that are representative of all possible variation within the category and based on specified stimulus characteristics are chosen. Quota-based stimulus sampling was used to ensure that there was equal representation of the three design characteristics (colorway – primary color, number of colors, and midsole color; strap feature; and ankle coverage) and their variations. Basketball shoes from several different brands, including but not limited to Nike, Adidas, Under Armour, Reebok, Fila, Puma, Air Jordan, and Ewing were included as these brands range in popularity within the basketball shoe market, and the logos were removed to avoid the influence of brand name on the proposed relationships. The final pool of stimuli was reduced from 135 to 100 to include the images that were most representative of the variance within the basketball shoe market.

2.2 Procedure

Data was collected online by providing the intramural coordinators with web links to distribute to males registered for intramural basketball for the spring 2018 semester. Respondents viewed 7 randomly assigned full-color images of basketball shoes out of a pool of 100 stimuli and rated them individually on a five-point semantic differential scale with 19 items using bipolar Kansei words. Shieh and Yeh's [4] 19 Kansei word pairs for sport shoes were used to gain understanding of the respondents' impressions of the product form of basketball shoes. These researchers used 20-word pairs, but the word pair of like-dislike was removed for the current study as it was determined to reflect product likeability, which was reflected in a later additional measurement. These word pairs were determined through expert interviews and an environmental scan of the sports shoe market, which provides evidence for validity and is the suggested method by Nagamachi [5], the creator of the Kansei engineering method. Respondents then answered several demographic questions related to age, race, socioeconomic status, university affiliation, how long they have been playing basketball, self-perceived basketball skill level, purchase behaviors of basketball shoes, and several other measures for data collection outside the scope of this paper. Respondents were given an unlimited amount of time for the questionnaire to facilitate careful consideration of the stimuli. Following completion of the questionnaire, respondents were thanked for their participation, provided with the contact information of the researcher, and were given the option to provide their email address to enter to win one of 50 $20 gift cards.

3   Results

3.1   Kansei/Affective Design Types

An exploratory factor analysis (EFA) using Principal Components Analysis extraction and Varimax rotation was used to determine how the 19 Kansei bipolar word pairs would load together. Stimulus-level analysis instead of respondent-level analysis resulted in a sample of 1281 valid cases, with around 10 responses per stimulus on average. The EFA revealed four factors based on the 19-word pairs. The researchers determined conceptual names for each factor, which represent an affective design typology for basketball shoes. Cronbach's *alpha* was used to test the reliability of each factor; the first factor was the sole factor considered to be statistically reliable ($\alpha > .70$), despite the conceptual validity of all four factors.

The first factor, the *Boldness Factor* ($\alpha = .811$), consisted of eight-word pairs that seem to describe the level of eye-catching or attention-getting design of the shoes. Word pairs such as vivacious:quiet, bright:dull, striking:mediocre, complex:simple, and obtrusive:modest highlight the "wow factor" of basketball shoes, focused on attractive (or unattractive) aesthetic design. Interestingly, the word pair expensive:cheap loaded higher on the *Boldness Factor* than the *Formality Factor*; this finding is noteworthy and original as it indicates that for basketball shoes, the perception of an expensive-looking shoe is more related to striking, modern, and complex aesthetic designs than formality or elegance. In addition, the negative loading of the mature:young word pair on the Boldness Factor suggests that youthfulness is associated with bold and striking shoes, whereas maturity is linked to traditional, simple, or modest shoes.

The second factor was renamed the *Structural Factor* ($\alpha = .641$) as it consisted of word pairs that were related to construction perceptions of the shoes. The creation of a factor based on word pairs such as thick:thin, sturdy:fragile, and rough:delicate is compelling as it illustrates the ability of athletes to infer structural attributes of a shoe solely based on aesthetics, without the need to feel or wear the shoe. Similarly, the third factor was conceptualized as the *Ergonomic Factor* ($\alpha = .526$) because its included word pairs were associated with comfort, pliability, and safety-related aesthetics, such as comfortable:tight, compliant:rigid, and safe:dangerous. Athletes were able to make inferences related to the ergonomic properties of the shoe without even trying them on, and these inferences were based solely on the shoes' designs.

Finally, the last factor was renamed the *Formality Factor* ($\alpha = .450$) and was focused on the level of elegance or casual nature of the shoes' designs through word pairs such as formal:casual, elegant:unrefined, and mature:young. As was previously noted, it is interesting that the expensive:cheap word pair did not load the highest on the *Formality Factor*, despite price being logically associated with formal or elegant design. This compelling finding allows one to draw connections between extremely high prices and highly formal basketball shoe designs currently being sold by some luxury brands in the market; it is not the formal or elegant designs of these luxury basketball shoes that elicits the feeling of "expensive," therefore providing evidence that the brand name is the sole reflector of the shoes' high cost. In the minds of athletes, expensive-looking shoes are not those which appear polished or refined, but those which audaciously grab attention.

3.2   Design Characteristic Linkages to Kansei/Affective Design Types

A multivariate analysis of variance (MANOVA), using the five design characteristics and their associated levels as independent variables and the four Kansei/affective design types as dependent variables, revealed significant multivariate and univariate effects of various design characteristics on the design types. There were significant multivariate interaction effects of primary color and number of colors [$F(12, 3204) = 1.81$, $p = .042$], primary color and strap feature [$F(24, 4225) = 2.16$, $p = .001$], and primary color and ankle coverage [$F(44, 4634) = 2.66, p < .001$]. Only the main effects of primary color [$F(24, 4225) = 6.53, p < .001$], number of colors [$F(8, 2422) = 2.996, p = .002$], and ankle coverage [$F(8, 2422) = 6.45, p < .001$] were significant; the main effects of midsole color and strap feature were found to be insignificant. To ascertain which of the dependent variables contributed to the overall significant multivariate main effects, univariate analyses of variance were analyzed. For the *Structural Factor*, the interactions between primary color and number of colors [$F(3, 1214) = 2.71$, $p = .044$], primary color and strap feature [$F(6, 1214) = 4.45$, $p < .001$], and primary color and ankle coverage [$F(11, 1214) = 3.40, p < .001$] were significant. In addition, the interaction between primary color and ankle coverage on the *Boldness Factor* was significant [$F(11, 1214) = 4.57, p < .001$]. There were also main effects of primary color on all four factors [*Boldness Factor*: $F(6, 1214) = 18.60$, $p < .001$; *Structural Factor*: $F(6, 1214) = 2.21$, $p = .039$; *Ergonomic Factor*:

$F(6, 1214) = 2.56$, $p = .018$; *Formality Factor:* $F(6, 1214) = 4.53$, $p < .001$], number of colors on the *Boldness Factor* [$F(2, 1214) = 5.96$, $p = .003$], and ankle coverage on both the *Boldness* [$F(2, 1214) = 6.26$, $p = .002$] and *Structural Factors* [$F(2, 1214) = 17.11$, $p < .001$].

4 Conclusion

As proposed, athletes do experience different feelings related to the design of basketball shoes; the results of this study show that, when controlling for brand name, perceptions of shoes differ based on design characteristics such as primary color, number of colors, and ankle coverage. Using Kansei engineering methods we were able to quantitatively analyze the feelings of the athletes when viewing various designs of basketball shoes, resulting in an affective design typology of basketball shoes based on elicited emotions and their associated design elements. The use of this method is novel for the field of apparel design, contributing to the literature and providing novel theoretical implications that can be continued to be used in design research of all kinds. Our findings have critical implications for athletic footwear product development, marketing, and mass customization as well; mass customization toolkits such as NikeID are often difficult to use for consumers, leading to increased perceptions of product risk. One way to simplify the co-design process is by limiting the overwhelming amount of choices to a smaller number of Kansei/affective design types for athletic shoes, each consisting of a subset of all available design characteristics. This strategy could minimize confusion during co-design and result in increased user satisfaction and sales. In effect, the results of this study will help to ensure that the customized shoe design elicits the exact 'feel' that the consumer intends, without requiring the consumer to choose the correct combination of design characteristics to create the 'feel'. Future research should replicate our processes to determine if the effects of strap feature and midsole color become significant with a larger sample, as well as test our newly-created design typology of basketball shoes to determine its external validity with athletes.

References

1. Feifer, J.: Shoes made for the man. Fast Company **191**, 52–54 (2014)
2. Engvall, N., Edler, B., Bengtson, R.: A beginner's guide to sneaker terminology. Complex (2014). http://www.complex.com/sneakers/2012/09/a-beginners-guide-to-sneaker-terminology/
3. Wang, C.H.: Integrating Kansei engineering with conjoint analysis to fulfill market segmentation and product customization for digital cameras. Int. J. Prod. Res. **53**, 2427–2438 (2015)
4. Shieh, M.D., Yeh, Y.E.: A comparative study on perceptual evaluations of sport shoe exterior colors in Taiwan. Color Res. Appl. **40**, 178–193 (2015)
5. Nagamachi, M.: Kansei engineering: a new ergonomic consumer-oriented technology for product development. Int. J. Indust. Eng. **15**, 3–11 (1995)

6. Branthwaite, H., Chockalingam, N.: What influences someone when purchasing new trainers? Footwear Sci. **1**, 71–72 (2009)
7. Lam, W.K., Kam, K., Qu, Y., Capio, C.M.: Influence of shoe colour on perceived and actual jumping performance. Footwear Sci. **9**, S3–S5 (2017)
8. Orth, U.R., Malkewitz, K.: Holistic package design and consumer brand impressions. J. Market. **72**, 64–81 (2008)

Music Retrieval and Recommendation Based on Musical Tempo

Masashi Murakami[1]([✉]), Takashi Sakamoto[2], and Toshikazu Kato[3]

[1] Doctral Program in Industrial and Systems Engineering, Chuo University,
1-13-27, Kasuga, Bunkyo-ku, Tokyo, Japan
masashi.murakami.indsys@gmail.com
[2] National Institute of Advanced Industrial Science and Technology,
1-1-1 Umezono, Tsukuba, Ibaraki, Japan
takashi-sakamoto@aist.go.jp
[3] Department of Industrial and System Engineering, Chuo University,
1-13-27, Kasuga, Bunkyo-ku, Tokyo, Japan
kato@indsys.chuo-u.ac.jp

Abstract. In this study, we offered an automatic recommendation technique that retrieves based on musical tempo and physiological index for the purposes of relaxation and refreshment As an physiological index, we used LF/HF ratio for our retrieval system. As a reason that using LF/HF ratio, we think LF/HF ratio has the merit to measure easily with mobile device, and can apply the result for the retrieval compared with the other several method to measure mental conditions.

Keywords: Human factors · Music recommendation · LF/HF ratio

1 Introduction

Recently, many music distribution services have emerged, each offering a personalizable music retrieval and recommendation system [1].

Often, we use such a system when we need relaxation or refreshment. It provides a personalizable playlist based on genre, songwriter, composer, or audience, allowing us to subjectively tailor the music selection so that we hear music that corresponds to our checked selections. The system also makes recommendations, based on our preferences. Music distribution services provide a vast array of songs and music including new music, unpopular music and not well-known music that does not correspond to our checked selections. Therefore, highly personalized preferences become a problem for music distributors who handle such a wide variety of artists.

Currently, we lack techniques to automatically retrieve a tailored mixture of music styles from a huge database not correspond to subjective index. By definition, such an automatic music retrieval and recommendation systems have no need of manpower support, which could otherwise index and provide not only new music, but also unpopular or not well-known music that fits our preferences. Automating this technique is expected to provide merit and benefits for both listeners and music companies that handle music composition, sales, and distribution.

© Springer International Publishing AG, part of Springer Nature 2019
S. Fukuda (Ed.): AHFE 2018, AISC 774, pp. 362–367, 2019.
https://doi.org/10.1007/978-3-319-94944-4_39

2 Related Work

2.1 User-Based and Content-Based Retrieval

Music retrieval researchers have developed the two types of retrieval method. One is based on metadata including user profile, social tags, and correlation between users [2, 3]. The music retrieval based on the metadata help to get a personalized result, however, it is difficult to model of users and many social tags and user profiles can be the noise for the results.

The other is based on auditory content [4]. These studies focused to investigate what musical feature such as melody, pitch, harmony, rhythm have the main effect of music. However music similarity is ambiguous even in a genre, it is difficult to extract the musical features classify the music itself. Moreover, these content-based studies used MIDI formatted files to evaluate the features, so it is difficult to evaluate the music itself.

As the similarity of music cannot be obtained by these studies, it can't be said that the retrieval system users have been always satisfied with the result of the retrievals.

2.2 Relaxation Effect of Music

Recently, there were various study about relaxation effect of music. These studies examined to investigate what musical structures, playback devices, listening situation have a main effect of relaxation. These studies used a physiological information of participants as an objective index. As an index for relaxation, the ratio of low- to high-frequency heart rates ratio (LF/HF), brain wave, cortisol, and salivary amylase activity is used in these studies. We use the LF/HF ratio as a physiological index for our retrieval system.

2.3 Mesurement LF/HF Ratio from Devices

To measure the heart rate, we use the photoplethysmography (PPG) with Android smartphone camera [5] (Fig. 1). As use of application of PPG, we calculate the LF/HF ratio with the mobile phone.

From the electrocardiograph data, we detected the peak time of the R waveform and calculated the RR intervals. RR interval data were resampled with 1.2 Hz via cubic spline interpolation. We used Fast Fourier Transform (FFT) for resampled RR interval data and calculated power spectral density. Finally, we calculated the integrated value of Low Frequency (0.05 Hz–0.15 Hz) and High Frequency (0.15 Hz–0.4 Hz) and the LF/HF ratio as the index of the sympathetic nervous system.

Therefore, in this study, we suggest the retrieval system based on analysis of music by the attributes of beat per minute (BPM) from the viewpoint of the audiological psychology. Especially, we suggest the retrieval for relaxation effect. With this approach, it is possible to establish the music retrieval system by the attribution of auditory perception without using a complicated model.

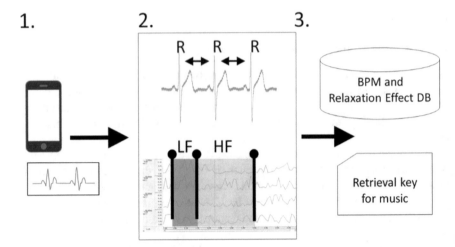

Fig. 1. Measurement method of LF/HF ratio using the photoplethysmography with mobile device. At first, we measure the cardiograph with mobile devie. Then, calculate the RR interval and LF/HF ratio. The data is given to retrieval key and to music DB to adapt BPM.

3   Music Retrieval System Based on BPM

We offer an automatic recommendation technique that retrieves a tailored mixture of music based on musical tempo and physiological index for the purposes of relaxation and refreshment (Fig. 2).

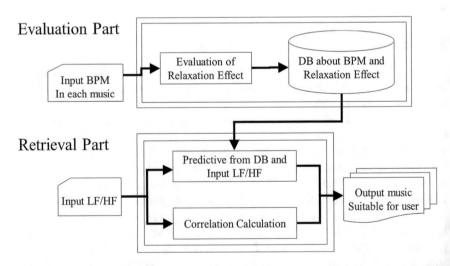

Fig. 2. Suggested music retrieval system and evaluation based on evaluation examination [6–8].

Having examined musical factors related to relaxation, we apply an assumed daily scenario in which a user wishes to listen to music and relax after hours of desk-work. This study examined the effect of background music on the LF/HF ratio of participants during rest with listening music after a mental stress task. To investigate the difference of genre, we used Classical music and Jazz music. The results showed that the LF/HF ratio during rest with a Classical music environment and Jazz music environment was lower than initial-before mental stress task-environment, and that especially in Classical music environment, LF/HF ratio was more lower than Jazz music environment. Therefore, we suggest that the Classical music and Jazz music has the effect of reducing the LF/HF ratio.

We know that musical tempos and melodies promote relaxation using physiological index [6–8]. Also, Okamatsu [9] showed that ambient music sustaining 60 beats-per-minute (BPM) provides the best relaxation effect. Thus, we consider a music retrieval technique that uses LF/HF ratio and the songs' BPM as the retrieval key to improve and ensure the relaxation effect

3.1 Evaluation Part

Evaluation part is constructed based on the result of past evaluation experiments [6–8]. As use of the result, we established the musical tempo database adapted via relaxation effect. Each collected music is measured BPM and weighted by weighting coefficient for the relaxation effect.

3.2 Retrieval Part

To retrieve the relaxing music suitable for users, we use the LF/HF ratio data for the retrieval key of system. After having received the LF/HF data, the system presume the user's mental condition and collate the music that BPM is suitable for user from BPM and relaxation effect database.

4 Discussion

There are several method and physiological information to measure the mental condition. However, using LF/HF ratio as a physiological index have the merit to measure the mental condition easily. Also, using the LF/HF for the retrieval key help users to retrieve music easily without any search form or check they need to listen.

Generally, human stress is an individual problem. We must refresh and relax ourselves by resting only after work or during a short break. Among stress reduction methods, music is the most effective. Thus, many users search and listen to playlists for this purpose [10]. A BPM-based music retrieval method should be suitable for the needs of relaxation.

Moreover, we think BPM is the important factor for the motivation, work efficiency, and the behavior in commercial space. Thus, BPM-based music retrieval method have the capability to satisfy the needs that many users such as audience, students, and consumers have. We require a system to measure and predict BPM correctly, and that is our problem to solve in this study.

Our suggested system allows distributors to provide a variety of music that have never been retrieved before because our BPM-based retrieval can find the music existing retrieval had not find. Also, suggested system help audience with a tailored retrieval than before.

5 Conclusion

In this study, we offered an automatic recommendation technique that retrieves based on musical tempo and physiological index for the purposes of relaxation and refreshment As an physiological index, we used LF/HF ratio for our retrieval system. As a reason that using LF/HF ratio, we think LF/HF ratio has the merit to measure easily with mobile device, and can apply the result for the retrieval compared with the other several method to measure mental conditions.

We focused on a musical tempo, BPM for the retrieval key. That is based on the past evaluation experiment that an assumed daily scenario in which a user wishes to listen to music and relax after hours of desk-work. As a result of evaluation, we investigated musical tempo have the effect of relaxation.

Our suggested system is useful in the situation in that user must relax and refresh in a short time, or user need to be motivated, consumers to be stimulated. Moverover, BPM-based retrieval have the capability to find the music existing retrieval have never find before.

Acknowledgements. This work was partially supported by JSPS KAKENHI grants (No. 25240043) and TISE Research Grant of Chuo University, "KANSEI Robotics Environment.".

References

1. HyperBot.com.: CONTEXT: the future of music streaming and personalization? January 2015. http://www.hypebot.com/hypebot/2015/01/context-the-future-of-music-streaming-and-personalization.html
2. Keiichiro, H., Kazunori, M., Naomi, I.: Personalization of user profiles for content-based music retrieval based on relevance feedback. In: Proceedings of the Eleventh ACM International Conference on Multimedia, pp. 110–119 (2003)
3. Paul, L.: Social tagging and music information retrieval. J. New Music Res. **37**(2), 101–114 (2008)
4. Casey, M.A., Veltkamp, R., Goto, M., Leman, M., Rhodes, C., Slaney, M.: Content-based music information retrieval: current directions and future challenges. Proc. IEEE **96**(4), 668–696 (2008)
5. Fernández, F.G., Ferrer-Mileo, V., Fernandez-Chimeno, M., García-González, M.A., Ramos-Castro, J.: Real time heart rate variability assessment from Android smartphone camera photoplethysmography: postural and device influences. In: Proceeding of 37th Annual International Conference of the IEEE Engineering in Medicine and Biology Society (EMBC), pp. 7332–7335 (2015)

6. Masashi, M., Takashi, S., Toshikazu, K.: Effect of classical background music on the arithmetic calculation task – psychological and physiological evaluations. In: Proceeding of the International Symposium on Affective Science and Engineering 2016 (2016)
7. Masashi, M., Takashi, S., Toshikazu, K.: Effect of classical background music tempo on a mental stress task: physiological evaluations. In: Proceeding of the International Symposium on Affective Science and Engineering 2017 (2017)
8. Masashi, M., Takashi, S., Toshikazu, K.: Study of the effect of background music tempo on a rest. In: Proceeding of the 7th International Conference on Kansei Engineering & Emotion Research 2018 (2017)
9. Keita, O., Makoto, F., Kazuhisa, M.: A 'Healing' effect by a tempo of music psychological evaluation by a simple sound made from acoustical property if 'Healing' music. J. Jpn Soc. Kansei Eng. 7(2), 237–242 (2007)
10. Linnemann, A., Ditzen, B., Strahler, J., Doerr, J., Nater, U.M.: Music listening as a means of stress reduction in daily life. Psychoneuroendocrinology 60, 82–90 (2015)

Affective Value and Kawaii Engineering

Affective Taste Evaluation System Using Sound Symbolic Words

Tomohiko Inazumi, Jinhwan Kwon[✉], Kohei Suzuki,
and Maki Sakamoto

Graduate School of Informatics and Engineering,
The University of Electro-Communications, Chofugaoka, Chofu, Tokyo, Japan
{i1410011, s1630068}@edu.cc.uec.ac.jp,
{kwonjh, maki.sakamoto}@uec.ac.jp

Abstract. Language and emotions play an important role in affective computing and advancement of artificial intelligence. Sound-symbolic words (SSWs) have become increasingly important for meaningful descriptions of perceptual experiences as a detailed and reliable vocabulary. In this research, we constructed a system for quantifying texture/taste impressions expressed by SSWs. This system decomposes input SSW into phoneme elements and refers to the category quantity of each phoneme element for each adjective scale referring to a quantitative rating database. Then, the system displays the evaluation values calculated by impression-rating predictive model. We anticipate that this system will be able to support food development and recommendation of food name which expresses the impression of goods.

Keywords: Affective system · Human-systems integration · Taste
Sound-symbolic words

1 Introduction

Language and emotions play an important role in affective computing and advancement of artificial intelligence [1–3]. Synesthetic associations between linguistic sounds and sensory experiences have been demonstrated over the decades and previous studies have reported that many languages have a word class that speech sounds are linked to visual or haptic experiences [4–11]. For example, English words starting with "sl-" such as "slime," "slush," "slop," "slobber," "slip," and "slide" symbolize smooth or wet impression [12]. In addition, "sound-symbolic words" (SSWs) have become increasingly important for meaningful descriptions of perceptual experiences as a detailed and reliable vocabulary [13–15]. Köhler (1929) reported the relationship between non-words and object shapes, revealing that participants tend to match some nonsense words (e.g., "maluma") with curvy rounded shapes and other (e.g., "takete") to spiky angular shapes [4]. Recent research suggests that this process, referred to as the "bouba/kiki effect", operates in a similar way to the correspondence between sound symbolism and visual perception [16, 17].

In recent years, the synesthetic associations called sound symbolism have been reported in a wide range of research field [18–21]. For example, Gallace et al.

© Springer International Publishing AG, part of Springer Nature 2019
S. Fukuda (Ed.): AHFE 2018, AISC 774, pp. 371–378, 2019.
https://doi.org/10.1007/978-3-319-94944-4_40

(2011) demonstrated that people reliably associate nonsense words (e.g., "maluma/takete" and "bouba/kiki") with real foodstuffs [18]. Ngo et al. (2011) also showed consistent cross-modal associations between the taste/flavor of chocolate and nonwords or shapes [19]. In particular, Sakamoto and Watanabe (2015) demonstrated that a hierarchical cluster analysis based on the relationship between linguistic sounds and taste/texture evaluations is linked to the structure of sensation categories [20]. In addition, Doizaki et al. (2017) proposed a system for estimating the fine impression of SSWs [21]. Specifically, the system can evaluate SSWs as quantitative adjectives by calculating subjective impressions of SSWs on the basis of the impressions evoked by each phoneme based on a quantitative rating database. Furthermore, the estimated ratings of SSWs enable us to visualize a tactile perceptual space.

In this research, we construct a system for quantifying the texture/taste impression expressed by SSWs. The system is characterized by outputting predicted impressions of SSWs on the basis of the impressions evoked by each phoneme based on a quantitative rating database.

2   System Construction

Figure 1 shows the structure of the system. The system consists of a user interface, a SSW analysis module, and a database. This system decomposes the input SSW into phoneme elements and refers to the category quantity of each phoneme element for each adjective scale from the database. Then, the system displays the evaluation values calculated by impression-rating predictive model.

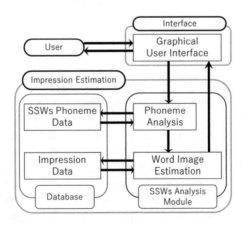

Fig. 1. Structure of the taste evaluation system.

2.1   Experiment to Build Sound-Symbolic Database

Stimuli. An experiment was conducted in order to prepare a data table of the connection between phoneme and impression evaluation. SSWs were presented to participants and the participants were asked to evaluate the impressions of each SSW using adjective scale on taste and texture. In the selection of the adjective scale, we used the 10 kinds of adjective measures, which Sakamoto and Watanabe used for evaluating drink tastes/textures [20]. To consider texture and taste, we prepared five evaluation adjectives expressing taste and five evaluation adjectives expressing texture. The adjective evaluation scales used in this study are shown in Table 1.

Table 1. List of adjective pairs used for evaluating taste and texture.

Taste-related scales	Texture-related scales
Sweet - Not sweet	Thick–light
Bitter - Not bitter	Sparkling–still
Salty - Not salty	Good texture–bad texture
Sour - Not sour	Smooth–rough
Delicious - Not delicious	Hot - Cold

Furthermore, it is necessary to select SSWs as experiment stimuli which are often used for expressing taste and texture. We collected 249 SSWs expressing taste and texture from taste/texture terms and the collected SSWs were used as experiment stimuli.

Experiment. We conducted experiments to rate the impression of each phonological form and element in order to construct a system for quantitatively evaluating the impression of SSWs. Therefore, participants evaluated the impression of SSWs and we obtained the evaluation data. 150 participants participated in this experiment and they are divided into 15 groups in which one group consists of 10 participants. Each group evaluated different SSWs. Therefore, we controlled that 10 participants responded to one SSW. Participants evaluated the impressions of each SSW presented on the questionnaire. The evaluation was made on a 7-point scale, for example, "sweet" as 6 and "not sweet" as 0 and "sparkling" as 6 and "still" as 0.

Analysis. We obtained a total of 24,900 response data (249 SSWs × 10 adjective pairs × 10 participants). In order to construct a database with high precision from the answer data, the variation of impression evaluation values among participants was investigated by calculating the standard deviation. The minimum value of the standard deviation was 0 and the maximum value was 2.80. Among them, we deleted 166 kinds of data with extreme variations such as standard deviation of 2.0 or more. The database of the system was constructed by taking the average value of the evaluation values with the standard deviation less than 2.0.

We then calculated the average rating value for each scale multiplied by each expression. By employing the average rating values of each sound-symbolic expression as the objective variables and variation of phonemes used in the expression as the predictor variables, we conducted mathematical quantification theory class I. This method for quantifying qualitative data uses a type of multiple regression analysis, calculating the degree to which each phoneme contributes to each rating scale.

2.2   Impression-Rating Predictive Model

On the basis of the hypothesis that taste impressions associated with sound-symbolic expressions can be determined by the sound symbolism of each expression, we used the impression-rating predictive model as follows [21]. Equation (1) predicts the value as a simple linear sum of the impact of each phoneme on the impression created by the expression, as a quantitative value.

$$Y = \sum_{i=1}^{13} X_i + Const. \tag{1}$$

Table 2 shows examples of the analysis results for each scale. We created a database using the category quantity (the degree of impact of each phoneme on the predictive rating value) for each phoneme.

Table 2. Correspondence between variables and phonemes.

First mora	Second mora	Phonological characteristics	Phonemes
X_1	X_7	Consonants	/k/, /s/, /t/, /n/, /h/, /m/, /y/, /r/, /w/ or absence
X_2	X_8	Voiced sounds/ p-sounds	Presence or absence
X_3	X_9	Contracted sounds	Presence or absence
X_4	X_{10}	Vowels	/a/, /i/, /u/, /e/, /o/
X_5	X_{11}	Semi-vowels	/a/, /i/, /u/, /e/, /o/ or absence
X_6	X_{12}	Special sounds	/N/, /Q/, /R/, /Li/ or absence
	X_{13}	Repetition	Presence or absence

Here, Y represents a predictive rating value of SSWs on a particular rating scale. X_1 – X_{13} represent the category quantity (the degree of impact of each phoneme on the predictive rating value) for each phoneme. X_1 – X_6 represent the consonant category, voiced/semi-voiced, palatalized, lowercase vowel, vowel and medial indicator for the first mora (the smallest sound unit in Japanese), respectively, and X_7 – X_{12} represent the consonant category, voiced/semi-voiced, palatalized, lower case vowel, vowel, and end of a word indicator for the second mora, respectively. X_{13} represents the presence or absence of repetition. The detailed relationships between variables and phonemes are shown in Table 2.

We show an example of the analysis results for each scale. From Eq. (1), the rating values for each sound-symbolic expression can be determined by totaling the category values for each phoneme in the expression. For example, the expression "huwa" is composed of the first mora /hu/ (/h/, /u/) and the second mora /wa/ (/w/, /a/). Therefore, the value of the "sweet–not sweet" scale is estimated by the following equation. For example, because evaluated values are rated on a 7-point scale, an estimated value of 4.2439 indicates that "huwa" is associated with the impression of sweet taste.

$$\hat{Y} = /h/ + /u/ + /w/ + /a/ + \text{Const.}$$
$$= /h/(X_1) + \text{semi} - \text{voiced sound}(X_2) + \text{absence}(X_3) + /u/(X_4) + \text{absence}(X_5)$$
$$+ \text{absence}(X_6) + /w/(X_7) + \text{absence}(X_8) + \text{absence}(X_9) + /a/(X_{10})$$
$$+ \text{absence}(X_{11}) + \text{absence}(X_{12}) + \text{absence}(X_{13}) + \text{Const.}$$
$$= (0.2799) + (0.1174) + (-0.0195) + (0.1114) + (-0.0007) + (0.0225) + (0.9087)$$
$$+ (0.0493) + (-0.1635) + (0.0006) + (-0.0304) + (0.1670) + (2.8012)$$
$$= 4.2439.$$

2.3   User Interface

User interface consists of input part of the SSW and output part of the result as impression-ratting values. User interface displays the impression evaluation values for each adjective calculated in the SSW analysis part. Figures 2 and 3 show examples of results obtained from the constructed taste impression evaluation system. As shown in Fig. 2, the system evaluates the impression associated with arbitrary SSW entered in the input form at the upper left of the screen, and presents the result to the user. 10 kinds of evaluation scale related to taste/texture are arranged in the center, and evaluation values from 0 to 1 are presented by the bar. In addition, the phonemes and form of SSW are displayed on the upper right of the screen.

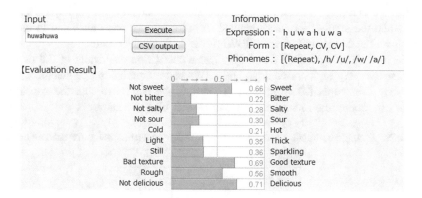

Fig. 2. Evaluation results for "huwa-huwa"

Figure 2 shows the result when "huwa-huwa" is input to the system. The "huwa-huwa" strongly expresses the impressions of *sweet*, *texture* and *delicious*, but weakly corresponded to the scales of *bitter* and *hot*. On the other hand, Fig. 3 shows the output result from "iga-iga". The result is symmetrical with the result of "huwa-huwa." The result shows the scales of *bitter* and *hot* are strongly corresponded to "iga-iga", indicating that sweetness is not felt.

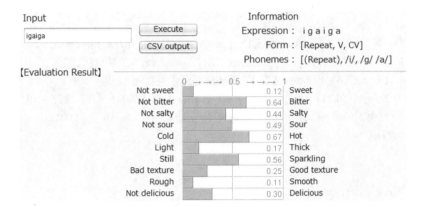

Fig. 3. Evaluation results for "iga-iga"

3 Validity Evaluation

In order to evaluate the validity of the constructed taste impression evaluation system, we compared the impression values evaluated by participants for database construction and the predicted impression values calculated by the system using multiple regression analysis. Table 3 indicates the multiple correlation coefficients values of each adjective scale.

The multiple correlation coefficient in the 8 kinds of evaluation scale was 0.6 or more in the 10 types of evaluation scale. Furthermore, the correlation coefficient with Pearson's correction between the measured values obtained from participant's experiments and the predicted values obtained from the system was calculated in order to investigate the correlation of each SSW. Table 4 shows the results.

The 146 SSWs out of all 249 SSWs had a correlation coefficient of 0.8 or more, which shows very high correlation. The SSWs had a correlation coefficient of 0.6 or more was 193 words (77%) in total. Therefore, we confirmed that the constructed system can predict the impression evaluation with high accuracy.

Table 3. Multiple correlation coefficient of impression evaluation value for each adjective.

Adjective scale	
Not sweet - Sweet	0.711***
Not bitter - Bitter	0.704***
Not salty - Salty	0.549***
Not sour - Sour	0.557***
Cold - Hot	0.628***
Light - Thick	0.792***
Still - Sparkling	0.765***
Bad texture - Good texture	0.801***
Rough - Smooth	0.714***
Not delicious - Delicious	0.806***

***p < 0.001

Table 4. Correlation coefficient for each SSW

Correlation coefficient	Number of SSW	Correlation coefficient	Number of SSW
1.0–0.8	146	0.4–0.2	17
0.8–0.6	47	0.2–0	13
0.6–0.4	26		

4   Conclusion

In this research, we constructed a system for quantifying the texture/taste impression from SSWs. The system is characterized by outputting predicted impressions of SSWs on the basis of the impressions evoked by each phoneme from a quantitative rating database. We anticipate that this system will be able to support food development and recommendation of food name which expresses the impression of goods. As future challenges, we aim to construct a database capable of dealing with new SSWs and multiple languages.

Acknowledgments. This work was supported by Grant-in-Aid for Scientific Research (A) Grant Number 15H01720), a Grant-in-Aid for Scientific Research on Innovative Areas "Shitsukan" (Grant Number 23135510 and 25135713), and JSPS KAKENHI Grant Number 15H05922 (Grant-in-Aid for Scientific Research on Innovative Areas "Innovative SHITSUKSAN Science and Technology") from MEXT, Japan.

References

1. Cambria, E.: Affective computing and sentiment analysis. IEEE Intel. Syst. **31**(2), 102–107 (2016)
2. Ren, F.: From cloud computing to language engineering affective computing and advanced intelligence. Int. J. Adv. Intel. **2**(1), 1–14 (2010)
3. Calvo, R.A., D'Mello, S., Gratch, J., Kappas, A. (eds.): The Oxford Handbook of Affective Computing. Oxford Library of Psychology, New York (2015)
4. Köhler, W.: Gestalt Psychology. Liveright Publishing Corporation, New York (1929)
5. Sapir, E.: A study of phonetic symbolism. J. Exp. Psychol. **12**, 225–239 (1929). https://doi.org/10.1037/h0070931
6. Newman, S.S.: Further experiments in phonetic symbolism. Am. J. Psychol. **45**, 53–75 (1933)
7. Werner, H., Wapner, S.: Toward a general theory of perception. Psychol. Rev. **59**, 324–338 (1952)
8. Wertheimer, M.: The relation between the sound of a word and its meaning. Am. J. Psychol. **71**, 412–415 (1958)
9. Taylor, I.K.: Phonetic symbolism re-examined. Psychol. Bull. **60**, 200–209 (1963)
10. Spence, C.: Crossmodal correspondences: a tutorial review. Atten. Percept. Psychophys. **73**(4), 1–25 (2011)
11. Schmidtke, D.S., Conrad, M., Jacobs, A.M.: Phonological iconicity. Front. Psychol. **5**, 80 (2014). https://doi.org/10.3389/fpsyg.2014.00080
12. Bloomfield, L.: Language. Henry Holt, New York (1933)

13. Osgood, C.E.: The nature and measurement of meaning. Psychol. Bull. **49**, 197–237 (1952). https://doi.org/10.1037/h0055737
14. Bhushan, N., Rao, A.R., Lohse, G.L.: The texture lexicon: understanding the categorization of visual texture terms and their relationship to texture images. Cogn. Sci. **21**, 219–246 (1997). https://doi.org/10.1207/s15516709cog21024
15. Guest, S., Dessirier, J.M., Mehrabyan, A., McGlone, F., Essick, G., Gescheider, G., et al.: The development and validation of sensory and emotional scales of touch perception. Atten. Percept. Psychophys. **73**, 531–550 (2011). https://doi.org/10.3758/s13414-010-0037-y
16. Ramachandran, V.S., Hubbard, E.M.: Synaesthesia—a window into perception, thought and language. J. Conscious. Stud. **8**, 3–34 (2001)
17. Ramachandran, V.S., Hubbard, E.M.: Hearing colors, tasting shapes. Sci. Am. **288**, 43–49 (2003)
18. Gallace, A., Boschin, E., Spence, C.: On the taste of "Bouba" and "Kiki": an exploration of word-food associations in neurologically normal participants. Cogn. Neurosci. **2**, 34–46 (2011)
19. Ngo, M.K., Misra, R., Spence, C.: Assessing the shapes and speech sounds that people associate with chocolate samples varying in cocoa content. Food Qual. Pref. **22**, 567–572 (2011)
20. Sakamoto, M., Watanabe, J.: Cross-modal associations between sounds and drink tastes/textures: a study with spontaneous production of sound-symbolic words. Chem. Sens. **41**(3), 197–203 (2015)
21. Doizaki, R., Watanabe, J., Sakamoto, M.: Automatic estimation of multidimensional ratings from a single sound-symbolic word and word-based visualization of tactile perceptual space. IEEE Trans. Haptics **10**(2), 173–182 (2017)

Affective Evaluation While Listening to Music with Vibrations to the Body

Ryohei Yamazaki[✉] and Michiko Ohkura

Graduate School of Engineering and Science, Shibaura Institute of Technology,
Toyosu, Koto-ku, Tokyo 135-8548, Japan
ma17125@shibaura-it.ac.jp,
ohkura@sic.shibaura-it.ac.jp

Abstract. Although wearable vibration devices have recently been developed for experiencing the effect of low frequency while listening to music, few studies have actually evaluated the physical sensitivity. It remains unclarified what kind of music enhances such sensibilities as the excitement or the interest felt by users. We focused on two musical genres that have a large bass sound pressure. By comparing the waveforms of the bass components of rock and electronic dance music, we identified the differences in the change of the maximum sound pressure of the bass component in each music genre and experimentally evaluated the effect of two types of music whose bass components have different maximum sound pressures when participants listened to music with vibrations and identified their affective differences.

Keywords: Music · Vibration · Affective evaluation

1 Introduction

In recent years, a vibration device has been developed that applies vibrations to the body in accordance with music's bass components while a person is listening to music [1]. However, few studies have evaluated such sensitivities as the excitement or the interest felt by users. Previous research [2] focused on music's dynamic range (DR) and targeted a range from 4.4 kHz to 360 Hz. The effect of increasing such sensitivity feelings as interesting and fierce was expected to be enhanced when the music's DR was large. On the other hand, our research emphasizes music's bass component (100 Hz or less) and investigates how music with vibrations affects the sensitivities of users when they are listening.

We focused on the changes in the bass components of rock music and electronic dance music (EDM) whose bass elements are strong. We calculated the DR of ten songs in each of our two musical genres. Rock's average DR was 10 (standard deviation: 1) and EDM's average DR was 6 (standard deviation: 1). There was a statistically significant difference at the 1% level. By comparing the bass component waveforms of rock music (less than 100 Hz) (Fig. 1) and EDM (Fig. 2), we identified the following bass components:

© Springer International Publishing AG, part of Springer Nature 2019
S. Fukuda (Ed.): AHFE 2018, AISC 774, pp. 379–385, 2019.
https://doi.org/10.1007/978-3-319-94944-4_41

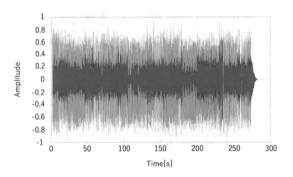

Fig. 1. Example of bass components of rock music. Title: *Zenzenzense* (movie. ver). Artist: RADWIMPS

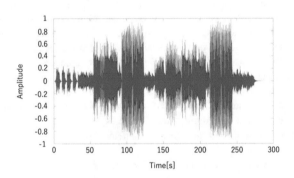

Fig. 2. Example of bass components of EDM music. Title: Done with Love. Artist: ZEDD

Rock: large DR, but monotonous change in the maximum sound pressure;
EDM: small DR, but large change in the maximum sound pressure.

   Therefore, we focused on the differences of the changes in the maximum sound pressure of the bass components of these two types of music and experimentally evaluated user sensibilities while they listened to music and physically experienced vibrations.

2   Experiment Method

2.1   Experiment Overview

We prepared the following three songs:

- One song selected by the participants as exciting;
- One rock song;
- One EDM song.

   We performed an evaluation experiment using a 7 Likert scale measure on eight evaluation items (interesting, exciting, pounding, annoying, relaxing, expectation, thrilling, standing goose bumps). After listening to the three songs, we asked whether the participants were familiar with each song and to identify the most exciting one.

2.2 Experimental System

Figure 3 shows the configuration of our experimental system that consists of the following components:

- Vibro Transducer (Acouve Laboratory, Vp7 series)
- Techtile Amp (TECHTILE toolkit [3])
- Air Sofa (INTEX, Empire chair)
- Headphones (SONY, MDR-900st)
- PC (Macintosh, MacBook Pro)

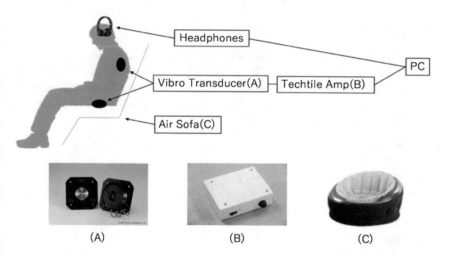

(A)   (B)   (C)

Fig. 3. Experimental system

2.3 Experimental Procedure

The following is our experiment's procedure:

I. The experimenter explained the experiment to the participants.
II. They listened to the music.
   (i) They listened to one song with vibrations.
   (ii) Then they answered questionnaire about the eight evaluation items.
      Next they repeated steps (i) and (ii) three times.
III. They answered whether they were familiar with each song and to identify the most exciting one.

Figure 4 shows the experimental landscape.

Fig. 4. Experimental landscape

2.4   Questionnaire

Participants answered a questionnaire about the participant's emotions while listening to music with physical vibrations. We conducted a questionnaire every time the participants finished listening to a song. In addition, we measured the following eight evaluation items on a 7 Likert scale where three is the highest score (strongly agree) and one is the lowest (strongly disagree):

- interesting
- exciting
- pounding
- annoying
- relaxing
- expectation
- thrilling
- standing goose bumps

We also asked the participants whether they were familiar with the songs and to identify the most exciting one.

3   Experimental Results

Our participants were eight students (six males and two females) over 20 years old. Figure 5 shows the average values for each item of all participants for the questionnaire. As a result of a one-way analysis of variance with music as a factor, we identified the main effects on music in four items, such as pounding. The subordinate test results are shown below:

pounding: EDM > rock ($p < 0.01$)
pounding: exciting music > rock ($p < 0.05$)
relaxing: rock > EDM ($p < 0.05$)
thrilling: EDM > rock ($p < 0.05$)

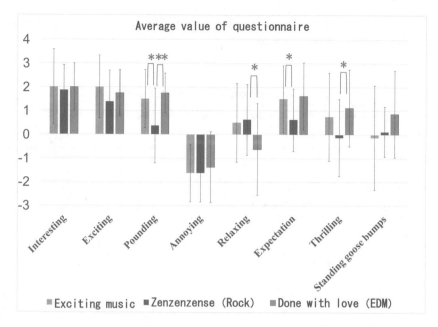

Fig. 5. Questionnaire results

When asked to identify the most exciting musical genre, four participants answered exciting music, no participant answered rock, and four said EDM (participant's exciting music is rock). All participants admitted that they had never heard EDM before. In addition, rock's DR was the largest of all the music used in our experiment.

A correlation analysis between the evaluation items for each kind of music revealed that the relation was different for each type of music:

Rock: significant positive correlation at 5% level between interesting and pounding; EDM: significant positive correlation at 5% level among interesting and expectation, interesting, and thrilling.

4 Discussion

From the result of our questionnaire analysis, we found no main effect of the music regarding interesting and exciting items.

For pounding, the rock score was significantly smaller than exciting music and EDM. Because rock's DR was the largest among all the music, its DR magnitude did not seem to affect the pounding experience of the participants. Regarding relaxing, the rock score was significantly larger than EDM. This suggests that the monotony of the maximum sound pressure of rock's bass component might have relaxed the participants more than the large change of EDM. Regarding thrilling, the EDM value was significantly larger than rock, suggesting that the large change of the maximum sound pressure of EDM's bass component might have thrilled participants more than the monotony of rock (Tables 1 and 2).

Table 1. Correlation coefficient between items in rock

	Interesting	Exciting	Pounding	Annoying	Relaxing	Expectation	Thrilling	Standing goose bumps
Interesting								
Exciting	0.665							
Pounding	0.706*	0.474						
Annoying	−0.645	−0.321	−0.269					
Relaxing	0.208	−0.119	−0.099	−0.541				
Expectation	0.597	0.546	0.369	−0.380	−0.135			
Thrilling	0.505	0.551	0.460	−0.040	−0.485	0.507		
Standing goose bumps	0.352	0.327	0.649	0.256	−0.605	0.304	0.817**	

Table 2. Correlation coefficient between items in EDM

	Interesting	Exciting	Pounding	Annoying	Relaxing	Expectation	Thrilling	Standing goose bumps
Interesting								
Exciting	0.516							
Pounding	0.603	0.701*						
Annoying	0.167	0.281	0.731*					
Relaxing	−0.194	0.518	0.137	−0.124				
Expectation	0.887**	0.756*	0.669	0.349	0.006			
Thrilling	0.696*	0.340	0.397	0.226	−0.415	0.680		
Standing goose bumps	0.667	0.335	0.473	0.165	−0.516	0.756*	0.892**	

Correlation analysis results revealed that different items correlated with the interesting item in rock and EDM. Perhaps its factor depends on the differences in the maximum sound pressure of music's bass components.

A previous study [4] reported that the brain activities of participants increased during their favorite songs or for songs that they knew. But in our study, all four participants who selected the rock music genre as exciting music answered that EDM is the most exciting music (even though they weren't familiar with that genre). When brain activity increases and exciting are considered synonymous based on experimental results and when listening to music with vibrations, we established the hypothesis that even if the change in the maximum sound pressure of the bass component is monotonous and even if participants have never heard of EDM, that musical genre increases brain activity more than rock.

5   Conclusions

Focusing on the change of the maximum sound pressure of the bass component, we performed an experiment for effective evaluation that compared rock and electronic dance music (EDM) when participants listened to music and physically experienced vibrations. Our analysis result suggests that because the change in the maximum sound pressure of music's bass components is different, perhaps there is a big difference in the sensations of pounding, relaxing, and thrilling. The factors that are deemed interesting may also differ between rock and EDM. Further studies are needed to clarify whether these results are caused by the maximum sound pressure of music's bass component.

Acknowledgements. We thank the students at Shibaura Institute of Technology who participated in our experiment.

References

1. Yamazaki, Y., Mitake, H., Hasegawa, S.: Tension-based vibroacoustic device for music appreciation. In: EuroHaptics, pp. 273–283 (2016)
2. Sugiya, K., Koushi, K., Koga, H., Oyama, Y.: Karada ni tsuketa shindou motor ni yoru ongaku jouhou dentatsu to kansei. ITE Technical Report, 24–51, pp. 33–40 (2000)
3. Minamizawa, K., Kakehi, Y., Nakatani, M., Mihara, S., Tachi, S.: TECHTILE toolkit: a prototyping tool for design and education of haptic media. In: Proceedings of the 2012 Virtual Reality International Conference, p. 26. ACM (2012)
4. Esther, R., Mittal, V.K.: Effect of different music genre: attention vs. meditation. In: Proceedings ACII2017 Workshops (2017)

Evaluation of Kawaii Feelings Caused by Looking at Stuffed Animals Using ECG

Takahumi Tombe$^{(\boxtimes)}$, Kodai Ito, and Michiko Ohkura

Shibaura Institute of Technology,
3-7-5, Toyosu, Koto-ku, Tokyo 135-8548, Japan
{mal7078, nb15501}@shibaura-it.ac.jp,
ohkura@sic.shibaura-it.ac.jp

Abstract. The Japanese word kawaii, which represents an element of Japanese culture, is used all over the world to express emotional values. We have been conducting research on its measurement and classification by biological signals. We previously identified exciting and relaxing kawaii images. However, no experiments have targeted actual objects. Therefore, we performed a new experiment using stuffed animals and obtained new knowledge on kawaii that produces relaxing experiences.

Keywords: Kawaii · Stuffed animals · Heartbeat · ECG

1 Introduction

The word kawaii, which evokes Japanese culture, is used all over the world to express emotional values. It is usually translated as cute, loveable, or adorable. However, in recent years, its definitions have become more diverse. One work argued that its definition is subjective based on personal taste and various situations [1]. Other aspects of kawaii have been discussed in psychology, sociology, Kansei engineering, and so on. Systematic studies on kawaii objects and attributes have been conducted in Kansei engineering [2]. Another study revealed that the physiological responses evoked by kawaii feelings can be measured in heartbeats [3]. Since these results have been adopted in industrial products [4], scientific research on kawaii feelings is critical for effective industrial applications.

In our previous study, we detected heartbeat modulations while participants looked at photographs. We also verified our hypothesis that the feelings, evoked while looking at photographs, can be classified as either excited or relaxed kawaii feelings. Exciting kawaii photographs significantly enhanced heart rates, but their relaxing counterparts induced smaller heart rate increases, validating our hypothesis. However, the photographs were not actual 3D objects. Therefore, we experimented with stuffed animals and evaluated the heart rate of kawaii feelings. Based on our experiment results, we report our new knowledge on kawaii feelings that are relaxing.

© Springer International Publishing AG, part of Springer Nature 2019
S. Fukuda (Ed.): AHFE 2018, AISC 774, pp. 386–391, 2019.
https://doi.org/10.1007/978-3-319-94944-4_42

2 Experimental Method

2.1 Experimental Outline

We previously performed an experiment with a woman in her 40's and measured her heartbeat modulation when she experienced kawaii feelings while she looked at photos [5, 6]. This heartbeat modulation was sharp when kawaii photos for the stimuli were subjectively selected by the subject herself. Therefore, in this experiment, we use three stuffed animals owned by the participant herself and three stuffed animals that she had never seen before.

We experimentally clarified the changes in heartbeat caused by kawaii feelings when she looked at the stuffed animals and the effect on her heartbeat by comparing when she looked at stuffed animals that she had never seen before and her own stuffed animals. This time, we used pulse wave to measure heartbeat [7].

2.2 System and Content

We used NEXUS-10 Mark II (manufactured by MIND MEDIA), which can measure various biological signals. In addition, we used a pulse wave sensor that can measure the heartbeat by simply being attached to the left index finger to lessen the psychological burden on participants. Heartbeat can be monitored in real time. Figure 1 shows our experimental system.

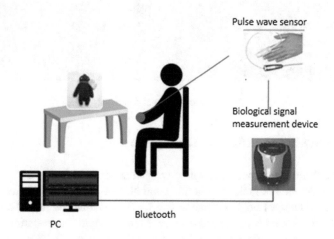

Fig. 1. Experimental system

2.3 Experimental Procedure

The following is the procedure of our experiment:

I. We explained our experiment and obtained informed consent from our participant.
II. We placed the pulse wave measurement equipment on her.

III. The experiment began.
  (i)   We measured the pulse waves.
  (ii) In the rest period, she didn't watch anything (30 s).
  (iii) During the task, she looked at a stuffed animal (30 s).
  (iv) (ii) and (iii) were repeated six times.
  (v)   During the rest period, she didn't watch anything (30 s).
  (vi) Finally, we ended the pulse wave measurement.
IV. We removed the measurement equipment.

Figure 2 shows the stuffed animals in the task and their presentation order.

Fig. 2. Stuffed animals in task

Figure 3 shows a measurement screenshot. The vertical axis shows the heart rate, and the horizontal axis shows the time.

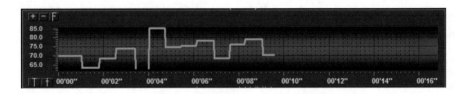

Fig. 3. Measurement screenshot

3   Experimental Result and Discussion

We again performed an experiment with a woman in her 40's. A 60-s trial was divided into two periods: 30 s of looking at nothing and 30 s of looking at a stuffed animal. In each trial, we calculated her heart rate at rest for 30 s and during the 30-second tasks based on the RR intervals of the pulse wave. Figure 4 shows an experimental scene.

Figure 5 shows the heart rate calculated for each trial, where the vertical axis shows the heart rate and the horizontal axis shows the rest and task periods in each trial.

From Fig. 5, we found the following.

  I. Except for Tasks 1 and 3, her heart rate was lower while looking at the stuffed animals that she had never seen before than at her own stuffed animals.
 II. She felt relaxed in the stuffed animal condition. Her heart rate was lower while she looked at a stuffed animal that she had never seen before (Tasks 1, 3, 5) than at her own stuffed animals (Tasks 2, 4, 6).

Fig. 4. Experimental scene

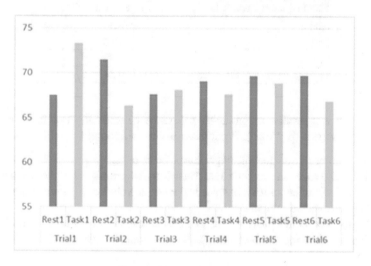

Fig. 5. Change in heart rate

III. She felt relaxed rather than excited when looking at a stuffed animal.

IV. Perhaps the heart rate was highest in Task 1 because she felt uplifted rather than relaxed because her own stuffed animal appeared as she expected.

V. Perhaps the heart rate was lowest in Task 2 because she reached a relaxed spot while looking at a stuffed animal that she had never seen before.

We confirmed the above speculation about her thoughts after the experiment. In addition, the differences in the heart rate in each trial were higher when she looked at a stuffed animal that she had never seen before than looking at her own stuffed animal. For each trial, we define the rest period (30 s) as the heart rate difference between the heart rates, which was assumed to be a baseline, during the task period and just after it (30 s). Table 1 shows the heart rate difference when she looked at her own stuffed animals. Table 2 shows the heart rate difference when she looked at stuffed animals that she had never seen before. For each stuffed animal condition, we averaged the heart rate differences of the three trials and called them the average heart rate difference. We conducted a one-sided test, based on them.

Table 1. Heart rate differences while she looked at her own stuffed animals

A stuffed animal owned by participant	Heart rate difference
Trial 1	5.8
Trial 3	0.45
Trial 5	−0.81
Average heart rate difference	1.81

Table 2. Heart rate difference when she looked at a stuffed animal she had never seen before

A stuffed animal which have never watch	Heart rate difference
Trial 2	−5.14
Trial 4	−1.46
Trial 6	−2.82
Average heart rate difference	−3.14

From the test results, a significant difference between the two types of stuffed animals was shown ($p < 0.05$). From those results, we found the following.

I. A relaxing kawaii feeling might lower heart rates.
II. The differences in the heart rate in each trial were higher while our participant looked at stuffed animals that she had never seen before than when she looked at her own stuffed animals.
III. By measuring her pulse waves, perhaps we can detect a relaxed kawaii feeling.
IV. When measuring her heartbeat, pulse waves can be substituted.

4   Conclusion

Our previous research found that heartbeat responses while looking at kawaii images are modulated differently depending on the types of kawaii experiences. In this study, we experimentally clarified the change in heartbeat caused by kawaii feeling when looking at stuffed animals and the effect on heartbeat of the differences between types of stuffed animals. We analyzed our experimental results and reached the following conclusions. Based on heartbeat evaluations, stuffed animals provide relaxing kawaii experiences and might lower heart rates. The effect on the heart rate was also higher when our participant looked at stuffed animals that she had never seen before than her own stuffed animals. However, since we didn't consider the experimental order effect, further verification is necessary. In addition, because we only got results from one participant, we must increase the number of participants in future work.

Acknowledgement. We thank Tomomi Eguchi (TBS) for preparing stuffed animals and participating in our experiment.

References

1. Yomota, I.: "Kawaii" Ron; Chikuma Shobo, Japan (2006)
2. Ohkura, M., Aoto, T.: Systematic study for "Kawaii" products. In: Proceedings of the 1st International Conference on Kansei Engineering and Emotion Research 2007 Sapporo, October 2007
3. Ohkura, M., Goto, S., Higo, A., Aoto, T.: Relation between Kawaii feeling and biological signals. J. Jap. Soc. Kansei Eng. **10**(2), 109–114 (2011)
4. Japan Society of Kansei Engineering; Kawaii kansei design award. http://kawaii-award.org/
5. Yanagi, M., Yamazaki, Y., Yamariku, Y., Takashina, T., Hirayama, Y., Horie, R., Ohkura, M.: Relation between Kawaii feeling and heart rates in watching photos. In: 15th Japan Society of Kansei Engineering Conference and General Meeting, B12, Tokyo (2013)
6. Horie, R., Yanagi, M., Ikeda, R., Yamazaki, Y., Yamariku, T., Takashina, T., Hirayama, Y., Ohkura, M.: Event-related potentials caused by Kawaii feeling in watching photos selected by heart beat modulation. In: EMBC 2014 Proceedings, TD18.19, Chicago (2014)
7. Dinh, A., Luu, L., Cao, T.: 6th International Conference on the Development of Biomedical Engineering in Vietnam

Cuteness in Japanese Design: Investigating Perceptions of *Kawaii* Among American College Students

Dave Berque[1], Hiroko Chiba[1(✉)], Ayako Hashizume[2],
Masaaki Kurosu[3], and Samuel Showalter[1]

[1] DePauw University, Greencastle, USA
{dberque,hchiba,samuelshowalter_2018}@depauw.edu
[2] Tokyo Metropolitan University, Tokyo, Japan
hashiaya@tmu.ac.jp
[3] The Open University of Japan, Chiba, Japan
masaakikurosu@spa.nifty.com

Abstract. Japanese products and pop culture, such as Hello Kitty, Pokemon, J-pop, and Anime, have gained global popularity, including in the United States. As a result, Japanese *Kawaii* (cute) design has also spread. It is not clear, however, how American perceptions of *Kawaii* compare to Japanese perceptions. In previous work by the third and fourth authors, Japanese college students rated 225 images with respect to *Kawaii* and other characteristics. This work compared perceptions between male and female Japanese college students. In the current paper, we report on a cross-cultural study of perceptions of *Kawaii* between Japanese students and American students. The study uses data we collected from American students who rated most of the images from the prior study that involved Japanese students.

Keywords: Cross-cultural studies · *Kawaii* · *Kawaii* design

1 Introduction

Japanese cuteness, described as *Kawaii*, is ingrained in Japanese contemporary society in many forms. As Yomota reports in the book "*Kawaii* Ron", the word *Kawaii* stemmed from the word kawayushi that appeared in Makura no Sōshi (The Pillow Book) in classical Japanese literature [1]. At this time, the word *Kawaii* meant pitiful, shameful, or too sad to see. During the course of Japanese history, the meaning of the word started to describe the small, weak, and someone or something that invokes the feeling of "wanting to protect" [1].

The meaning of *Kawaii* has been extended to the concept of "Japanese cuteness" in contemporary society. Indeed, in their book "Cuteness Engineering: Designing Adorable Products and Services" Marcus, Kurosu, Ma and Hashizume confirm that *Kawaii* is the closest Japanese word to the English word cute [2]. However, the authors also acknowledge that the words are not perfect translations for each other and they note that *Kawaii* can also take on aspects of the English words "pretty" and

© Springer International Publishing AG, part of Springer Nature 2019
S. Fukuda (Ed.): AHFE 2018, AISC 774, pp. 392–402, 2019.
https://doi.org/10.1007/978-3-319-94944-4_43

"interesting" [2]. In the context of cuteness, the notion of *Kawaii* is pervasive in Japan and ranges from Hello Kitty products to road signs to posters created by the Japanese government, just to name a few examples. Japanese products are consciously tailored to accommodate widely preferred "cuteness." Therefore, when designing a product, it is important to understand how specific groups of target users perceive *Kawaii*.

Japanese products and pop culture, such as Hello Kitty, Pokemon, J-pop, and Anime, have gained popularity around the globe, including in the United States. As a result, Japanese *Kawaii* design has also spread to some extent. For example, a recent survey of 25 college students at an American University revealed that more than 70% of the respondents were familiar with the concept of *Kawaii*. It is not clear, however, how closely American students' perceptions of *Kawaii* match their Japanese counterparts [3].

In previous work, the third and fourth authors studied the extent to which perceptions of *Kawaii* in 225 specific photographs differ between male and female Japanese college students [4]. The photographs were divided into subgroups including products, objects, foods, geometric shapes, animals, characters and people. Gender differences were identified, depending on the subgroup of object studied.

2 Details of the Current Study

2.1 Participants

Participants were recruited from the student body at DePauw University, which is an undergraduate, residential, liberal arts college in the Midwestern United States. A general invitation to participate was posted on an electronic bulletin board and email invitations were sent to a variety of student groups. Due to University Institutional Review Board requirements, in order to be eligible for the study, participants had to be at least 18 years of age. In order to limit the study to students who would view the images primarily through an American cultural lens, we excluded participants who had not been raised primarily in the United States and we further excluded participants who had taken classes in Japanese language or culture.

In total, 47 students participated in the study at DePauw University. All 47 of these participants were raised primarily in the United States and 37 of the participants reported that they had never lived outside of the United States. One participant had lived in Ghana for 7 years, one had lived in England for 2 years, and 9 had lived outside of the United States for at most one year, likely during college study abroad programs that typically last between 2 and 4 months. In the remainder of this paper, for convenience, we will refer to these 47 participants as "American" participants since they were raised primarily in the United States, even though we did not ask them about citizenship.

The American participants ranged in age from 18 to 22 with a mean age of 20.29. Of the 47 American participants, 23 identified as female and 22 identified as male. The remaining two American participants self-reported a non-conforming gender identity. Data from these two participants were excluded from consideration when comparing results of male and female participants.

Data from the American participants are compared to the data that were previously collected from 89 university students in Japan, who we will refer to as the Japanese participants. These participants ranged in age from 18 to 24 with a mean age of 20.07. The Japanese participants included 54 males and 34 females [4].

2.2   Materials and Setup

In previous work by the second and third authors, Japanese participants had been asked to rate each image in a series of 225 photographs that depicted objects, people, scenes, small animals, sweets, geometric shapes, female celebrities, etc. [4]. In order to receive expedited Institutional Review Board Approval, we excluded eight images with potentially sensitive material (for example naked people) from consideration in the current study. Thus, the American participants viewed 217 images, which were a subset of the 225 images viewed by the Japanese participants.

The current study took place in a computer laboratory in an academic building at DePauw University. Data were collected in three sessions over a four-day period and each session had between 11 and 19 student participants.

2.3   Procedure

The procedure for the current study consisted of six parts.

- Participants were welcomed to the computer laboratory, shown to a seat and given an informed consent form and a participant number.
- After all the participants arrived, one of the experimenters read the informed consent form while the participants followed along. Each participant then signed an informed consent form.
- Students navigated to a Google Form where they entered their subject number and answered some basic demographic questions.
- The experimenters then presented a series of 217 images one by one on a screen at the front of the computer laboratory. Each image was shown for approximately fifteen seconds. While each image was presented, students used a Google Form to respond to several questions about the image.
- After the final image was presented, participants answered an additional question about their perceptions of cuteness and beauty as well as an additional question about their overall experience during the study.
- Finally, the students were debriefed.

Figure 1 shows the survey questions that were presented after each image.

Do you know this person or thing? For a human, character, or picture, if you know the item's name, answer yes otherwise answer no. For animals or plants, if you have seen the item or you know the species or category, answer yes, otherwise answer no. *

○ Yes

○ No

Indicate your level of agreement with the following statements with regard to the person or thing shown in the photograph. *

	1. Strongly disagree	2. Disagree	3. Neither agree nor disagree	4. Agree	5. Strongly agree
This is cute.	○	○	○	○	○
This is beautiful.	○	○	○	○	○
I like this.	○	○	○	○	○

What would a suitable expression or adjective for this person or thing be? (For example: cute, gross, annoying, scary, eerie, etc.) *

[   ]

Fig. 1. Questions presented after participants viewed each image.

3   Results

A list of the descriptive titles of each image used in this study can be found in [4]. Where we provide image titles in this paper, the titles are often taken verbatim from [4]. The actual images are not displayed in this paper due to copyright limitations. However, Fig. 2 provides an example of an image, purchased from iStock with rights for re-use, similar to one of the images (Image 22) used in the study.

Fig. 2. Japanese sweets

Five pair-wise comparisons were made for each of the images: (a) Japanese males were compared to Japanese females, (b) American males were compared to American females, (c) Japanese males were compared to American males, (d) Japanese females were compared to American females, and (e) all Japanese participants were compared to all American participants.

Tables 1, 2, 3, 4 and 5 report mean results for one of the images of Japanese Sweets (Image 22) used in the study. See Fig. 1 to review the full questions summarized in the tables. When participants responded that they were familiar with the image (Know) the response was recorded as 1, otherwise, the response was recorded as 0. For the remaining questions (Cute, Beauty, Like) answers were recorded on a five-point Likert scale with 1 meaning strongly disagree through 5 meaning strongly agree. A t-test was computed to test for differences between populations. Significant differences at the p < .01 level are marked with ** and significant differences at the p < .05 level are marked with *.

Table 1. Japanese males versus Japanese females for Japanese sweets (Image 22.)

	Know	Cute	Beauty	Like
Japanese males	0.63	3.18	3.76	3.60
Japanese females	0.91	3.50	4.12	4.03
Significance level	**			

Table 2. American males versus American females for Japanese sweets (Image 22.)

	Know	Cute	Beauty	Like
American males	0.27	3.50	3.41	3.82
American females	0.22	4.09	3.39	3.96
Significance level		*		

Table 3. Japanese males versus American males for Japanese sweets (Image 22.)

	Know	Cute	Beauty	Like
Japanese males	0.63	3.18	3.76	3.60
American males	0.27	3.50	3.41	3.82
Significance level	**	**		

Table 4. Japanese females versus American females for Japanese sweets (Image 22.)

	Know	Cute	Beauty	Like
Japanese females	0.91	3.50	4.12	4.03
American females	0.22	4.09	3.39	3.96
Significance level	**		*	

Table 5. All Japanese versus all American for Japanese sweets (Image 22.)

	Know	Cute	Beauty	Like
All Japanese	0.74	3.30	3.90	3.76
All American	0.26	3.81	3.40	3.89
Significance level	**	**	*	

As shown in the Table 1, Japanese females were more familiar (Know) the image than Japanese males (p < .01) but there were no differences in ratings for Cute/*Kawaii*, Beauty, or Like between these groups. Table 2 indicates that American females rated the image more highly than American males with respect Cute/*Kawaii* (p < .05) but there were no other differences between these groups.

As shown in Table 3, Japanese males were more familiar with the image than American males (p < .05) but American males rated the image more highly with respect to Cute/*Kawaii* (p < .05). Similarly, as shown in Table 4, Japanese females were more familiar with the image than American females (p < .01). While there were no significant differences in Cute/*Kawaii* ratings between these groups, Japanese females rated the image more highly than American females with regard to Beauty (p < .05).

Finally, Table 5 compares responses between all Japanese participants (combining males and females) and all American participants. As reported in the table, Japanese participants were more familiar with the image than American participants (p < .01) but American participants rated the image more highly than the Japanese participants with regard to Cute/*Kawaii* (p < 0.1), while the Japanese participants rated the image more highly with regard to Beauty than the American participants (p < .05).

A similar set of 5 tables were produced for each of the 217 images used in the study, which resulted in 1085 tables in total. Space limitations prevent printing all of the tables.

We used the tables and Excel conditional formulas to identify images that had a Cute/*Kawaii* rating of 4.0 or greater for at least one population (Japanese males, Japanese females, American males, and American females). Using this set of images as a base, we then identified images for which the Cute/*Kawaii* ratings differed significantly between a pair of groups. In particular:

- There were significant differences between Japanese males and Japanese females in the Cute/*Kawaii* rating for 16 of these images.
- There were significant differences between American males and American females in the Cute/*Kawaii* rating for 6 of these images.
- There were significant differences between American females and Japanese females in the Cute/*Kawaii* ratings for 20 of these images.
- There were significant differences between American males and Japanese males in the Cute/*Kawaii* ratings for 18 of these images.
- There were significant differences between all Americans and all Japanese in the Cute/*Kawaii* ratings for 27 of these images.

Excel conditional formulas were used to help us identify additional patterns in the data. These patterns are discussed in the next section, particularly with respect to cross-cultural differences.

4   Discussion

We analyzed the data with a focus on understanding differences in perceptions of Cute/ *Kawaii* between American participants and Japanese participants. Key findings are presented in the remainder of this section.

Table 6 shows the number of images for which each group's mean Cute/*Kawaii* rating was 4.0 or higher. These images represent the notion of Cute/*Kawaii* for the respective group. The table also shows the mean rating for those images that were rated 4.0 or higher. For example, as indicated in the first row of the table, Japanese males rated 11 different images with a mean Cute/*Kawaii* score of 4.0 or higher. The mean rating provided by Japanese males for these 11 images was 4.16.

Table 6. Number of Cute/*Kawaii* images and mean ratings for these images.

	Number of images rated Cute/*Kawaii*	Mean rating for these images
Japanese males	11	4.16
American males	22	4.37
Japanese females	36	4.28
American females	27	4.31
All Japanese	20	4.19
All Americans	22	4.33

As shown in Table 6, Japanese males rated fewer images to be Cute/*Kawaii* than any other group. In particular, they rated half as many images to be Cute/*Kawaii* as American males, and they rated about one-third as many images to be Cute/*Kawaii* as Japanese females. Japanese males also had lower mean ratings for those images they did judge to be Cute/*Kawaii*. The lower Cute/*Kawaii* ratings provided by Japanese males largely accounts for the difference in ratings between the All Japanese group and the All American group.

Table 7 shows the number of images that had a mean Beauty rating of 4.0 or higher for each group. These images represent beauty for the respective group. Japanese males rate fewer images as Beautiful than any other group. However, for those images they do find beautiful, the mean Beauty rating for Japanese males is similar to other groups. The difference in the number of images all Japanese participants find beautiful, compared to all American participants, is primarily due to the differences between Japanese males and other groups.

Table 7. Number of beautiful images and mean ratings for these images.

	Number of images rated beautiful	Mean rating for these images
Japanese males	8	4.24
American males	19	4.21
Japanese females	18	4.24
American females	21	4.20
All Japanese	10	4.22
All Americans	18	4.19

Table 8 shows the number of images that had a mean rating of 4.0 or higher with regard to "Like" for each group. Once again, Japanese males rate fewer images as likeable than any other group. However, for those images they do like, the mean rating for Japanese males is similar to that of other groups. The difference in the number of images all Japanese participants liked, compared to all American participants, is primarily due to the differences between Japanese males and other groups.

Table 8. Number of liked images and mean ratings for these images.

	Number of images rated like	Mean rating for these images
Japanese males	7	4.22
American males	42	4.25
Japanese females	22	4.14
American females	30	4.19
All Japanese	9	4.15
All Americans	31	4.21

In summary, when compared to all other groups, Japanese males rate fewer images at 4.0 or above in terms of Cute/*Kawaii*, Beauty and Likeability. For those images they do rate at a 4.0 or above, Japanese males seem to have lower ratings than other groups with regard to Cute/*Kawaii* but not with regard to Beauty or Likeability.

There were only four images that had mean Cute/*Kawaii* ratings of 4.0 or higher for each group (Japanese males, Japanese females, American males, American females). These images are described in Table 9. Three of these images depict animals and one image (Pikachu) is an animation character that represents a small animal.

Table 9. Images with a Cute/*Kawaii* Rating >= 4.0 for each Group.

Image number	Description	Japanese male rating	Japanese female rating	American male rating	American female rating
52	An animation character of an animal (Pikachu)	4.02	4.35	4.45	4.30
67	A cat (American Shorthair)	4.11	4.29	4.45	4.35
148	A dog (Pomeranian)	4.30	4.52	4.55	4.61
181	A baby dog (Mame-shiba a small kind of Shiba Inu)	4.40	4.48	4.68	4.69

In addition to the four images listed in Table 9, there were seven images that Japanese males assigned a mean Cute/*Kawaii* rating of at least 4.0. These images are listed in Table 10. Note that each of the 11 images listed in Tables 9 and 10 depict animals or adult humans with the exception of the Pikachu image, which is an animated animal character. Japanese males did not assign any other types of images a mean Cute/ *Kawaii* rating of 4.0 or more.

Table 10. Additional images Japanese males assigned a Cute/*Kawaii* Rating >= 4.0.

Image number	Description	Japanese male rating	Japanese female rating	American male rating	American female rating
62	A female Japanese model	4.00	4.15	3.59	3.48
138	A female model with fair skin and hair	4.20	3.28	3.64	3.60
165	A Japanese actress (Yui Aragaki)	4.10	4.44	3.95	3.83
167	A cat (Koyuki, Scottish Fold and American Shorthair)	4.40	4.16	3.77	3.48
178	A sparrow	4.10	4.44	4.09	3.78
194	An illustration (dogs painted on a picture scroll)	4.10	3.72	4.14	4.39
205	An American actress (Scarlett Johansson)	4.00	2.64	3.68	3.00

Table 11. Comparison of Cute/*Kawaii* between Japanese and American females.

Image	Description	Japanese females	American females	Sig. Level
18	**An ordinary Japanese baby**	3.74	4.60	**
49	A British actress (Audrey Hepburn)	4.00	3.22	**
62	**A female Japanese model**	4.15	3.48	**
107	**A small Chinese figure**	3.06	4.04	**
136	A Japanese girl group (Perfume)	4.12	3.30	**
140	**A baby that is breast feeding**	4.00	3.09	**
145	**Little monsters**	3.36	4.38	**
153	A character (Gru and friends, despicable me)	4.36	3.83	*
157	A character (Baymax, Big Hero 6)	4.36	3.78	*
158	**A character (Chinese red cats)**	2.32	4.00	**
161	A Japanese idol group (Momoiro Clover Z)	4.00	2.96	**
162	A mascot character (Peko from Fujiya)	4.24	3.61	*
165	A Japanese actress (Yui Aragaki)	4.44	3.82	*
178	A sparrow	4.44	3.78	*
180	Sweets (chocolates covered with colored sugar)	3.04	4.13	**
194	An illustration (dogs painted on a picture scroll)	3.72	4.39	*
196	A female Japanese model	4.24	3.48	**
203	A character (Disney version of Alice's Adventures in Wonderland)	4.56	3.87	**
224	A character (Gudetama)	4.16	3.48	*

The 19 images shown in Table 11 received a mean Cute/*Kawaii* rating of 4.0 or greater from at least one female group (Japanese females or American females) and also had ratings that differed significantly between these groups.

The 9 images shown in Table 12 received a Cute/*Kawaii* rating of 4.0 or greater for at least one male group (Japanese males or American males) and also had ratings that differed significantly between these groups.

Table 12. Comparison of Cute/*Kawaii* between Japanese and American males.

Image	Description	Japanese males	American males	Sig. level
18	**An ordinary Japanese baby**	2.64	4.04	**
23	A panda	3.80	4.64	**
33	A hedgehog	3.82	4.50	*
40	A dog (Pug)	3.54	4.41	**
45	A baby girl model	3.93	4.32	**
57	A baby boy model	3.51	4.41	**
62	**A female Japanese model**	4.00	3.59	*
68	A koala	3.51	4.55	**
124	A baby boy with a rabbit	3.71	4.68	**

In Tables 11 and 12, bold rows indicate images that appear in both tables. In each table a * indicates significance at the $p < .05$ level while a ** indicates significance at the $p < .01$ level.

For 13 of the 19 images in Table 11, Japanese females rated the image significantly higher than American females with regard to Cute/*Kawaii*. For the remaining 6 images, American females rated the image significantly higher than Japanese females with regard to Cute/*Kawaii*.

For 8 of the 9 images in Table 12, American males rated the image significantly more highly then Japanese males with regard to Cute/*Kawaii*. For the remaining image (a female Japanese model), Japanese males rated the image more highly with regard to Cute/*Kawaii* than American males.

For the images presented in this study, the differences between Japanese females and American females seem more varied with regard to perceptions of Cute/*Kawaii* as compared to the differences between Japanese males and American males. For images that at least one male group finds cute (mean rating of at least 4.0) the groups only differ for images representing animals or people. Additionally, as shown in Table 12, when the male groups differ, American males are more likely to give higher Cute/*Kawaii* ratings (with an apparent exception when they are rating adult females).

As shown in Table 11, ratings of Cute/*Kawaii* differ more frequently for female groups. In addition to images of animals and people, image of characters, sweets and toys were rated significantly differently with respect to Cute/*Kawaii* by Japanese females versus American female. Additionally, American males and females seem less likely than Japanese males and females to use "cute" to describe adults.

Evaluating an image in the context of *Kawaii* for Japanese females may not be just a matter of observing the characteristics of the two-dimensional photo. Rather, it is possible that Japanese females look at familiar "*Kawaii*" photos in richer contexts. For example, the group of Japanese female participants highly rated a pop group, Perfume, with respect to *Kawaii*, while Japanese males did not. Perfume is a well-known celebrity group active in music and TV programs. Japanese females might have associated the photo of the group with other attributes, which brought a wider sense of cuteness.

5   Future Work

Given the increasing global popularity, including popularity in the United States, of Japanese products that exhibit *Kawaii*, it would be interesting to perform a cross-cultural comparison between American participants and Japanese participants with regard to ratings of Cute/*Kawaii* in physical representations of commercial products such as toys, school gear, sweets, clothing, home decor and electronic goods.

It would also be interesting to determine the specific attributes of products (color, size, shape, decoration, etc.) that lead to feelings of Cute/*Kawaii*. Previous work by Laohakangvalvit, Achalakul and Ohkura [5] has developed a preliminary model for *Kawaii* feelings based on ratings of pictures of various proposed spoon designs. Additional studies reported by Ohkura [6] have reported on affective values of *Kawaii* in physical products. All of this work would inform the proposed cross-cultural study of commercial products. Results of such a study could be useful to manufacturers who are interested in designing products that appear to consumers outside of Japan.

References

1. Yomota, I.: Kawaii Ron. The Theory of Kawaii Tokyo, Chikuma Shobō (2006)
2. Marcus, A., Kurosu, M., Ma, X., Hashizume, A.: Cuteness Engineering Designing Adorable Products and Services. Springer, Cham (2017)
3. Berque, D., Chiba, H.: Evaluating the use of LINE software to support interaction during an American Travel Course in Japan. In: Rau, P.L. (ed.) Cross-Cultural Design CCD 2017. LNCS, vol. 10281, pp. 614–623. Springer, Cham (2017)
4. Hashizume, A., Kurosu, M.: The gender difference of impression evaluation of visual images among young people. In: Kurosu, M. (ed.) HCI 2017, Part II. LNCS 10272, pp. 664–677. Springer, Cham (2017)
5. Laohakangvalvit, T., Achalakul, T., Ohkura, M.: A proposal of model of Kawaii feeling for spoon designs. In: Kurosu, M. (ed.) HCI 2017. Part I, LNCS 10271, pp. 687–699. Springer, Cham (2017)
6. Ohkura, M.: "Kawaii" Engineering: Asakura Publishing Co., Ltd, Tokyo, Japan (2017)

Evaluation of Feeling of the Like in Watching Pictures by Physiological Signals

Takuma Hashimoto[1], Kensaku Fukumoto[2], Tomomi Takashina[2],
Yoshikazu Hirayama[2], Michiko Ohkura[1], and Ryota Horie[1(✉)]

[1] Graduate School of Engineering and Science,
Shibaura Institute of Technology, 3-7-5, Toyosu, Koto-ku,
Tokyo 135-8548, Japan
mal6079@shibaura-it.ac.jp,
{ohkura,horie}@sic.shibaura-it.ac.jp
[2] Nikon Corporation, 471, Nagaodai-Cho, Sakae-ku, Yokohama-city,
Kanagawa 244-8533, Japan
{Kensaku.Fukumoto,Tomomi.Takashina,
Yoshikazu.Hirayama}@nikon.com

Abstract. We usually use the word 'like' when we see good pictures, e.g. in social networks. When we choose a picture that evokes the feeling of 'like' out of many pictures, we evaluate each picture based on our subjective and intuitive preference. We hypothesized that physiological responses arise when participants watch pictures that evokes the feeling of 'like'. In this study, we examined whether we can measure physiological responses when participants feel 'to like' a picture they choose among many pictures. We measured electroencephalogram, electrooculogram, electrocardiogram, and respiration of the participants while these are choosing the best picture from a set of pictures which had different setting of depth of field. We found that EEG responses appears when the participant watched the best pictures which evokes the feeling of 'like'.

Keywords: Feeling of the like · Pictures · Physiological signals
Electroencephalogram · Electrooculogram · Electrocardiogram
Respiration

1 Introduction

Recently, we use the word 'like' when we see good pictures, as used in social networks. There are many situations in which we choose a particular picture to which we feel 'like' from many pictures. For example, when we take a picture or edit a picture, we choose a picture by setting parameters for taking or editing the photo to appropriate values. In choosing a picture by setting the parameters, we have to evaluate each picture based on our subjective and intuitive preference. When options in choosing a picture are many, we have high cognitive load. Therefore, our goal is to develop a system that assists our choosing a picture which evokes the feeling of 'like'.

However, it is hard for us to define criteria for choosing pictures based on subjective and intuitive preference explicitly. In our previous studies, we found that

© Springer International Publishing AG, part of Springer Nature 2019
S. Fukuda (Ed.): AHFE 2018, AISC 774, pp. 403–408, 2019.
https://doi.org/10.1007/978-3-319-94944-4_44

physiological responses, in both heart rate and brain waves, arises when participants feel 'kawaii' in watching photos [1–6]. Similarly, we hypothesized that physiological responses arise when participants watch pictures which evokes the feeling of 'like'.

In this study, we examined whether we can measure physiological responses when participants feel 'to like' a picture among many pictures. In our experiment, the participants are supposed to choose the best picture that evokes the feeling of 'like' from a set of pictures with different settings of depth of field. We measured multimodal physiological signals in the participants, that is, electroencephalogram (EEG), electrooculogram (EOG), electrocardiogram (ECG), and respiration, while these are making the choice of the best picture. We report how changes of the physiological signals arose when the participants watched the best picture which evokes the feeling of 'like'.

2 Materials and Methods

2.1 Photographing

Six volunteers in their twenties participated in the experiment. For photographing, the participants arranged toy figures in a light tent and determined a camera angle and a focus position. Pictures were taken by using a digital single-lens reflex camera (Nikon, D5200). The participant chose one lens from three lenses, AF-S NIKKOR 50 mm f/1.8G, AF-S DX Micro NIKKOR 85 mm f/3.5G ED VR, and AF-S DX NIKKOR 18–55 mm f/3.5–5.6G VR. The camera was fixed to a tripod, and the experimenter obtained 21 images with different depth of field by continuous shooting while changing the lens diaphragm in 1/3 increments. Degree of picture blurring was changed depending on the depth of field. This trial was repeated five times. In each of the trials, the participants changed the arrangement of the toy figures, the camera angle and a focus position. Each participant prepared five set of 21 pictures. A snapshot of conducting the photographing was shown in Fig. 1. Examples of pictures which have different depth of field are shown in Fig. 2.

Fig. 1. A snapshot of conducting the photographing.

Fig. 2. Examples of pictures which have different depth of field.

2.2   Experiments

Pictures were displayed on a 27 in. display (BenQ, GW 2760). By manipulating a controller, pictures with different depth of field was presented in sequential order. The participant chose the best picture which evoked the feeling of the 'like' from a set of 21 pictures. During the participant was choosing the best picture, physiological signals were measured. The participant's head was set on a chin rest at a position 60 cm away from the display. For each set of the 21 pictures, the measurement was repeated three times. The measurement was carried out 15 times in total across the five sets of pictures.

2.3   Recording Physiological Signals

We developed an experimental system to measure multimodal physiological signals from a participant during the participant is choosing the best picture. A system configuration of the experimental system is shown in Fig. 3. ECG, EOG and respiration were measured by a single board computer (PLUX, BITalino [7]).

ECG was measured by the 3-lead system. Horizontal EOG was measured by electrodes placed on the left and right temples. Vertical EOG was measured by electrodes attached above and beneath the dominant eye. Silver-silver chloride disposable electrodes were used. The ECG signal was amplified in a frequency band 0.5–40 Hz at a gain of 1100. The EOG signals were amplified in a frequency band 0.05–40 Hz at a gain of 2000. The respiration was measured by a band to which a piezoelectric film is attached. The band was attached to the abdomen. The signals were sampled continuously at 1000 Hz in 10bits, transferred from the single board computer to a PC via Bluetooth, and recorded by a homemade application running in the PC. The homemade application was developed by using API distributed by PLUX.

Simultaneously, EEG signals were measured by a compact EEG recorder (Emotiv, EEG headset). Electrodes of the compact EEG recorder were attached at the sites, AF3, AF4, F3, F4, F7, F8, FC5, FC6, T7, T8, P7, P8, O1, and O2, of the international 10–20 system. All of the electrodes were referenced to an average of CMS and DRL. The EEG signals were amplified in a frequency band 0.16–43 Hz, sampled continuously at 128 Hz in 16bits, transmitted to the PC wirelessly and recorded by an application running in the PC (Emotiv, TestBench).

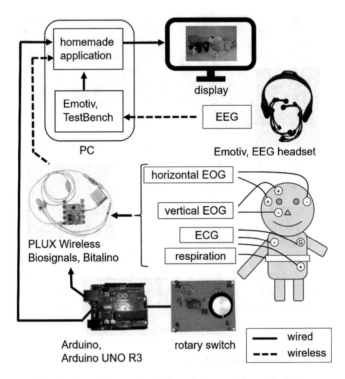

Fig. 3. A system configuration of the experimental system.

2.4   Presenting Pictures

The pictures were presented on the display by the homemade application running in the PC. The application was developed by using Python and PyQt. A participant changed a picture presented on the display by rotating a rotary switch. A homemade circuit board detected direction of the rotation. When the rotary switch is rotated, pulse signals having one of two different phases were output according to the direction. The signals were input to a single board computer (Arduino, Arduino Uno), which were connected to the PC by a USB cable. The single board computer discriminated the direction from the input signal, sent presence or absence of rotation to BitAlino digitally, and sent one-byte signal of the direction of rotation to the PC. The BitAlino recorded the digital signal of presence or absence of rotation along with recording the physiological signals. The one-byte signal was transferred to the homemade application. The homemade application changed a picture displayed on a monitor according to the one-byte signal. The homemade application sent one-byte signal representing an index of the picture to TestBench. In the TestBench, the one-byte signal was recorded as triggers along with recording the EEG signals.

2.5   Pre-processing of Physiological Signals

The EEG signals and the triggers were upsampled to 1000 Hz, which is a sampling frequency of the BitAlino. Then, the EEG signals and the triggers were merged with the ECG, EOG and respiration signals with synchronizing the time by matching between the triggers recorded in the TestBench and the signal of presence or absence of rotation recorded in the homemade application.

Heart rates were calculated from R–R intervals in the ECG signals. The EEG signals were passed through each of band pass filters in increments of 1 Hz from 1 Hz to 40 Hz. Instantaneous amplitude of EEG signal at each of the frequency band was obtained as a positive envelope of the filtered signal.

3   Results and Discussion

A typical EEG response during a participant was choosing the best picture within one trial was shown in Fig. 4. The horizontal axis is time [s] and the vertical axes are indices of pictures (the top panel), heart rates [bpm] (the second panel), respiration (the third panel), horizontal EOG (the fourth panel), and vertical EOG (the fifth panel). Units of the respiration, the horizontal EOG, and the vertical EOG are defined by BitAlino. The last two panel shows instantaneous amplitudes of EEG signals at each of the frequency band from 5 Hz to 15 Hz at the electrodes AF4 and F4. The figure shows that the physiological signals were successfully measured.

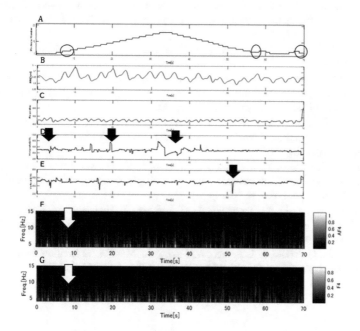

Fig. 4. A typical EEG response within one trial.

The circles on the graph of the indices of pictures indicates an index of the best picture. Black arrows in the graphs of the horizontal EOG and the vertical EOG indicates time points at which eye artifacts occurred. In the graphs of instantaneous amplitudes of EEG signals, there are several time points at which amplitudes increase in the wide frequency bands. White arrows in the graphs of instantaneous amplitudes of EEG signals points to EEG responses appeared at the timing of watching the best picture but not appeared at the timing of artifacts. We considered that the EEG responses might be caused by watching the picture which evokes the feeling of the 'like'.

4 Conclusions

In this study, we examined whether we can measure physiological responses when participants feel 'to like' a picture among several ones. We measured EEGs, EOGs, ECG, and respiration during the participants' choice of the best picture from a set of pictures with different settings of depth of field. We found that EEG responses appears when the participant watched the best picture which evokes the feeling of 'like'. We will conduct statistical evaluation of the results in the future works.

References

1. Ohkura, M., Aoto, T.: Systematic study of Kawaii products: relation between Kawaii feelings and attributes of industrial products. In: Proceedings of the IDETC/CIE 2010 (DETC 2010-2818), pp. 587–594 (2010)
2. Ohkura, M., Goto, S., Higo, A., Aoto, T.: Relation between Kawaii feeling and biological signals. Trans. Jpn. Soci. Kansei Eng. 10(2), 109–114 (2011)
3. Ohkura, M., Komatsu, T., Tivatansakul, S., Settapat, S., Charoenpit S.: Comparison of evaluation of Kawaii ribbons between gender and generation. In: Ji, Y.G. (ed.) Advances in Affective and Pleasurable Design, pp. 59–68 (2012)
4. Takashina, T., Yanagi, M., Yamariku, Y., Hirayama, Y., Horie, R., Ohokura, M.: Toward practical implementation of emotion driven digital camera using EEG. In: Proceedings of the Augmented Human (2014)
5. Yanagi, M., Yamasaki, Y., Yamariku, Y., Takashina, T., Hirayama, Y., Horie, R., Ohkura, M.: Physiological responses caused by Kawaii feeling in watching photos. In: Proceedings of the AHFE 2014, pp. 839–847 (2014)
6. Yanagi, M., Yamasaki, Y., Yamariku, Y., Takashina, T., Hirayama, Y., Horie, R., Ohkura, M.: Differences in heartbeat modulation between excited and relaxed Kawaii feelings during photograph observation. IJAE 15(2), 189–193 (2016)
7. da Silva, H.P., Guerreiro, J., Lourenço, A., Fred, A.L., Martins, R.: Bitalino: a novel hardware framework for physiological computing. In: PhyCS, pp. 246–253 (2014)

Author Index

© Springer International Publishing AG, part of Springer Nature 2019
S. Fukuda (Ed.): AHFE 2018, AISC 774, pp. 409–410, 2019.
https://doi.org/10.1007/978-3-319-94944-4

Printed in the United States
By Bookmasters